Archimedes

New Studies in the History and Philosophy of Science and Technology

Volume 70

Archimedes has three fundamental goals: to further the integration of the histories of science and technology with one another; to investigate the technical, social and practical histories of specific developments in science and technology; and finally, where possible and desirable, to bring the histories of science and technology into closer contact with the philosophy of science.

The series is interested in receiving book proposals that treat the history of any of the sciences, ranging from biology through physics, all aspects of the history of technology, broadly construed, as well as historically-engaged philosophy of science or technology. Taken as a whole, Archimedes will be of interest to historians, philosophers, and scientists, as well as to those in business and industry who seek to understand how science and industry have come to be so strongly linked.

Submission / Instructions for Authors and Editors: The series editors aim to make a first decision within one month of submission. In case of a positive first decision the work will be provisionally contracted: the final decision about publication will depend upon the result of the anonymous peer-review of the complete manuscript. The series editors aim to have the work peer-reviewed within 3 months after submission of the complete manuscript.

The series editors discourage the submission of manuscripts that contain reprints of previously published material and of manuscripts that are below 150 printed pages (75,000 words). For inquiries and submission of proposals prospective authors can contact one of the editors:

Please find on the top right side of our webpage a link to our *Book Proposal Form.*

Fabian Kröger

From Automated to Autonomous Driving

A Transnational Research History on Pioneers, Artifacts and Technological Change (1950–2000)

Springer

Fabian Kröger
Equipe d'Histoire des Techniques
Paris 1 Panthéon-Sorbonne University
Paris, France

ISSN 1385-0180 ISSN 2215-0064 (electronic)
Archimedes
ISBN 978-3-031-49880-0 ISBN 978-3-031-49881-7 (eBook)
https://doi.org/10.1007/978-3-031-49881-7

This work was supported by Bundesministerium der Verteidigung

This Springer imprint is published by the registered company Springer Nature Switzerland AG
The registered company address is: Gewerbestrasse 11, 6330 Cham, Switzerland

Paper in this product is recyclable.

Acknowledgments

First of all, I would like to thank Ernst Dieter Dickmanns (UniBwM) for his encouragement to write this book, for providing me with important sources and most importantly for his patience. He was also successful in obtaining the funding of this project from the University of the German Federal Armed Forces/University of Bundeswehr Munich (UniBwM). I am also very much indebted to Sadayuki Tsugawa (MEL) whom I had the chance to interview in Tokyo in May 2018. Without his support, the chapter on Japan could not have been written like it is now. He also provided excellent images of the Intelligent Vehicle that are published here for the first time. I would also like to thank Chuck Thorpe and Todd Jochem (CMU) who gave very useful comments on an early version of the US chapter. Todd was also helpful to provide good images of research vehicles. Steven Shladover (PATH) is one of the practitioners in the field who included always the historical dimension of vehicle automation in his papers. I would like to thank him for his advice that improved the book considerably. He also warned me that a book about the research history would be a very challenging project, a prediction that revealed to be absolutely true. I also have to thank Michel Parent who showed me the research facility of INRIA in Roquencourt near Paris and provided research reports of the French participants of the PROMETHEUS program. It was the first time that I saw guidance cables by myself in the pavement of a road. I also have to thank Walter Zimdahl (VW) for sending me the original manuscript of his dissertation and a copy of Super 8 films. Catherine Fargeon provided important insights and photos for the chapter on the autonomous vehicle of the French military and I thank her for that. Reinhold Behringer kindly provided a photo of UniBwM research vehicles. I am also grateful to William (Red) Whittaker, Brian Staszel and Debra Tobin from the CMU Robotics Institute for providing images of their robot vehicles.

Many other persons provided sources and helpful comments and I like to thank them: Javier Ibañez-Guzmán (Renault) was very helpful discussing machine vision during an online meeting. The economist Yvelyne Lecler and Jeehoon Ki from the Korean Academy of Science and Technology made valuable remarks that improved the chapter on Japan. The economist Gerhard Prätorius provided me with a source that was difficult to find. Holger Meinel (Daimler) helped to distinguish early infrared and radar devices. Gunter Schaub (Airbus) indicated the exact dates of the PRIMUS program. Nils Fehlhaber and Hans Stibbe provided me with material from the corporate archives of Continental and MAN. Markus Henrichs from VDI-Nachrichten discussed the Optopilot of Volkswagen with me. Many thanks go also to Gery Deffontaines and Tommaso Pardi from GERPISA (Groupe international d'études et de recherche permanent sur l'industrie et les salariés de l'automobile) for inviting me to present my research in April 2019 at ENS Paris-Saclay; to the Fondation France-Japon (FFJ) de l'EHESS for their invitation in June 2019 to a workshop on the future use of AVs at PSL University in Paris and to Christian Long for his invitation to the Robomobile Workshop in September 2020. I also wish to thank the editor Jed Z. Buchwald of Caltech for including my book in the Archimedes series. Many thanks go also to the publishing editor Chris Wilby at Springer and Solomon George and Divya Rajasekaran of Straive in India for the book production. Without you, this book would not exist.

Above all, I want to thank my wife Reakka for her patience with me during this project that went sometimes on a collision course with our family life, especially when the Covid-19 pandemic started to affect our daily routines. I dedicate this book to her and to our sons Aaron and Carl who will perhaps one day take a ride in an autonomous vehicle.

Photographic Acknowledgements

We wish to express our thanks to everybody who provided images and gave the permission to reproduce them in this book. While every effort has been made to identify and credit all copyright holders, we would like to apologize to anyone who has not been formally acknowledged. All claims remain unaffected.

Contents

Chapter 1
Introduction

Since its invention at the end of the nineteenth century, the automobile remained a machine that had to be controlled by a human driver. Without human control of steering wheel, gas pedal and brakes none of the billions of miles could have been traversed by the billions of cars in the world: A car always needed the driving skills of a human to fulfill its function. Without a driver, it would have been only a static artifact, condemned to stand still in its parking lot. In the early years of motoring, this necessity of a human driver was not seen as a hindrance. Manual driving promised to fulfill the human dream of individual mobility and freedom, of self-guidance, of autonomy. But with the emerging mass automobility in the first decades of the twentieth century, the negative effects of human agency behind the steering wheel – accidents for example – became a serious topic of concern. It is no surprise that the phantasy of a self-driving car, a car that can navigate without a human driver, can be dated to this period.

Automated driving has a long history that is mostly unknown to a wider public. It received huge public attention at the World's Fairs in NY in 1939–40 and 1964–5, thanks to the GM Futurama displays at both fairs. Disney also highlighted it for their amusement parks. It became a big topic of the media discourse after the turn of the century when DARPA held its robotic challenges in the Mojave desert (2004, 2005 and 2007). These races were surrounded with much publicity. But it is another event that left a deeper impression in the collective memory: In 2009 the technology company Google announced to develop "self-driving" cars. Today, many people think indeed that Google started the development of autonomous driving. Some years later, a second name became associated with the topic: In 2014, the electric car company Tesla offered a self-driving feature called Auto-Pilot, including autonomous steering, braking and parking functions. Investments in neural networks triggered a new hype around artificial intelligence. Expectations of the finance markets fueled the hype – venture capitalists invested massively in the field – $27,5 billion

F. Kröger, *From Automated to Autonomous Driving*, Archimedes 70,
https://doi.org/10.1007/978-3-031-49881-7_1

in 2018 alone.[1] Autonomous driving became an icon of the future of mobility, at the same time menacing and rejuvenating the old car industry.

The coverage of these developments in media reports created the distorted impression that the idea of a vehicle that is able to find its way without a human driver is a recent development that started somewhere in the Silicon Valley between DARPA, Google and Tesla. Autonomous driving seemed to have no history and its introduction seemed only some years away. This book has the aim to correct this impression. "Hype hides history" as Michael Mahoney observed.[2] We will see that the techno-optimism of the last decades was only possible with a lack of historical knowledge. Like it is often the case when new technologies are developed, unrealistic timelines and overly optimistic promises started to fall back on the protagonists. The peak of excitement was reached in 2019 when some automotive companies admitted that they had overestimated the possibility of fast market introduction. This seems to be a good moment to take some distance and look back at the long-term research history of this vision.

1.1 Between Fiction and Science

This book is a contribution to the history of technology, as many other publications. But its topic is in several ways different from other technologies that were examined by historians: First, autonomous driving is situated between Fiction and Science, second its mass deployment remained until today 20 years in the future and third, it has not yet become a technology-in-use. The next paragraphs will detail these hypotheses.

Throughout the twentieth century, automated and autonomous driving unfolded in two complementary domains: Fiction and Science, cultural dreams and engineering practices. Driverless vehicles were a field of fiction before they became a field of engineering – a feature that they share with robots.[3] Before they became material artifacts and prototypes, they were powerful symbols of a technological sublime, promising a better future. This can be called the *sociotechnical imaginary* of autonomous driving in the twentieth century.

Fiction and Science are very different fields, but Bruno Latour argued that they should not be understood as contradiction.[4] Their relation can be conceptualized like a circle: Beneath other factors, the power of this imaginary pushes research institutions around the world to pursue their work. And the realm of invention also

---

[1] Statista. 2020. By 2030, one in 10 vehicles will be self-driving globally, Press release, February 6, 2020, https://mailchi.mp/statista/autonomous_cars_20200206?e=145345a469.

[2] Mahoney, Michael S. 2005. The Histories of Computing (s). *Interdisciplinary Science Reviews* 30 (2):120.

[3] The word robot was first used in the theater piece R.U.R. of Karel Čapek.

[4] Latour, Bruno.1992. *Aramis ou l'amour des techniques*, p. 28. Paris.

produces fictions, as pointed out by Lucien Sfez.[5] By promoting visions of a better future fixed by technology, the hard sciences and engineering disciplines participate in the creation of this powerful technoscientific imaginary.

The author of this book explored the cultural side of this imaginary in several publications since 2012. My publications, situated at the crossroads of technical and cultural history, focused mostly on the popular imaginations of driverless cars in film and literature.[6] Throughout the twentieth century the cultural logic of the self-driving car developed in the space between the weird and the wonderful. Literary metaphors, pictorial imaginings and technological concepts of autonomous vehicles evolved in mutual interplay, sparking off each other, but never evolving in step. The cultural phantasy of an autonomous vehicle emerged for the first time in literary and cinematographic sources of the 1930s.[7] The idea of automated vehicles guided by wires in the pavement can also be dated back to the 1930s.[8] As serious technical developments only started after World War II, until the 1950s the cultural phantasies were one step ahead of the technical concepts.

When my publications came to the attention of the Daimler and Benz-Stiftung they integrated me in a Fellow-program with several experts, giving me the chance to examine the cultural and societal dimensions of autonomous driving further. In 2015 the results were published in an anthology in German language.[9] Having heard about this publication, Ernst Dieter Dickmanns contacted me and asked for the reasons that I did not mention the achievements of the University of Bundeswehr Munich (UniBwM) in my article. I explained that my focus privileged those research projects that remained in a direct connection with popular imaginations. As the whole picture of the technical research was not in my focus, I had to skip many projects of the engineering domain in my article. Agreeing that his pioneering works on machine vision were of great importance for autonomous driving, I modified the

---

[5] Sfez, Lucien. 2002. La technique comme fiction, *Revue européenne des sciences sociales* XL-123, published December 3, 2009, accessed on August 23, 2020, online in: http://journals.openedition.org/ress/612.

[6] Kröger, Fabian. 2012. Fahrerlos und unfallfrei. Eine frühe automobile Technikutopie und ihre populärkulturelle Bildgeschichte. In *Technology Fiction, Technische Visionen und Utopien in der Hochmoderne*, ed. Uwe Fraunholz and Anke Woschech, pp. 93–114. Bielefeld. Kröger, Fabian. 2014. Die Zukunft, die nie eintrat, Über die Nützlichkeit filmischer Zukunftsbilder des autonomen Fahrens für die Prospektive. *Technology Assessment – Theory and Practice (TATuP)*, N° 1/2014, Verkehrszukünfte – Visionen jenseits aktueller öffentlicher Aufmerksamkeit, p. 3–6. Kröger, Fabian. 2016b. Le long cheminement de l'imaginaire cinématographique de la voiture autonome. In *Choc de Mobilités*, ed. Mathieu Flonneau and Stéphane Levesque, pp. 55–67. Paris.

[7] Werner Illing, Utopolis, 1930, David H. Keller, The Living Machine, 1935, GM, The Safest Place, 1935 (see Kröger 2016a, pp. 45–47).

[8] Murtfeldt, E. W. 1938. Highways of the future. *Popular Science*, 132/5, pp. 27–29 and 118–119.

[9] Kröger, Fabian. 2015. Das automatisierte Fahren im gesellschaftsgeschichtlichen und kulturwissenschaftlichen Kontext. In *Autonomes Fahren, Technische, rechtliche und gesellschaftliche Aspekte*, ed. Maurer, M., Gerdes, J.C., Lenz, B., Winner, H., pp. 41–67. Heidelberg. Note: The findings of the article were cited in 2015 in a book of the Daimler and Benz Foundation that summarizes the project (Minx, Eckard, and Rainer Dietrich. 2015. *Autonomes Fahren, Wo wir heute stehen und was noch zu tun ist*, pp. 38–55. Ladenburg).

English version of the chapter that was published in 2016.[10] After finalizing my PhD-thesis on the knowledge history and the image history of car accidents, I came back to him and proposed to start an inquiry on the technical concepts and engineering efforts of automated and autonomous driving. Dickmanns agreed to start a book project and got a funding of the University of Bundeswehr Munich (UniBwM).

1.2 Predictions: 20 Years in the Future

With its position between Fiction and Science, autonomous driving has a special relationship to time: Autonomous driving is an "Utopia that always hangs 'out there' in the future somewhere", observes Richard Bishop, former program manager of the US-Department of Transportation.[11] Other experts agree that the deployment of these systems has always been "over the horizon, just out of sight and reach."[12]

The historian Jameson M. Wetmore discovered that the promise of automated vehicles was not only situated *somewhere* in the future, it was bound to the very precise promise that it will become a reality in 20 years.[13] This time span is indeed a timeless estimation that can be found in many different sources, especially from the automotive industry, highway authorities and engineers who developed the technology of these systems. One of the first events where it had been promoted was the Futurama-Exhibition of General Motors at the World's Fair 1939 in New York: "Strange? Fantastic? Unbelievable? Remember, this is the world of 1960!"[14] In the following decades this time span was often repeated, especially by the car industry: The General Motors-film *Key to Future* promised in 1956 that automated cars would become a reality between 1960 and 1976. Joseph Bidwell from GM stated in 1959 that "completely automatic control for highway operation will be here in 1980."[15] A report of the Michigan Department of State Highways predicted in 1966 that

[10] Kröger, Fabian. 2016a. Automated Driving in its Social, Historical and Cultural Contexts. In *Autonomous Driving, Technical, Legal and Social Aspects*, ed. Maurer, M., Gerdes, J.C., Lenz, B., Winner, H., pp. 41–68. Heidelberg.

[11] Bishop, Richard, Whatever Happened to Automated Highway Systems (AHS)? Originally published in the August–September 2001 issue of Traffic Technology International, online in: https://staff.washington.edu/jbs/itrans/bishopahs.htm (accessed on July 3, 2019).

[12] Shladover, Steven E., Jane Lappin, Robert P. Denaro, and Bryant Walker Smith. 2018. Introduction: The Transportation Research Board's 2013 Workshop on Road Vehicle Automation. In *Road Vehicle Automation*, ed. Gereon Meyer and Sven Beiker, p. 8.

[13] Wetmore, Jameson M. 2003. Driving the Dream, The History and Motivations Behind 60 Years of Automated Highway Systems in America. In *Automotive History Review*, p. 14.

[14] General Motors. 1940. Futurama, Exhibition brochure, p. 8. USA.

[15] Bidwell, J. B., General Motors Research Laboratories, Vehicles and Drivers, 1980, SAE Annual Meeting, Detroit, 1959.

"electronic controls and automated highways are 'just around the corner'. They will be a reality within the next two decades, on selected freeways."[16]

The estimation of a fast market introduction was also shared by engineers that worked in the field. For example, Ernst Dieter Dickmanns estimated in 1986 that "the introduction of autonomous guidance systems is one or two decades away"[17] and "mass production systems may be more than one vehicle generation away".[18] In 1993, he said that market introduction could be expected in "ten or even 20 years" at the earliest.[19] In 2002, he claimed to have "no doubt that the average performance level in road vehicle guidance of humans will be achievable in a few decades by technical systems based on vision and some additional signals from more conventional sensors."[20] And in 2018, he said in an interview that truly autonomous vehicles are still a decade or two away.[21]

The timelessness of the 20 years-prediction appears like an inverse version of Moore's law with its self-fulfilling prediction of an ever increasing computer power. Wetmore explained recently why 20 years are a "perfect timeframe" for predictions: "It seems close enough that it will happen in many of our lifetimes, and therefore it is worth working toward. At the same time, it is far enough away that the person, company, or government making the prediction probably won't be held responsible if time passes and the promises aren't met."[22]

However, the crucial point is not whether these predictions were accurate or not. The historian Peter Bowler argues in his history of the future that we should not "judge the accuracy" of predictions as they are "extrapolated from an element". For instance, from the 1970s on, predictions had been made with regard to the development of digital microprocessors and machine vision. In the 1980s and 1990s, they had been made with regard to successful demonstrations of autonomous research vehicles. The successes in these *research contexts* inspired rhetoric's of a "sublime technological future"[23] and pushed many actors to imagine short timelines until

[16] Lutz, Randolph B. 1966, April. *The Motor Vehicle of the Future*. Michigan Department of State Highways, Technical Report Number 2, p. 4. It should be mentioned that the report states also that it is "unlikely that there will be complete automation of the freeway systems, even by the year 2000." It would more be a "gradual evolution." (Ibid., p. 28).

[17] Dickmanns, Ernst Dieter. 1986. Computer Vision in Road Vehicles – Chances and Problems. *ITCS-Symposion on "Human Factors Technology for Next-Generation Transportation Vehicles"*, p. 1. Amalfi, Italy.

[18] Ibid., p. 16.

[19] Der Lotse geht an Bord, Interview with Ernst Dieter Dickmanns. In *Hochschulkurier* 1993, p. 7.

[20] Dickmanns. 2002. The Development of Machine Vision for Road Vehicles in the Last Decade. *Intelligent Vehicle Symposium*, p. 275.

[21] Delcker, Janosch. 2018. The man who invented the self-driving car. *Politico*, July 19, 2018, online in: https://www.politico.eu/article/delf-driving-car-born-1986-ernst-dickmanns-mercedes/.

[22] Wetmore, Jameson M. 2020. Reflecting on the Dream of Automated Vehicles, Visions of Hands Free Driving Over the Past 80 Years. *Technikgeschichte* 87 (1): 70.

[23] Carey, James, and John Quirk. 2009. The History of the Future. In *Communication as Culture: Essays on Media and Society*. New York, London. See also Packer, Jeremy. 2008. Automobility

market introduction. This proved wrong as the successful introduction of a new technology into a consumer market is a challenge that is very different from successes in basic research and technology development.

1.3 Autonomous Driving Was Never a Technology-in-Use

The observation that autonomous driving developed between Science and Fiction and that it remained always 20 years in the future is important for the theoretical framing of our study. Until now, historians of technology often explored *technologies-in-use*, technologies that are integrated in the everyday practices of millions of people in the world – like the automobile, the washing machine, mobile phones, vacuum cleaners etc. But autonomous vehicles remained always in the *antechamber of history*. They never became items of individual consumption on the market.

Today, no fully autonomous vehicle is available for purchase, for sale. Therefore they have no users and no user cultures. Without users, sources about any social appropriation of these vehicles are missing. While historians of the automobile have the advantage to access a great amount of sources from the ubiquitous, worldwide deployment of what the sociologist John Urry called the "system of automobility"[24] with thousands of possible case studies, the history of autonomous driving must be written without information about the everyday use of driverless cars. There are no communities of driverless car enthusiasts, no clubs, meetings, no fandom, and no specialized newspapers. There are no tinkering practices, no individualization of different models, no national or cultural differences of hypothetical driverless car cultures around the world. We do not know anything about the use of autonomous vehicles in different regions of the world. For that reason, writing about its history requires another methodological approach than for example a book about the history of automobile assistance systems that are now used by millions of people each day.

Autonomous vehicles are also very different from the histories of other automobile subspecies – the electric car for example. Even when the combustion engine became the dominant propulsion concept, electric cars existed in niches – as delivery vans, taxis, airport vehicles, municipal services, fire brigades, buses, postal services, as Gijs Mom has shown.[25]

In contrast to the electric vehicle, autonomous driving was never a technology-in-use, but a *technology in the making*: The only group who used autonomous

---

and the Driving Force of Warfare: From Public Safety to National Security. In *The Ethics of Mobility, Rethinking Place, Exclusion, Freedom and Environment*, ed. Sigurd Bergmann and Tore Sager, p. 39f. Hampshire, Burlington.

[24] Urry, John, The 'System' of Automobility. In *Theory, Culture and Society* Vol. 21, No 4/5, p. 25–39.

[25] See Mom, Gijs. 2004. *The Electric Vehicle, Technology and Expectations in the Automobile Age*. Baltimore, London.

vehicles were engineers in different countries who drove the first test vehicles. Only on special occasions – test drives and demos – autonomous vehicles were allowed to leave their fenced proving grounds.[26] It seemed that the quest for autonomous driving was an "eternal research project, one built of steady, incremental advances that would produce a steady stream of PhD dissertations" as Alex Davies resumes.[27] Therefore it is too early to write a *general* history of autonomous driving – its market introduction is just starting. What can be written today is a history of the *research* on autonomous driving, a history of a technology that remained for long time in the domain of basic research.

In contrast to other technologies of historical interest, automated driving is part of the very recent history, part of the contemporary world. We are writing about history "while it's still smoking"[28] as Barbara Tuchman said. What does this mean for the future? Will autonomous driving remain a dream forever? We should be aware that the history of technology is not only full of predictions about what *will be realized* at a given moment in the future, but also predictions on the opposite side, predictions of what *will never exist* in the future. And they also proved to be often wrong as the example of Wilbur Wright shows, one of the early flight pioneers. He said to his brother Orville in 1901, that "man would not fly for fifty years."[29] According to his biographer he even said that man would not fly in a thousand years.[30] But already two years later, in December 17, 1903, the first successful motorized flight of the Wright brothers took place in Kitty Hawk, North Carolina.

The 20 years-prediction proved wrong on behalf of market introduction and mass deployment. But focusing our attention on technical breakthroughs, we will see that autonomous driving had several of these Kitty Hawk-moments – for example in 1994 at the final demo of PROMETHEUS when the twin Mercedes of UniBwM and Daimler drove autonomously on three-lane highways with speeds up to 130 km/h. Even a twofold lane change has been performed autonomously as we will discover later in this book. In this perspective, autonomous driving already exists since over 20 years – but only in the domain of research.

Today, the time spans that are mentioned for market introduction are becoming much shorter.[31] We should be aware that in the last decade, more and more prototypes of autonomous vehicles circulate on our streets and in the last years, the development efforts advanced further into useable systems.

---

[26] An exception was VaMoRs that – after five years of testing as mentioned above – was allowed to drive on any type of public road in Germany since 1992.

[27] Davies, Alex. 2021. *Driven, The Race to Create the Autonomous Car*, p. 20. New York, London, Toronto, Sydney, New Delhi.

[28] Tuchman, Barbara. 1982. In Geschichte denken, Essays, p. 31. Düsseldorf. (Thanks to Hannah Zindel for letting me know about this citation).

[29] Rescher, Nicholas. 1998. *Predicting the Future: An Introduction to the Theory of Forecasting*, p. 2. New York.

[30] Kelly, Fred C., ed. 2002. *Miracle at Kitty Hawk, The Letters of Wilbur & Orville Wright*, p. 42. New York.

[31] Wetmore 2020, p. 70.

Waymo offers since December 2018 a level 4 autonomous ride-hailing service with 600 mini-vans to a selected group of 1000 "early riders" in Phoenix, Arizona, United States.[32] The taxis drive in a precisely defined, "geofenced" geographical area, operating only if the weather is good (no rain, no sand storm). In fall 2019, Waymo even started to operate a handful of their Chrysler Pacifica vehicles without safety drivers.[33] However, the vehicles are surveilled in real-time by Waymo employees, meaning that this type of "autonomy" depends heavily on human labor. Nevertheless there are three reasons why the experiment in Phoenix is an important rupture in the history of autonomous driving: The vehicles are now available on public streets. They operate without drivers. And the users are not only engineers. For the first time, this technology left the realm of research only and slowly becomes a technology-in-use.

1.4 Terms for Characterizing Automated and Autonomous Driving

Before we will present the method that had been followed in the inquiry, we should define our topic more precisely. What exactly is autonomous driving? A definition can be started with four terms that are used today in an interchangeable way in popular media discourses but also in many engineering articles. *Automated driving* and *autonomous driving* focus on the task (driving) and most of the time these terms are not further distinguished. Two other terms are centered on the vehicle, describing it as *self-driving vehicle*[34] or classifying it with an even more popular term *driverless vehicle*. The following section will differentiate these terms and show that they designate different phenomena.

The Society of Automotive Engineers (SAE International) coined in 2014 six levels of vehicle *automation* that became an official standard for the automobile industry. This often cited classification is based on the amount of driver attention needed to operate the vehicle.[35]

---

[32] Ohnsman, Alan. 2018. At Waymo, It's Launch Time for Google's Biggest Moonshot. *Forbes*, November 13, 2018, online in: https://www.forbes.com/sites/alanohnsman/2018/11/30/10-million-miles-later-waymo-ceo-says-its-self-driving-cars-are-ready-for-prime-time/#222ddf675e7e.

[33] Hawkins, Andrew J. 2019. Waymo's Driverless Car: Ghost-Riding in the Back Seat of a Robot Taxi. *The Verge*, December 9, 2019, online in: https://www.theverge.com/2019/12/9/21000085/waymo-fully-driverless-car-self-driving-ride-hail-service-phoenix-arizona. Accessed September 3, 2020.

[34] Interestingly, Nuro, Waymo and Uber use the term "self-driving" for their vehicles (Wetmore 2020, p. 70).

[35] SAE International. 2019. SAE J3016, Levels of Driving Automation. https://www.sae.org/news/press-room/2018/12/sae-international-releases-updated-visual-chart-for-its-"levels-of-driving-automation"-standard-for-self-driving-vehicles. Accessed July 2, 2019.

Level 0 means a manually operated vehicle, including warning systems.[36] In level 1 (driver assistance or hands on), the human driver shares control with an automated system that controls longitudinal *or* lateral movements. An example is the adaptive cruise control (ACC): The driver controls the lateral movements of the car (steering) while the longitudinal control (acceleration, brake) is automated. A more recent technology is the lane keeping assistant.

In level 2 (partial automation or hands off), the automated system controls *both* lateral and longitudinal movements but the human driver must monitor the process and must intervene *immediately* if required. This is also called semi-automated driving. It is available on the market today, for instance with the Tesla *AutoPilot* that follows lanes and keeps the distance to the vehicle in front. It can also do automated lane changes under certain conditions. The name *AutoPilot* had been criticized because it can give the impression that a supervision of the system by a driver is no longer mandatory. Many fatal accidents with Teslas are known today, involving the AutoPilot. The first fatality occurred in May 2016 when a trailer made a left turn in front of a Tesla in Williston, Florida – the system failed to brake and the driver who failed to monitor the system was killed. Semi-automated driving revealed to be a dangerous concept as the driver does not have to drive but must be able to drive at every moment.[37] If the system inspires confidence, the way to distract attention is open and can lead to new types of accidents.

In level 3 (conditional automation or eyes off), the human driver can concentrate his attention to other tasks than the surveillance of the automated system. The driver is for instance allowed to watch a movie or read a newspaper. Control is shifted from the driver to the vehicle – the system now monitors the driving environment and can handle actions that need immediate responses like emergency braking. It can make informed decisions like overtaking other vehicles.

Nevertheless, the human passenger must be prepared to intervene like in level 2 – he only has more time to take over this task. How many seconds would be appropriate for the machine-to-human handover is currently a topic of discussion (around ten seconds). Level 3 is today imagined as limited to specific controlled environments like highways. But if we remember that a vehicle driving with 120 km/h advances 33 meters in one second, this handover may present considerable risks even on highways. In the worst case, a driver would not retake control. For that reason, Volvo decided in 2017 not to develop level 3 vehicles in the future.[38] Audi developed a level 3 traffic jam pilot for the A8 but had to cancel its introduction in

---

[36] A new version of the SAE-norm (2019) mentions "automatic emergency braking" in level 0. It is an assistance system that first warns the driver, asking for action. Without human reaction, the system brakes automatically – independently from the driver. But as it does not perform sustained control of the vehicle and does not change the role of the driver, it is not placed in level 1.

[37] Torchinsky, Jason. 2019. *Robot, Take the Wheel: The Road to Autonomous Cars and the Lost Art of Driving*, 96ff. New York.

[38] Volvo, Volvo Cars CEO urges governments and car industry to share safety-related traffic data, Press release, April 3, 2017, in: https://www.media.volvocars.com/global/en-gb/media/pressreleases/207164/volvo-cars-ceo-urges-governments-and-car-industry-to-share-safety-related-traffic-data. Accessed July 3, 2019.

2020 because of a missing legal framework.[39] At this time, it was explicitly prohibited to take the hands off the steering wheel.[40]

Level 4 (high automation or mind off) is a more advanced version of level 3. Now the human driver is becoming a passenger, he can even go to sleep or leave the driver's seat. Driver attention is not required. This level is limited to certain environments like highways or car parks. It resembles automated shuttles at the airport. Waymo, Uber, Baidu and MOIA are working on this type of automation. Peugeot claimed not to develop cars beyond the level three because this would add too much extra-costs to the price of the vehicle.[41]

Level 5 (full automation) is "probably a myth", says John Krafcik, CEO of Waymo.[42] It means would be capable of driving under all conditions in which humans are capable of driving. A human driver is never required to take control, he becomes a passenger who doesn't even need to have a driver's license. In contrast to level 4 the vehicle is able to drive in all environments, in all landscapes and countries and under all meteorological and traffic conditions. As it can even operate without a human passenger, it is not only a self-driving car, but can be called with the more popular term "driverless" vehicle. The first three levels of automation cannot be called driverless: The role of the driver changes but a driver – a person that should be able to intervene – is required.

It is intriguing that even a level 5 car would not be able to decide where it should go if it is not given a destination by a human or an algorithm. It also is not able to

---

[39] Schmidt, Herbie. 2020. Audi steckt beim autonomen Fahren zurück. In *Neue Zürcher Zeitung*, April 28, 2020, online in: https://www.nzz.ch/mobilitaet/auto-mobil/autonomes-fahren-stufe-3-audi-verzichtet-im-a8-auf-staupilot-ld.1553933. Accessed May 24, 2020.

[40] On July, 21, 2021 Germany passed a law that legalized the operation of autonomous vehicles with level 4 in public road traffic. Requiring a local limitation to a specified operating area, it concerns passenger and freight transport providers – shuttles on defined routes, transport of goods between distribution centers or during delivery missions (Gesetz zur Änderung des Straßenverkehrsgesetzes und des Pflichtversicherungsgesetzes – Gesetz zum autonomen Fahren. In Bundesgesetzblatt Jahrgang 2021, Teil I, Nr. 48, ausgegeben zu Bonn am 27. Juli 2021, p. 3108–3114). Since August 5, 2022 operating a level 3 autonomous vehicle is legal in Europe (COMMISSION IMPLEMENTING REGULATION (EU) 2022/1426). On roads without pedestrians, equipped with lanes that are separated physically from incoming traffic, the driver is no longer required to keep his hands on the steering wheel and it is not mandatory to supervise the car. But the speed is limited to 60 km/h – this indicates clearly that traffic jams on the highway are the main use case. From January 1, 2023 on, a speed of 130 km/h is allowed if the vehicle is able to make lane-changes without human intervention.

[41] Feitz, Anne. 2019. Premiers coups de frein sur la voiture autonome. *Les Echos*, March 26, 2019, online in: https://www.lesechos.fr/tech-medias/intelligence-artificielle/premiers-coups-de-frein-sur-la-voiture-autonome-1003606. Accessed on July 3, 2019.

[42] Ohnsman, Alan. 2018. At Waymo, It's Launch Time for Google's Biggest Moonshot. *Forbes*, November 13, 2018, online in: https://www.forbes.com/sites/alanohnsman/2018/11/30/10-million-miles-later-waymo-ceo-says-its-self-driving-cars-are-ready-for-prime-time/#222ddf675e7e.

decide autonomously when it has to start. For that reason Jason Torchinsky proposed to complete the SAE-scale with a sixth level.[43]

Interestingly, the SAE-scale only mentions different degrees of *automation*, avoiding the term *autonomy*. The historian of technology Martina Heßler explains the important difference between both terms: *Automated* technologies function without human intervention, but they are programmed by humans and are therefore predictable.[44] *Autonomous* technologies go one step further. They learn by themselves and can adapt to new situations. Based on great amounts of data from different sources, autonomous systems *create* algorithms to solve a problem, to reach a goal. The most important point of *autonomous* technologies is that they have *decision-making* capabilities that depend on actual sensor data; they may or may not have been programmed by a human. Therefore it is not predictable how an autonomous system comes to a decision, emphasizes Heßler.

An example may clarify the difference further: An *automated* system is provided with an input, an *autonomous* system decides by itself, given the sensor data. For instance, modern *automated* washing machines execute a set of pre-programmed instructions that must be chosen in advance by a human operator (tissue, temperature, time). When the machine is switched on, it functions without the help of these outside actors. An automated machine cannot adapt itself to changes in the environment.[45]

If a washing machine would be able to check the laundry itself and take the decision when to wash what kind of clothes (color, tissue), how long and with what kind of temperature, it would be an autonomous washing machine. It would be able to take complex decisions and to adapt itself to the environment. While an *automated* vehicle can execute a procedure or task without human intervention, the driving functions of an *autonomous* vehicle are done by a *self-governed* technology, independently from the human capability of decision making. The vehicle will for instance take *itself* the decision to effectuate lane changes. As it is capable to adapt to changes in the environment it could even *refuse* to function autonomously in certain situations.

From this background we can come back to the SAE-levels and the more popular media terms mentioned above, resuming that the lower levels of automation (level 2–3) can be addressed with the terms self-driving and automated, the higher levels with the terms autonomous (level 4–5). As already explained above, only level 5 corresponds to the term driverless.

The degree of human intervention is not the only method to separate automated from autonomous driving. Other definitions explain the difference with relation to the infrastructure: According to Tsugawa, *automated* driving depends on a type of

[43] Torchinsky, p. 181f.

[44] Heßler, Martina. 2019. Technik und Autonomie, Kulturhistorische Bemerkungen zu einem komplexen Verhältnis. In *Autonome Systeme und Arbeit*, ed. Hartmut Hirsch-Kreinsen and Anemari Karacic, p. 259. Bielefeld.

[45] Tessier, Catherine. 2015. Autonomie : enjeux techniques et perspectives. In *Drones et killer robots: faut-il les interdire?*, ed. Ronan Doaré, Didier Danet, and Gérard de Boisboissel, p. 2. Rennes.

infrastructure that is built specially for this purpose (guidance cables, magnetic markers etc.).[46] Therefore he calls it cooperative driving. In contrast, Tsugawa understands *autonomous* driving as independent from these infrastructures, only using GPS and lane markers on the road, two elements that are conceived for more purposes than autonomous driving only. Other engineers share this definition: Graefe and Kuhnert define a fully autonomous robot in the sense "that no special installations on the road will be provided (e.g. buried cables or radio beacons) and that no external source of computing power" will be needed.[47] The robot will "only utilize those markings and traffic signs that exist" already and must coexist with manually driven vehicles.

Autonomous systems can also be differentiated according to their autarchy or self-sufficiency, i.e. how much they have to rely on other facilities and services. Most of today's civilian visual systems for vehicle guidance are network-based, they only function with GPS and continuously updated highly precise maps. As they depend on systems that go beyond the vehicle, they are not self-sufficient. If a system can do its job without these elements, it is self-sufficient. We will see later that UniBwM used machine vision to create autonomous systems that are as self-sufficient as possible.

1.5 State of the Art

Until the turn of the century, autonomous driving was not an important research topic for historians and was neglected in the historiography of the automobile. The first historical overviews were published by engineers and practitioners who worked in the field of vehicle automation.[48] Dissertations and technical papers in engineering journals often introduce their findings with a brief summary of the state of the art, mentioning what practitioners see as historical landmarks.

One of the first historical studies about Automated Highway Systems (AHS) in the United States was published in 2003 by the historian Jameson M. Wetmore.[49] In

[46] E-mail from Sadayuki Tsugawa to the author, August 30, 2019.

[47] Graefe, Volker, and Klaus-Dieter Kuhnert. 1992. Vision-Based Autonomous Road Vehicles. In *Vision-Based Vehicle Guidance*, ed. Ichiro Masaki, p. 2. New York.

[48] See for example Fenton, Robert E. 1994. IVHS/AHS: Driving into the Future. *IEEE Control Systems Magazine* 14 (6): 13–20. Shladover, Steven. 2007. PATH at 20 – History and Major Milestones. *IEEE Transactions on Intelligent Transportation Systems* 8 (4): 584–592. Tsugawa, Sadayuki, 2008. A History of Automated Highway Systems in Japan and Future Issues, Institute of Electrical and Electronics Engineers. *International Conference on Vehicular Electronics and Safety*, Ohio, USA, September 22, 2008, p. 2–3. Dickmanns, Ernst Dieter, 2016. Summary of contributions to Vision-based Autonomous Vehicles. http://www.dyna-vision.de/main/Review2016%20of%20ContribVisBasAutVeh.html

[49] Wetmore, Jameson M. 2003. Driving the Dream, The History and Motivations Behind 60 Years of Automated Highway Systems in America. In *Automotive History Review*, Summer 2003, pp. 4–19.

2015, Marc Weber from the Computer History Museum in California published a paper on the history of different kinds of autonomous vehicles, including sea-based devices and rockets.[50]

Since 2017, several books addressed the advent of autonomous driving, but mostly from a Silicon Valley-business perspective.[51] Book publications with a historical interest are rare, they are mostly popular accounts with short narrations and no interest in detailed presentations of the research concepts: In 2018 the author Michael Fallon published a book about the history of self-driving cars, giving a quick overview about some milestones.[52] Curiously, the research efforts of Ford are heavily featured in the last chapters, giving the impression of a biased narration. The automotive historian Dan Albert wrote 2018 a popular book about the twentieth-century car culture heading to driverless cars.[53] In the same year the transport expert Samuel Schwartz mentioned some of the early historical achievements in his description of the dramatic revolution that self-driving cars will bring.[54] In 2019, the author Malcolm Gladwell produced the documentary film *Autonomy*, directed by Alex Horwitz. The film contains interviews with some of the engineers that are also the main protagonists of this book. But the film is only available on Amazon Prime in the US. In his book *Robot, Take the Wheel*, Jason Torchinsky gave in 2019 a short overview about the numerous research vehicles and their implications for the future.[55] He also presents many insightful ideas about potential effects of self-driving cars in the future.

Two other books go more into details, but they limit their historical memory to the DARPA-challenges. Lawrence D. Burns, former R&D-Director of General Motors, published in 2018 a very interesting first-hand account of these

---

[50] Weber, Marc. 2014. Where to? A History of Autonomous Vehicles. http://www.computerhistory.org/atchm/where-to-a-history-of-autonomous-vehicles

[51] Jamthe, Sudha. 2017. 2030 The Driverless World: Business Transformation from Autonomous Vehicles, (independently published). Wadhwa, Vivek, and Alex Salkever. 2019. *The Driver in the Driverless Car: How Our Technology Choices Will Create the Future, Oakland* 2017. 2nd ed. Simoudis, Evangelos. 2017. *The Big Data Opportunity in Our Driverless Future*. Menlo Park. Eliot, Lance. 2018. Introduction to Driverless Self-Driving Cars: The Best of the AI Insider, (no place of publication mentioned). McGrath, Michael E. 2018. *Autonomous Vehicles: Opportunities, Strategies, and Disruptions* (independently published). Herrmann, Andreas, Walter Brenner, and Rupert Stadler. 2018. *Autonomous Driving: How the Driverless Revolution will Change the World*. Bingley. Sperling, Daniel. 2018. *Three Revolutions: Steering Automated, Shared, and Electric Vehicles to a Better Future*. Washington, DC. Vaid, Kamal. 2018. Selbst ist das Auto – automatisiertes und autonomes Fahren, Die Zukunft der Mobilität, Schriftenreihe der Northern Business School zur angewandten Wissenschaft, Band 10, Hamburg.

[52] Fallon, Michael. 2019. *Self-Driving Cars, the New Way Forward*. Minneapolis.

[53] Albert, Dan. 2018. *Are We There Yet?: The American Automobile Past, Present, and Driverless*. New York.

[54] Schwartz, Samuel I. 2018. *No One at the Wheel: Driverless Cars and the Road of the Future*. New York.

[55] Torchinsky, Jason. 2019. *Robot, Take the Wheel: The Road to Autonomous Cars and the Lost Art of Driving*. New York.

competitions.[56] Another book on the DARPA-Challenges, published in 2021 by Alex Davies, the editor of Wired, mentions some early pioneers of autonomous driving, but only in five of 300 pages.[57]

This review of the literature shows that until now, the long-term research history of autonomous driving – including the history of mobile robots – had not been explored with more details in a monograph. A more exhaustive in-depth study of the paths or trajectories that the different technological concepts of automated and autonomous driving followed in the last 50 years in different geographical spaces is missing so far. This book will hopefully be a starting point to fill this gap.

1.6 Historical and Geographical Delimitation and Main Research Question

The main objective of the following study is to present the most important contributions of the transnational research on automated and autonomous driving during the five decades from 1950 until the year 2000. Initially it was planned to focus the book only on autonomous driving and to start the inquiry with the 1970s. This timeframe revealed to be too limited. If all the early research on *automated* driving – like cable guidance-experiments – would have been ignored, the driving forces behind the quest for *autonomous* driving in the 1970s would have been invisible. Especially if the advent of mobile robots and computer vision in the 1960s would have been ignored, it would have been impossible to understand the historical genealogy, the sources and the origins of autonomous driving. Therefore the book now covers an extended historical period of 50 years. The geographical focus of our inquiry is centered on the United States, Japan and Europe as the most important projects were situated in these three major industrialized regions of the world. The aim of the three case studies is to describe the different approaches, experimental systems and technologies in a comparative perspective.

The inquiry is guided by the following lead question: *Who participated where and when with what kind of technical concepts and artifacts in the research of automated and autonomous ground vehicles?* Or in a more naïve form: *Who invented the driverless car?*

The notion *ground vehicle* is important because it contains a systematic limitation of our topic: Autonomous aerial vehicles (helicopters, planes, drones) underwater vehicles (torpedoes, submarines) and surface vehicles (ships) have a rich history on their own.[58] They are excluded from this book that focuses on ground vehicles

---

[56] Burns, Lawrence D. 2018. *Autonomy: The Quest to Build the Driverless Car – And How It Will Reshape Our World (Together with Christopher Shulgan).* London.

[57] Davies, Alex. 2021. *Driven, The Race to Create the Autonomous Car*, New York.

[58] One important element that had not been integrated in our study is the fact that autonomous systems were first introduced in airplanes. It would be recommended that future studies integrate the insights of this engineering history in their narration.

only. The term ground vehicles is also important because it signifies that we are not only interested in systems that were integrated in automobiles – mobile robots are also included.

The four sub-questions can be detailed further:

Who did the research? We decided to use the interplay between different academic, governmental, industrial and military actors in each of the three geographical spaces as one important *guideline* for our narration. A focus on different players shows that technology development is always influenced by specific interests, research strategies, funding policies, historical events, very different objectives and hopes and does not follow a predefined path. It should also be mentioned that most of the stakeholders that are presented here are public laboratories from state sponsored universities. The private industry is present but not in the center of the study. It may be exaggerated to claim that the automotive industry was "neither interested nor involved" in the research on autonomous driving.[59] In the early years, GM invested in the field, in the 1980s Volkswagen and Daimler started own projects. Other automotive players, especially in Japan, may have been also active in the field. But the restrictive publishing policy of the automotive industry is a hindrance for historical research as important sources are not available.

It is also important to note that the engineers and practitioners from different knowledge communities and academic disciplines (electrical engineering, artificial intelligence, robotics, automotive industry and defense research) play a decisive role in the book. In a certain way, they are the main actors. This highlights that it was not only "technology" but social forces – ideas, hopes and aspirations of engineering communities – that formed the material conditions for the different concepts of automated and autonomous driving.

When did the research start? In historical studies, dates are important as they allow *situating* concepts, experiments, inventions and events in a historical context. They allow to highlight contacts between different geographical zones and connections to other research teams. But they can also prove that some inventions were made at the same time by different actors at different places without knowing from each other. Finally they allow proving that a pioneer applied a specific concept, developed a distinct technology or organized an event at a certain moment in history *for the first time*. It can be criticized that some historians of technology fetishize dates and over personalize research processes. From our perspective it seems however important to identify who did things first and when. This does not mean to conceive engineering as a process driven by geniuses working isolated in hidden laboratories, having brilliant ideas without communicating with the outside world. Attributing ideas to single researchers should not occult that they are influenced from colleagues, research networks, information from other areas and engineering communities. Researchers in the three regions

[59] Davies, p. 19.

exchanged knowledge on international conferences and via scientific publications.

If an engineering team claims to be the first who invented a certain system, it is important to examine how they justified their own design as the first. Ann Johnson emphasizes that "any machine could be claimed as "the first" if the definition of that machine were peculiar enough."[60] The first autonomous vehicle moving through a laboratory is followed by the first autonomous vehicle driving outdoors with analog computers on-board, followed by the first vehicle with digital computers on-board and so on. That means firsts are always at the same time seconds, thirds, thirties and so on, depending on the context where they are placed.

Being interested in firsts does not mean that a first approach to a question is always the ultimate solution. Not only firsts but also *lasts* are important – we will also analyze under which circumstances a concept had been *abandoned* because it was not seen as beneficial for the future. And we will also see that some concepts that seemed to be abandoned may reappear in the future.

Where did the research take place? As we have already seen above, the question of time is linked to the question of space: The experiments on vehicle automation – both, automated and autonomous – had been effectuated in different geographical regions. To understand the research efforts in the three regions, we had to address their different historical and societal settings. Each country had different motivations to invest in automation. Research was effectuated under different conditions which redirects to the first question as funding was provided by different local actors. It is also important to understand that the question of space is not limited to national and cultural entities, it also directs our attention to the very different places or environments of the research vehicles: laboratories, laboratory yards, test tracks, highways – and the desert.

Which concepts of automation with what kind of technology had been developed? Technical artifacts – the mechanical and electronic devices that were developed for automated and autonomous driving – served as the second *guideline* for our narration. As mentioned above, we decided to include two different types of research vehicles or prototypes in our presentation – automobiles and robots. On the one hand, modified automobiles, equipped with different kinds of electronic devices served as test vehicles. On the other hand, specially designed robots with advanced capabilities for perception and mobility were used for research purposes.[61] They are called mobile robots. Marc Weber resumes that a moving robot must first *sense* the environment, second *model* the outside world and make decisions. Third he must *react* appropriately to change its position in the environment. While sensing and reacting are "feasible with known technology", the real

[60] Johnson, p. 104. Johnson identifies for instance several first antiskid devices as their design answered different questions – a first commercially available system, a first production device, a first four wheel device, a first electronic device, a first integrated device… (Johnson, p. 134f.).

[61] Stationary robots, used for example in manufacturing environments – bolted at fixed positions and manipulating recurring tasks – are not considered in this publication.

challenge is the second step, the processing and machine intelligence needed for interpretation.[62]

Some of the specially equipped automobiles can also be called mobile robots because the tasks they have to achieve and the methods they use are similar to robots. It is important to understand that fully autonomous cars are no longer cars in the historical sense of the term – without the then historical figure of the human driver they will be "fundamentally different from automobiles" as Torchinsky pointed out.[63] The commercialization and the appearance of these robot vehicles in the public space will be the "first mass deployment of potentially lethal autonomous machines in human history."[64]

All of the vehicles presented in this book are mentioned in other sources, but mostly without a more detailed description of their abilities. This book has the aim to evaluate the capabilities of both – the mobile robots and the automobile research vehicles -, paying special attention to four factors of interest:

First, it is important if a vehicle was *self-contained* or if it depended on off-board computing.

Second, the *environment* is crucial, as already mentioned. Johnson emphasizes that the main loci of automobile testing are the laboratory and the track.[65] The inquiry will show that the availability of test tracks was an important element in all three regions. But these proving grounds are closed and very controlled environments. Therefore we propose to add a third space: Did the vehicles have been tested under *real-world conditions on public roads*? Did they even traverse all-terrain surfaces?

Third, the *speed* of the test vehicles is an indicator if the system worked well.[66]

Fourth, the *distance* that had been traversed between disengagements of the system is decisive – did the system only function with regular stops? A fifth question concerns especially the automobile research vehicles: When reports claimed that test runs were successful, we had to precise if longitudinal *and* lateral movements were automated or only one of them. And most importantly, we had to distinguish automated from autonomous vehicles.

As outcome of the detailed description of the actors, the historical and geographical or cultural situations and the technical capabilities we will be able to answer another question at the end of the inquiry: Is it possible to identify different research paradigms or historical periods of the research on automated and autonomous driving? Are these periods transnational or specific in each region and cultural context? An

---

[62] Weber 2014, in: http://www.computerhistory.org/atchm/where-to-a-history-of-autonomous-vehicles/

[63] Torchinsky, p. 109f.

[64] Torchinsky, p. 111.

[65] Johnson, p. 64.

[66] Often, contradictory speed indications can be found in the research literature – this had been rectified in direct contact with the researchers that participated in the projects.

important hypothesis that will be tested in this book is if a *shift from a guided to a seeing vehicle can be observed in the twentieth century*.[67] Finally we will try to answer the question if there are any lessons to be learned from history for the present development.

1.7 Sources and Methods

To answer these questions, the inquiry is based on a great amount of *primary sources* – scientific articles, monographies and films – from the engineers and practitioners of the laboratories that participated in the research.

These primary sources were retrieved by five methods: Our inquiry was started searching the online archive of the Institute of Electrical and Electronics Engineers (IEEE), one of the most important professional associations of engineers working in this field worldwide. Primary sources – articles in scientific journals – were found in a first step with the search keywords "Automated driving", "autonomous driving" "Unmanned ground vehicles" and "vehicle automation". In the second step, further articles were found by entering the names of engineers that made contributions to the field, for instance Ernst Dieter Dickmanns, Robert Fenton, Michel Parent, Steven Shladover, Chuck Thorpe, Sadayuki Tsugawa, Walter Zimdahl. Third, additional sources were found by online searches in databases, journals, books and the websites of the institutes. Fourth, further sources were retrieved by exploring the references cited in these articles. Fifth, important sources were sent directly to the author by practitioners who worked for several decades on the development of these technologies. Especially grey literature, for instance research reports were made available by these personal contacts.

Drawing on these primary sources it was our aim to present the concepts and functions of the different technological devices in a detailed but comprehensive form to assure accessibility for a broad audience. The publication should be interesting for scholars of history of technology, Science and Technology Studies, engineers working in the field, executives of the automotive industry, but also for political decision makers and media workers who are following the actual development.

[67] To avoid a common misunderstanding of the topic, we have to emphasize that vision is not only meant as optical vision in analogy to the human eye but as a process based on sensing, perception. In this definition, a seeing car is not only based on cameras, but also on sensors like radar, lidar, infrared detectors, ultrasonic range sensors and odometry.

1.8 Theoretical Background

The history of the research on automated driving can be examined from different perspectives. The next section has the aim to situate my research in the history of technology field and to highlight its theoretical background.

1.8.1 Artifacts and the Stairway-Model of Progress

Until the 1960s, the historiography of technology had been dominated by *internalistic* approaches from engineers who explored inventions, constructions, machines, material objects and scientific discoveries.[68] It was a history "from technicians for technicians" as the historian Günther Oetzel highlights.[69] A *deterministic* view of technology was dominant, conceptualizing society as determinated by technology. This purely *technical* history of technology was only interested in artifacts as they seemed to form a chain of progress, emphasizes the sociologist Otto Ulrich.[70] The historian Wolfgang König reveals that this notion of progress has a religious origin. It came from Christianity and signified a conversion to god. In modern times it had been secularized and meant mastering nature – and therefore it was associated to the development of technology.[71]

Often, this conventional historiography had the aim to glorify individual scientists and engineers. *The heroic figure of great men* who invented masterpieces of humankind were in the center of these studies.[72] The French philosopher of technology Pierre Musso called this conception of the creative genius the "romantic vision" of innovation.[73]

The history of computing can serve as an example for this kind of historiography as it was mostly written by participants of the computer community, not by historians. Computer scientists were mostly interested in establishing the firsts – the first digital computer, the first microprocessor etc. Another example is the history of the automobile that also was too often written as a history of progress, powered by a

---

[68] König, Wolfgang. 2009. *Technikgeschichte, Eine Einführung in ihre Konzepte und Forschungsergebnisse*, p. 46. Stuttgart.

[69] Gleitsmann, Rolf-Jürgen, Kunze, Rolf-Ulrich, Oetzel, Günther, *Technikgeschichte*, Konstanz 2009, p. 183.

[70] Ullrich, Otto. 1980. Der Charakter des Fortschritts moderner Technologien, Zur Geschichtsphilosophie der Technikgeschichtsschreibung. In *Technologie und Politik*, Issue 16, p. 47. Hamburg.

[71] König, p. 100.

[72] Oetzel, p. 220, König, p. 20, 40.

[73] Musso, Pierre, Laurent Ponthou and Éric Seuillet. 2007. *Fabriquer le futur 2, L'imaginaire au service de l'innovation*, p. 6. Paris.

succession or chain of precisely dated key inventions of great men as the historian of technology Kurt Möser complains.[74]

Since the 1960s, the history of technology made by engineers had been replaced by a history of technology made by historians.[75] Now, the succession of technical artifacts had not been seen only as result of technical research but economic, political and social influences. Causes and effects of technology had been considered in a broader view. The cultural context of technology became more and more important. Today, it is impossible to write a history of technology as path of great men and great innovations, as history of machines and technical objects only, emphasizes Bruno Jacomy. According to him, historical studies of technology should explore the big picture, they have to integrate the way of life, the social and historical context in their narration.[76] The genealogy of a technology and its use cases should be integrated in the research, emphasizes Günter Rohpol.[77]

The internalistic perspective with its over-simplifying stairway model of progress neglected the fact that technology can have very different outcomes and includes always different ways to use it. This means that historical events are contingent, open and not determined. According to Ulrich Wengenroth historians should privilege the notion of technological change over the notion of progress. Change is a less teleological and more open term.[78] For that reason we choose the term technological change in the title of our publication.

Computer historians like Michael S. Mahoney rejected the idea to unite all sources "into a single stream", reaching for example from Charles Babbage's analytical engine to George Boole's algebra as if their concepts are related.[79] Like Mahoney, Möser emphasizes that the history of the automobile and its technology cannot be written as a chain of innovations that are based on each other.[80] It would for instance be wrong to assert a genealogical connection between the ABS-devices of the 1930s and the 1970s.[81] The same statement can be made on behalf of guidance-cable (1950s) and machine vision concepts (1980s).

---

[74] Möser, Kurt. 2012. Innovationskulturen um das Automobil?, In *Innovationskulturen um das Automobil, Von gestern bis morgen, Stuttgarter Tage zur Automobil- und Unternehmensgeschichte*, ed. Rolf-Jürgen Gleitsmann, Jürgen E. Wittmann, Wiss. Schriftenreihe der Mercedes-Benz Classic Archive, Vol. 16, p. 26. Mühlacker/Mühlhausen.

[75] König, p. 54f.

[76] Jacomy, Bruno. 1990. *Une histoire des techniques*, p. 10. Paris.

[77] Rohpohl, Günter. 1999. *Allgemeine Technologie, Eine Systemtheorie der Technik*, München Wien, p. 56 (cit. in: König, p. 50).

[78] Wengenroth, Ulrich. 1997. Technischer Fortschritt, Deindustrialisierung und Konsum, Eine Herausforderung für die Technikgeschichte. In *Technikgeschichte* 64, p. 1–18.

[79] Mahoney, Michael S. 1988. The History of Computing in the History of Technology. *Annals of the History of Computing*, April–June 1988, 10(2): p. 5.

[80] Möser, p. 41.

[81] Johnson, Ann. 2009. *Hitting the Brakes, Engineering Design and the Production of Knowledge*, p. 25. Durham and London.

The historian of technology Anne-Françoise Garçon agrees with the idea that technological change is always driven by the intention to improve technology – but it should not be forgotten that what is right for an object or a specific procedure must not be right for the global evolution of technology that is influenced by regressive moments, useful steps back that may redirect the development in other directions.[82] David Edgerton has recently shown that the deployment of new technologies did not end the use of old technologies.[83]

Möser gives three examples why the stairway model of innovation is misleading: First, some innovations are introduced in the market but later they are discontinued.[84] Second, innovations can disappear and later reappear under different socio-technical conditions and user attentions (e.g. the curve light). As we will see, this mechanism can also be observed in the field of automated driving (e.g. guidance-cable). Third, innovations like the seat belt were known for a long time before they became standard. Resistance from users and the industry retarded their implementation.

If the stairway-model is not useful, how could we conceptualize instead the historical development of technology? According to Möser the interest of historians should be recentered on anonymous history opposed to the over personalized narration of inventions and inventors.[85] Instead of single inventions, "cultures of innovation", that means the reconstruction of long historical trends should be a main goal of historical research.

Further suggestions of theoretical and methodological frames for the analysis of innovation processes can be found in approaches of the Science and Technology Studies (STS). For example, the engineering studies scholar Michael S. Mahoney suggested that historians should look at the practices and artifacts of engineers. The STS-scholar Ann Johnson (1965–2016) published a case study – supervised by Mahoney – about the history of antilock braking systems (ABS), a technology that – like automated driving – had been developed by "many different actors, institutions, and artifacts embedded in an array of dynamic national, economic, and cultural contexts"[86] over several decades. Johnson decided to focus on the relationship between three elements – knowledge, artifacts and social communities.[87]

First, Johnson highlights that compared to scientific practices and knowledge, the work and knowledge production of engineers is "significantly understudied" and undertheorized.[88] The philosopher Davis Bairs showed in *Thing knowledge: A Philosophy of Scientific Instruments* (2004) that artifacts were mostly "ignored in

---

[82] Garçon, Anne-Françoise. 2012. *L'Imaginaire et la pensée technique, Une approche historique, XVIe-XXe siècle*, p. 17. Paris.

[83] Edgerton, David. 2008. *The shock of the old, Technology and global history since 1900.* London.

[84] For example, the Proconten system of Audi that retired the steering wheel from the chest of the driver was introduced in 1986 and discontinued in 1990.

[85] Möser, p. 50.

[86] Johnson, p. ix.

[87] Johnson, p. 14.

[88] Johnson, p. 6, p. 9.

epistemology".[89] For that reason, concepts from the history and philosophy of science cannot be directly applied to engineering. Johnson emphasizes the difference of scientific knowledge and engineering knowledge: "Engineering knowledge is a genuine kind of knowledge"[90] that "rarely takes the form of formal scientific theories"[91] – it often "resides in the artifacts" that are designed by engineers "or in the process of designing those artifacts". For that reason, a study of engineering "requires special attention to objects and the ways artifacts both bear knowledge and can constrain and shape knowledge."[92] According to Johnson, there are three types of knowledge that emerge around artifacts: "the knowledge that frames the initial steps of artifact design, the knowledge produced in the process of designing the artifact, and the knowledge needed for the artifact's use."[93]

Second, Johnson emphasizes the existence of a knowledge community in the research on ABS, meaning a "small informal community of practitioners that aggregates around a simultaneously developing question".[94] A knowledge community can be very diverse – "multilingual, multidisciplinary"[95] – but it needs a common language and needs to center on a common problem.

This theoretical perspective is very productive for our topic as Johnsons approach can be used to retrace the research history of autonomous driving that is also constituted over more than fifty years, focusing on concepts, prototypes, vehicles, hard- and software, driven by engineering communities around the world.

In the last decade, the history of technology field showed a renewed interest in the material culture, like Johnsons. Following this material turn, we decided to pay special attention to material objects, instruments and technical artifacts – guidance cables, magnetic nails, pick-up coils, odometers, actuators, speed measuring meters, computers, sensors (cameras, radar, lidar), but also *infrastructural things* like the Global Positioning System (GPS) and virtual things like software and maps. Things matter.

All these artifacts are nothing without being integrated and connected in mobile laboratories, research vehicles or mobile robots, a second category of artifacts that allow to traverse the space. With these laboratories on wheels, another spatial dimension is added to the experimentation – the test track.

In contrast to the field of ABS, the research on automated or autonomous driving has its origins in a much more diverse field. Research projects of electrical engineering, the automotive industry and military funded computer science and robotics institutes formed distinct knowledge communities. Adopting the concept of the knowledge community from Ann Johnson, it will be shown that at least three

[89] Johnson, p. 8.

[90] Johnson, p. ix.

[91] Johnson, p. 7.

[92] Johnson, p. 7.

[93] Johnson, p. 13.

[94] Johnson, p. ix.

[95] Johnson, p. 20.

distinct communities participated in the development of automated and autonomous driving: Highway engineers, roboticists and defense officials. Only in the mid-1980s a knowledge community on automated driving appeared. For example, actors like UniBwM did not know about the research efforts in Japan and in the US before this period. We will see that the different research fields converged at specific moments in history e.g. when Carnegie Mellon University participated in DARPA-funded AI robotics research, the Demo programs of the military from 1990 on and the automobile focused PATH demo 1997.

1.8.2 From Invention to Innovation

As the book is centered on technological inventions and innovations, a short discussion of these terms seems to be appropriate. Following a definition of David Edgerton, an invention is the *creation* of a new *idea* while an innovation is the *first use* of a new *device*.[96] A more precise definition of Joseph Schumpeter (1883–1950) differentiates both notions according to their capability to conquer a market.[97] An invention is the first description – in verbal form, as drawing, as model or prototype – of a new technology. It is always a solution for a problem. An innovation consists of three elements – the invention, the development of the technology until it is ready for the market and diffusion on the market. It can be resumed that an innovation is the introduction of new "things, ideas or ways of doing something" into the market, for example walkmans, videorecorders, copy-machines or the high-tensioned steel that allows the flexible shapes of modern cars.[98]

Inventions and patents alone cannot explain the dynamic of technology development – only a small part of them lead to innovations in the sense of products on the market. Therefore inventions have no longer a central place in the history of technology – most modern studies of technology put innovations in the center of their interest.[99]

The following study suggests differentiating this perspective: On the one hand it is true that the *inventions* made during 40 years of research on vehicle automation did not lead to the market introduction of fully autonomous vehicles. They did not lead to a *rupture innovation*, defined by the historian of technology Bertrand Gille as a new technological system, a new *sociocultural and economic ensemble* (like the internet). The research history that is presented in this book – mostly basic or fundamental research – was not driven by the goal to achieve immediately a

---

[96] Edgerton, p. ix.

[97] Schumpeter, Joseph A., *Konjunkturzyklen, Eine theoretische, historische und statistische Analyse des kapitalistischen Prozesses*, Göttingen 1961 (see also König, p. 57, 60).

[98] Nagano, Hiroshi, 2012. Science and Technology Policy in Japan. In *The Dynamics of Regional Innovation, the Policy Challenges in Europe and Japan*, ed. Lecler, Yveline, Yoshimoto, Tetsuo and Fujimoto, Takahiro, p. 33. Singapore.

[99] König, p. 60.

marketable innovation. The Intelligent Vehicle in Japan, the Navlab in the US and VaMoRs in Germany – to name only a few – were experimental devices. Their primary goal was not to be developed into products sold on the market. Their primary goal was a scientific contribution to basic research. In other words: Most of the artefacts that are described in this book are closer to inventions than to innovations.

But on the other hand, these inventions contributed to *incremental innovations* that improved the performance of existing devices like driver assistance systems. The combination of these incremental innovations could one day lead to the *rupture innovation* of an autonomous vehicle: It would not only improve the main elements of cars as they are known since over 100 years, it would create a completely different technical system.

1.9 Outline of the Study

After the presentation of the theoretical background we will present the outline of our inquiry: The book is structured in three chronologically ordered *parts* with different *chapters* that contain different *sections*. The interior structure of each part follows a more loosely chronological order because the concepts of different actors – academia and car industry working on automated highway systems, robotics and military actors researching unmanned ground vehicles – were gathered in separate subchapters. These subchapters then follow their own chronological order.

As most of the different approaches had been influenced by similar efforts in other countries, a systematic instead of a geographical and chronological structure of the book would also have been possible. The decision of a geographical structure of the publication had been chosen for practical reasons only. As a multitude of different projects had to be presented, it seemed to be useful to avoid jumps between different countries as much as possible. This should *not* give the impression that the research efforts have more national than transnational references. To highlight their connections and juxtapositions, it had been privileged to work with cross references to activities in other countries.

The **first part** of the book presents the research on vehicle automation in the United States. Following a chronological order, the early guidance cable approaches and the research on an Automated Highway System are presented. This is followed by a portrait of different mobile robots. After a detailed presentation of defense related projects, we will come back to the AHS-development of the 1990s.

As Japan followed quickly with its own projects, the **second part** focuses on the technology development in this country. Japan was one of the first countries in Asia who started experiments on automated driving. The first chapters introduce MITI, a main actor of the research policy in Japan. The increasing number of traffic accidents, a factor that motivated the research on vehicle automation, is also mentioned. The main part presents the first guidance-cable experiments of the Government Mechanical Laboratory during the 1960s. The next chapters retrace the shift to Machine Vision-concepts in the 1970s. At this time, the Mechanical Engineering

Laboratory worked on the first visually guided autonomous vehicle, presented in 1977. It recorded and processed (on-board) pictures of lateral guide rails at the side of the road via two vertically aligned cameras.

Europe is presented in the **third part** as the most important projects were started later in the twentieth century. Its emphasis is on visually guided vehicles with on board computing, developed by Ernst Dieter Dickmanns and his team at the University of the Federal Armed Forces in Munich in Germany. The first chapter details the early experiments on automated driving in Europe, focusing on guidance-cable research in England and Germany during the 1960s. The second chapter presents the shift to vision systems in Europe at the end of the 1970s, presenting the mobile robot HILARE in France and experiments with dynamic machine vision at UniBwM in Germany. Experiments of the automotive industry in visual guidance experiments are also presented (Volkswagen Opto-Pilot). The third chapter is dedicated to the concepts, research vehicles and milestones of UniBwM in Germany, especially the test-drives with VaMoRs and the 4-D approach of Ernst Dieter Dickmanns. The fourth chapter is concentrated on the PROMETHEUS-program of the European Union (1988–1994). It gives a portrait of the research vehicles VITA-II and VaMP. After a paragraph about the second generation vision-system of VaMoRs the performances of VaMP and VITA-II at the final PROMETHEUS-demo in Paris (1994) are detailed, followed by an overview about the state of the art of other research projects in France, the UK and Germany during this time. The fifth chapter is focused on the research efforts that were undertaken after PROMETHEUS (1995–2000), especially in France (Praxitèle, Cycab and La Route Automatisée). The next paragraphs will highlight two important demonstrations, VaMP's long-distance drive from Munich to Odense (1995) and the Mille Miglia in Automatico of the University of Parma in Italy (1998).The sixth chapter will go back in the chronology and present the defense related activities in Europe. Starting with the French military project DARDS, the experimental robotics-program PRIMUS in Germany is presented together with the AutoNav-agreement (1997–2003) and the third-generation vision system (EMS) of UniBwM.

Bibliography

Albert, Dan. 2018. *Are We There Yet?: The American Automobile Past, Present, and Driverless*. New York.

Bidwell, J.B. 1959. Vehicles and Drivers – 1980, General Motors Research Laboratories, Presented at SAE Annual Meeting, Detroit.

Bishop, Richard. 2001. Whatever Happened to Automated Highway Systems (AHS)? Originally published in the August–September 2001 issue of Traffic Technology International, online in: https://staff.washington.edu/jbs/itrans/bishopahs.htm. Accessed on July 3, 2019.

Burns, Lawrence D. 2018. *Autonomy: The Quest to Build the Driverless Car – And How It Will Reshape Our World (Together with Christopher Shulgan)*. London.

Carey, James, and John Quirk. 2009. The History of the Future. In *Communication as Culture: Essays on Media and Society*. New York, London.

Davies, Alex. 2021. *Driven, The Race to Create the Autonomous Car*, New York, London, Toronto, Sydney, New Delhi.
Delcker, Janosch. 2018. The man who invented the self-driving car. *Politico*, July 19, 2018, online in: https://www.politico.eu/article/delf-driving-car-born-1986-ernst-dickmanns-mercedes/.
Dickmanns, Ernst Dieter. 1986. Computer Vision in Road Vehicles – Chances and Problems. *ITCS-Symposion on "Human Factors Technology for Next-Generation Transportation Vehicles"*, Amalfi, Italy.
———. 1993. Interview, Der Lotse geht an Bord. *Hochschulkurier*, pp. 6–7.
———. 2002. The Development of Machine Vision for Road Vehicles in the Last Decade. *Intelligent Vehicle Symposium* 1: 268–281.
———. 2016. Summary of contributions to Vision-based Autonomous Vehicles. http://www.dyna--vision.de/main/Review2016%20of%20ContribVisBasAutVeh.html.
Edgerton, David. 2008. *The shock of the old, Technology and global history since 1900*. London.
Eliot, Lance. 2018. Introduction to Driverless Self-Driving Cars: The Best of the AI Insider, (no place of publication mentioned).
Fallon, Michael. 2019. *Self-Driving Cars, the New Way Forward*. Minneapolis.
Feitz, Anne. 2019. Premiers coups de frein sur la voiture autonome. *Les Echos*, March 26, 2019, online in: https://www.lesechos.fr/tech-medias/intelligence-artificielle/premiers-coups-de-frein-sur-la-voiture-autonome-1003606. Accessed on July 3, 2019.
Fenton, Robert E. 1994. IVHS/AHS: Driving into the Future. *IEEE Control Systems Magazine* 14 (6): 13–20.
Garçon, Anne-Françoise. 2012. *L'Imaginaire et la pensée technique, Une approche historique, XVIe-XXe siècle*. Paris.
General Motors. 1940. Futurama, Exhibition brochure, USA.
Gleitsmann, Rolf-Jürgen, Rolf-Ulrich Kunze, and Günther Oetzel. 2009. *Technikgeschichte*, Konstanz.
Graefe, Volker, and Klaus-Dieter Kuhnert. 1992. Vision-Based Autonomous Road Vehicles. In *Vision-Based Vehicle Guidance*, ed. Ichiro Masaki, 1–29. New York.
Hawkins, Andrew J. 2019. Waymo's Driverless Car: Ghost-Riding in the Back Seat of a Robot Taxi. *The Verge*, December 9, 2019, online in: https://www.theverge.com/2019/12/9/21000085/waymo-fully-driverless-car-self-driving-ride-hail-service-phoenix-arizona. Accessed September 3, 2020.
Herrmann, Andreas, Walter Brenner, and Rupert Stadler. 2018. *Autonomous Driving: How the Driverless Revolution will Change the World*. Bingley.
Heßler, Martina. 2019. Technik und Autonomie, Kulturhistorische Bemerkungen zu einem komplexen Verhältnis. In *Autonome Systeme und Arbeit*, ed. Hartmut Hirsch-Kreinsen and Anemari Karacic, 247–274. Bielefeld.
Jacomy, Bruno. 1990. *Une histoire des techniques*. Paris.
Jamthe, Sudha. 2017. 2030 The Driverless World: Business Transformation from Autonomous Vehicles, (independently published).
Johnson, Ann. 2009. *Hitting the Brakes, Engineering Design and the Production of Knowledge*. Durham and London.
Kelly, Fred C., ed. 2002. *Miracle at Kitty Hawk, The Letters of Wilbur & Orville Wright*. New York.
König, Wolfgang. 2009. *Technikgeschichte, Eine Einführung in ihre Konzepte und Forschungsergebnisse*. Stuttgart.
Kröger, Fabian. 2012. Fahrerlos und unfallfrei. Eine frühe automobile Technikutopie und ihre populärkulturelle Bildgeschichte. In *Technology Fiction, Technische Visionen und Utopien in der Hochmoderne*, ed. Uwe Fraunholz and Anke Woschech, 93–114. Bielefeld.
———. 2014. Die Zukunft, die nie eintrat, Über die Nützlichkeit filmischer Zukunftsbilder des autonomen Fahrens für die Prospektive. *Technology Assessment – Theory and Practice (TATuP)*, N° 1/2014, Verkehrszukünfte – Visionen jenseits aktueller öffentlicher Aufmerksamkeit, pp. 3–6.
———. 2015. Das automatisierte Fahren im gesellschaftsgeschichtlichen und kulturwissenschaftlichen Kontext. In *Autonomes Fahren, Technische, rechtliche und gesellschaftliche Aspekte*, ed. M. Maurer, J.C. Gerdes, B. Lenz, and H. Winner, 41–67. Heidelberg.

———. 2016a. Automated Driving in its Social, Historical and Cultural Contexts. In *Autonomous Driving, Technical, Legal and Social Aspects*, ed. M. Maurer, J.C. Gerdes, B. Lenz, and H. Winner, 41–68. Heidelberg.

———. 2016b. Le long cheminement de l'imaginaire cinématographique de la voiture autonome. In *Choc de Mobilités*, ed. Mathieu Flonneau and Stéphane Levesque, 55–67. Paris.

Latour, Bruno. 1992. *Aramis ou l'amour des techniques*, Paris.

Lutz, Randolph B. 1966, April. *The Motor Vehicle of the Future*. Michigan Department of State Highways, Technical Report Number 2.

Mahoney, Michael S. 1988, April–June. The History of Computing in the History of Technology. *Annals of the History of Computing* 10 (2): 113–125.

———. 2005. The Histories of Computing (s). *Interdisciplinary Science Reviews* 30 (2): 119–135.

McGrath, Michael E. 2018. *Autonomous Vehicles: Opportunities, Strategies, and Disruptions* (independently published).

Minx, Eckard, and Rainer Dietrich. 2015. *Autonomes Fahren, Wo wir heute stehen und was noch zu tun ist*, 38–55. Ladenburg.

Mom, Gijs. 2004. *The Electric Vehicle, Technology and Expectations in the Automobile Age*. Baltimore, London.

Möser, Kurt. 2012. Innovationskulturen um das Automobil?, In *Innovationskulturen um das Automobil, Von gestern bis morgen, Stuttgarter Tage zur Automobil- und Unternehmensgeschichte*, ed. Rolf-Jürgen Gleitsmann, Jürgen E. Wittmann, Wiss. Schriftenreihe der Mercedes-Benz Classic Archive, Vol. 16, 25–50. Mühlacker/Mühlhausen.

Musso, Pierre, and Laurent Ponthou, Éric Seuillet, ed. 2007. *Fabriquer le futur 2, L'imaginaire au service de l'innovation*. Paris.

Nagano, Hiroshi. 2012. Science and Technology Policy in Japan, In *The Dynamics of Regional Innovation, the Policy Challenges in Europe and Japan*, ed. Yveline Lecler, Tetsuo Yoshimoto, and Takahiro Fujimoto, Singapore.

Ohnsman, Alan. 2018. At Waymo, It's Launch Time for Google's Biggest Moonshot, in: *Forbes*, November 13, 2018, online in: https://www.forbes.com/sites/alanohnsman/2018/11/30/10-million-miles-later-waymo-ceo-says-its-self-driving-cars-are-ready-for-prime-time/#222ddf675e7e.

Packer, Jeremy. 2008. Automobility and the Driving Force of Warfare: From Public Safety to National Security. In *The Ethics of Mobility, Rethinking Place, Exclusion, Freedom and Environment*, ed. Sigurd Bergmann and Tore Sager, 39–64. Hampshire, Burlington.

Rescher, Nicholas. 1998. *Predicting the Future: An Introduction to the Theory of Forecasting*. New York.

Schmidt, Herbie. 2020. Audi steckt beim autonomen Fahren zurück. In *Neue Zürcher Zeitung*. April 28, 2020, online in: https://www.nzz.ch/mobilitaet/auto-mobil/autonomes-fahren-stufe-3-audi-verzichtet-im-a8-auf-staupilot-ld.1553933. Accessed May 24, 2020.

Schwartz, Samuel I. 2018. *No One at the Wheel: Driverless Cars and the Road of the Future*. New York.

Sfez, Lucien. 2002. La technique comme fiction, *Revue européenne des sciences sociales*, XL-123, published December 3, 2009, online in: http://journals.openedition.org/ress/612. Accessed August 23, 2020.

Shladover, Steven. 2007. PATH at 20 – History and Major Milestones. *IEEE Transactions on Intelligent Transportation Systems* 8 (4): 584–592.

Shladover, Steven E., Jane Lappin, Robert P. Denaro, and Bryant Walker Smith. 2018. Introduction: The Transportation Research Board's 2013 Workshop on Road Vehicle Automation. In *Road Vehicle Automation*, ed. Gereon Meyer and Sven Beiker, 1–12.

Simoudis, Evangelos. 2017. *The Big Data Opportunity in Our Driverless Future*. Menlo Park.

Sperling, Daniel. 2018. *Three Revolutions: Steering Automated, Shared, and Electric Vehicles to a Better Future*. Washington, DC.

Tessier, Catherine. 2015. Autonomie: enjeux techniques et perspectives. In *Drones et killer robots: faut-il les interdire?*, ed. Ronan Doaré, Didier Danet, Gérard de Boisboissel, Rennes.

Torchinsky, Jason. 2019. *Robot, Take the Wheel: The Road to Autonomous Cars and the Lost Art of Driving*. New York.

Tsugawa, Sadayuki. 2008. A History of Automated Highway Systems in Japan and Future Issues, Institute of Electrical and Electronics Engineers. *International Conference on Vehicular Electronics and Safety*, Ohio, USA, September 22, 2008, p. 2–3.

Tuchman, Barbara. 1982. In Geschichte denken, Essays. Düsseldorf.

Ullrich, Otto. 1980. Der Charakter des Fortschritts moderner Technologien, Zur Geschichtsphilosophie der Technikgeschichtsschreibung. *Technologie und Politik*, Issue 16, Hamburg, pp. 23–51.

Urry, John. 2004. The 'System' of Automobility. *Theory, Culture and Society* 21 (4/5): 25–39.

Vaid, Kamal. 2018. *Selbst ist das Auto – automatisiertes und autonomes Fahren*, Die Zukunft der Mobilität, Schriftenreihe der Northern Business School zur angewandten Wissenschaft, Band 10, Hamburg.

Wadhwa, Vivek, and Alex Salkever. 2019. *The Driver in the Driverless Car: How Our Technology Choices Will Create the Future,* Oakland 2017. 2nd ed.

Weber, Marc. 2014. Where To? A History of Autonomous Vehicles, May 8, 2014, in: http://www.computerhistory.org/atchm/where-to-a-history-of-autonomous-vehicles/.

Wengenroth, Ulrich. 1997. Technischer Fortschritt, Deindustrialisierung und Konsum, Eine Herausforderung für die Technikgeschichte. *Technikgeschichte* 64: 1–18.

Wetmore, Jameson M. 2003. Driving the Dream, The History and Motivations Behind 60 Years of Automated Highway Systems in America. *Automotive History Review*, Summer 2003, pp. 4–19.

———. 2020. Reflecting on the Dream of Automated Vehicles, Visions of Hands Free Driving Over the Past 80 Years. *Technikgeschichte* 87 (1): 69–94.

Chapter 2
History of the Research on Vehicle Automation in the United States

Following a chronological order, the first part of this book will present the historical context, the aims and different technical concepts of the research on automated and autonomous driving in the United States.

The first chapter starts with early experiments on Automated Highway Systems. The efforts of the electronics and automobile industry (Radio Corporation of America, General Motors) on cable-based vehicle guidance during the 1950s are presented. In the 1960s, public institutions like the Ohio State University joined the efforts to realize an Automated Highway System with a long-term research program. These experiments were dominated by approaches of electrical engineering. Another outcome was a Personal Rapid Transport-System with guidance-cables, developed by Bendix.

While the first chapter focuses on automobile guidance systems, the following chapters are centered on a very different type of vehicle – mobile robots that became much more relevant for *autonomous* driving than the guidance-cable experiments. The second chapter centers on academic research on mobile robots since the 1960s. It starts with a section about the beginning of artificial intelligence and a short historical portrait of the Defense Advanced Research Projects Agency (ARPA/DARPA) that funded most of the robotics research. The following sections present two mobile robots: The Stanford Cart, built by the Stanford Mechanical Engineering Laboratory, was conceived with the aim to experiment teleoperation for space missions. Later, the Stanford Artificial Intelligence Lab (SAIL) used the Stanford Cart for the exploration of computer vision and obstacle detection. This robot followed a bottom-up-approach – based on sensory data, it oriented itself in the world. Shakey, a robot developed by Stanford Research Institute followed the top-down-approach – it aimed to understand its environment by reasoning.

The third chapter presents a series of robot vehicles that had been developed at Carnegie Mellon University in Pittsburgh during the 1980s and 1990s. Beginning with the CMU Rover and the Terregator, it focuses on the different capabilities of the NavLab research vehicles (1986–1995).

F. Kröger, *From Automated to Autonomous Driving*, Archimedes 70,
https://doi.org/10.1007/978-3-031-49881-7_2

The next chapters will make a shift to defense related research: After a fourth chapter about the Autonomous Land Vehicle (ALV) of DARPA (1983–1989), the Demo I-III experiments of the US-Army Research Lab in the 1990s are presented in the fifth chapter.

The sixth chapter will go back to the AHS-development and discuss the Demo '97 in San Diego with its platooning and Free Agent demonstrations. The Navlabs 6–10 from CMU will be presented here.

2.1 Early Research on Automated Highway Systems

The idea of an automated highway system goes back to the Futurama exhibition of General Motors at the World's Fair 1939 in New York. The industrial designer Norman Bel Geddes designed a giant model landscape, demonstrating an automatic traffic system of tomorrow, guided by radio transmissions.[1] This was the first time that the idea of an Automated Highway System (AHS) – a highway with designated lanes for specially equipped vehicles that are operated under fully automatic control[2] – had been promoted publicly. In this system, the control of all driving functions is moved from the driver to a cooperation of vehicle and infrastructure. In 1940 Bel Geddes proposed in his book *Magic Motorways* to equip roads with cog tracks, moving forward cars equipped with cogwheels.[3] He also described a system of radio-controlled cars propelled by electromagnetic fields. But until the end of World War II, these ideas were not developed technically and remained abstract concepts.

The following chapter will present the first technical concepts of automated highway systems that had been developed by electrical engineers of the Radio Corporation of America (RCA) and General Motors in the 1950s.[4] It will then discuss the automatic steering and control experiments for a network of vehicles, undertaken at the Bureau of Public Roads (BPR) and the Ohio State University in the 1960s and 1970s.

[1] For a more detailed description see: Kröger, Fabian, 2016. Automated Driving in its Social, Historical and Cultural Contexts. In *Autonomous Driving, Technical, Legal and Social Aspects*, ed. Maurer, M., Gerdes, J.C., Lenz, B., Winner, H., 48–53. Heidelberg.

[2] Rillings, Jim. 1997, February. *The National Automated Highway System: Cooperative Program*. National Automated Highway System Consortium, p. 4.

[3] Bel Geddes, Norman, 1940. *Magic Motorways*, p. 100. New York.

[4] Zworykin, V.K., and L.E. Flory. 1958. Electronic Control of Motor Vehicles on the Highway. *Highway Research Board Proceedings* 37: 436–451.

2.1.1 *The First Guide-Wire Prototype of GM and RCA (1953–1960)*

The first technical concept for automated driving had been developed by the electronics and automotive industry. As highway automation was based on the idea that vehicles should receive and send information, it is not surprising that the first prototype was created by a radio company:

The television pioneer Vladimir Zworykin (1889–1982) from the Radio Corporation of America (RCA) Laboratories in Princeton, New Jersey, started to work on highway automation in 1953.[5] He was persuaded that *car-based* sensor systems like "anti-collision-radar, ultrasonic devices, and infrared-detectors" *alone* would not be sufficient to solve the problem of highway accidents and roadway congestion. The only system satisfying all requirements should be installed "in the road itself", the infrastructure.

Zworykin's approach was based on two components, a lead cable for lateral control and rectangular wire loops buried in the pavement for longitudinal control: Following a "cable carrying high-frequency current in the center of the lane" – a concept directly translated from railroad tracks – vehicles should assure lane-keeping automatically. Six by twenty feet long detector loops in the road should avoid collisions by sensing the presence and speed of a car and issue a warning signal to the following vehicle: "As a vehicle passes over any loop, the inductance of the loop is altered by the metal of the vehicle", explains Zworykin.[6] This signal will then be detected by an "electronic circuit" beneath the road. It will generate a "flying tail" of warning signals that will actuate accelerator and brakes of the next incoming vehicle.

Zworykin emphasizes that the cables in the road could also be used as "communication system". The highway authority could issue warnings directly to the driver, receive "a complete, instantaneous representation of traffic conditions" and measure if all vehicles respect the speed limitations.

Zworykin envisaged in 1959 that the system could be developed in three stages. First, only the road would be equipped. "Detectors actuate warning lights" about vehicles ahead when the visibility is obscured by a hill, a curve or fog (Fig. 2.1). In the next stage, the vehicles should be equipped to detect these warnings. Visual or auditory signals should be given to the driver. In the third stage, vehicles will automatically adjust the steering and speed-control-mechanisms. Zworykin emphasizes that these three stages could also "coexist for many years".[7]

---

[5] Zworykin, Vladimir K. 1959. The Human Aspect of Engineering Progress. The Fourth Clerk Maxwell Memorial Lecture, July 1, 1959, Cavendish Laboratory, Cambridge. *Journal of the British Institution of Radio Engineers*, September 1959, p. 537.

[6] Ibid., p. 538.

[7] Ibid., p. 539.

Fig. 2.1 An illustration of the warning system of RCA, 1961. (Credits: RCA. Image Source: Flory (1961), p. 272)

Directed by Zworykin, this electronic highway control system was developed by the RCA-engineers Leslie E. Flory, George W. Gray and Winthrop S. Pike.[8] In 1953, the RCA-team demonstrated for the first time a small-scale system of miniature cars (1:5) on the laboratory floor of the RCA Center. According to the historian James Wetmore, it was developed with General Motors (GM)[9] who started in 1956 under the direction of J. B. Bidwell to work on automation concepts. In 1956, GM's promotional show Motorama popularized the guide-wired highway with the Firebird II concept car. Soon, public authorities also became interested in the idea: According to the RCA-Journal *Electronic Age*, the traffic engineer Leland M. Hancock and his director, L. N. Ress, state engineer of the Department of Roads of the State of Nebraska became inspired by an article about the experiment in Colliers Magazine.[10] They decided to test the system in cooperation with RCA on a real highway.

The first complete road installation of the system was demonstrated on October 10, 1957 on a 400-foot (122 m) strip of a newly-constructed highway intersection of US Route 77 and Highway 2 on the outskirts of Lincoln, Nebraska.[11] Nearly one hundred state and federal officers, representatives of the automotive industry and media members attended the RCA demo.[12] As the car followed an electronic track, press coverage pointed to the similarities of the Nebraska and GM test runs to railroad principles.[13]

[8] author unknown, Highway of the future. In *Electronic Age (RCA- Journal)*, January 1958, Vol. 17, No. 1, p. 13.

[9] Wetmore 2003, p. 6.

[10] author unknown, Highway of the future. In *Electronic Age (RCA- Journal)*, January 1958, Volume 17, Number 1, p. 13.

[11] Zworykin 1959, p. 540, and author unknown, Highway of the future. In *Electronic Age*, January 1958, Volume 17, Number 1, p. 12f.

[12] Fallon, Michael. 2019. *Self-Driving Cars, the New Way Forward*, p. 30. Minneapolis.

[13] Mann, Martin, 1958 May. The Car That Drives Itself. *Popular Science*, 172, pp. 75–79 and 226–227.

During this experiment, only lateral control – lane-keeping – was automated with the guide-wire. This event did not show automatic longitudinal control – a driver had to adjust the acceleration manually. But "it did replace the human driver's need for vision" as Torchinsky observed.[14] According to *Electronic Age*, remotely operated warning lights issued visual and audible warnings to the driver when he followed another vehicle too closely. The windshield of the Chevrolet Bel Air test-vehicle was covered and the driver adjusted the speed according to a swinging needle on a M1 vehicle detector meter[15] in the interior, designed by Leland Hancock. A video from the test run confirms this configuration.[16] It is obvious that this experiment was dangerous: At one moment, a test driver had to disengage the system – he had forgotten to flip a switch and was about to crash into a car ahead of him.[17] But from our contemporary perspective the experiment is important because it was the first demonstration of an infrastructure-based driver assistance system in a real highway situation and an early version of a smart road. RCA was confident that automatic highways would be realized 20 years later, in 1975.

On February 14, 1958, General Motors tested the guidance function on a full-scale test-track at the Technical Center of GM in Warren, Michigan. Presented in a press release as the first "automatically guided automobile",[18] a 1958 Chevrolet followed a guidance cable of one mile at low speeds, up to 15.65 m/s (35 mph).[19]

On June 5, 1960, the system was demonstrated at the quarter-mile oval test track at the David Sarnoff headquarter of RCA in Princeton, New Jersey. Reporters were invited to take a ride in two Chevrolet Impala test-vehicles, one without a steering wheel but a joystick called Unicontrol (Fig. 2.2).[20] Beneath the track, GM presented also a scale model of the system (1:40).

A chain of transistorized detector circuits was installed along the roadside (see Fig. 2.3), each connected to rectangular wire loops in the road. This system could sense the position, direction and speed of any metallic vehicle on the surface. Two

---

[14] Torchinsky, Jason. 2019. *Robot, Take the Wheel: The Road to Autonomous Cars and the Lost Art of Driving*, p. 60. New York.

[15] Different vehicle detectors from this test run are now in the stock of the Nebraska State Historical Society (see Nebraska State Historical Society, Online Archive, online in: https://nebraskahistory.pastperfectonline.com/search?utf8=✓&search_criteria=%22automatic+driving%22&searchButton=Search (accessed on June 18, 2019)).

[16] Image Source: JournalStarNews (Youtube-account), Driverless Car (video title). In Youtube, uploaded on September 13, 2017, online in: https://www.youtube.com/watch?v=Eufcsa46gGk (accessed on June 13, 2019).

[17] Fallon, Michael, 2019. *Self-driving cars, The new way forward*, p. 30. Minneapolis.

[18] General Motors: "An automatically guided automobile cruised along a one-mile check road at General Motors technical Center today…", Press release, USA 1953.

[19] Fenton, Robert E., and Robert J. Mayhan. 1991 February. Automated Highway Studies at the Ohio State University – An Overview. *IEEE Transactions on Vehicular Technology* 40 (1), p. 107. (citing: Zworykin, Flory, 1958 and Gardels, K. 1960. *Automatic car controls for electronic highways*. General Motors Research Laboratories, Rep. GMR-276. Warren, MI).

[20] Ingraham, Joseph C. 1960. Electronic Roads Called Practical; New System of Guiding Cars Safely on Highways Is Shown at Princeton Fruit of 7 Years' Study R.C.A. and G.M. Jointly Conducted It – Full Use Seen 15 Years Away. *The New York Times*, June 6, 1960.

Fig. 2.2 RCA-demo with two Chevrolet Impala in Princeton on June 5, 1960. The Unicontrol-Joystick is positioned on the left hand near the driver's seat. (Credits: RCA. Image Source: Kurman 2017)

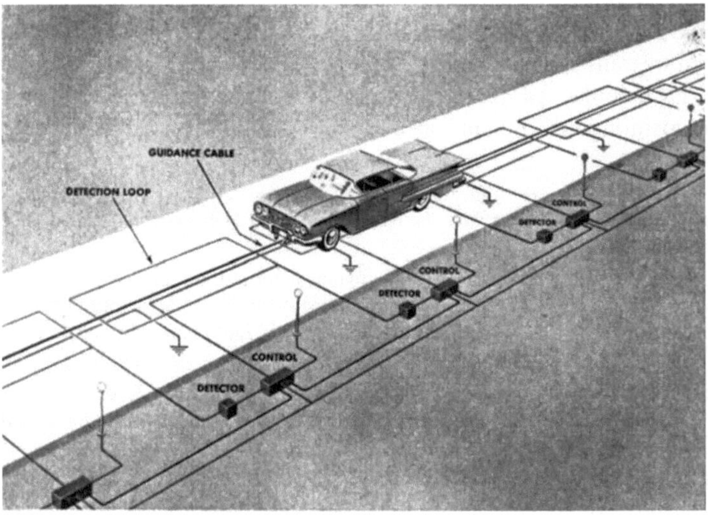

Fig. 2.3 Graphical illustration of the guidance cable system. (Credits & Image Source: Flory, L. E., Electronic Control of Highway Vehicles, May 1961, p. 271)

meters on the dashboard showed the speed and the distance to the preceding car. The press announced it as "smart road", a term that is in use again today.[21] According to Ümit Özgüner et al. the test vehicles of GM showed automated "lane tracking, lane changing, and car following functions".[22] But we have not been able to find any evidence for lane changing functions. The system was also very limited in two other aspects: First, James Wetmore points out that the driver was not completely taken out of the loop because the system of GM was only envisioned for highways. A human driver had to retake the steering wheel in rural areas and cities.[23] Second, the high investments that would be necessary to realize such a system were a hindrance: RCA officials estimated that the system would add about 10% to the cost of building the road – "$100.000 per mile over the complete course."[24] But at the beginning of the 1960's, this was state of the art in vehicle automation: At the same time, the Road Research Laboratory in Great Britain and the Government Mechanical Laboratory in Japan also started tests with guidance cables (see Sects. 3.2 and 4.1).

2.1.2 Vehicle Control Studies at Ohio State University (1964–1980)

The guide wire concept, first invented and tested by the electronics and automobile industry, became in the next decades a topic of academic research. From 1964–1980, an ambitious program on what is called today Automated Highway Systems (AHS) was conducted at Ohio State University (OSU) by the electrical engineer Robert E. Fenton under the direction of Robert L. Cosgriff, head of the university's Communication and Control Systems Laboratory.

The research efforts were motivated by the growing number of vehicles and car accidents in the United States. The principal aim of the program was the development of "controllers for both longitudinal and lateral vehicle control."[25] It was funded by the Ohio Department of Highways and the Bureau of Public Roads (BPR), later the Department of Transportation (DOT) and finally by the Federal Highway Administration (FHWA). It became part of a large AHS program that was started by BPR at the end of the 1960s.[26]

---

[21] Quigg, Doc. 1960, June 7. Reporter Rides Driverless Car. In *The Press-Courier*.

[22] Özgüner, Ümit, Tankut Acarman, and Keith Alan Redmill. 2011. *Autonomous Ground Vehicles*, p. 3. Boston, London.

[23] Wetmore 2020.

[24] Kilbon, Kenyon. 1960, Autumn. Tomorrow's Thruway is Here Today! *Electronic Age*, p. 29.

[25] Fenton, Mayhan, 1991, p. 100.

[26] Fenton, Robert E. 1994 December. IVHS/AHS: Driving into the Future. *IEEE Control Systems Magazine* 14 (6), p. 14.

The work of Fenton is regularly mentioned in the literature on automated driving but a more detailed presentation is missing so far.[27] For that reason, the following paragraph will give a chronological overview about the main technical concepts conceived at OSU.

The investigation of car-following systems started already in 1962.[28] The first control studies "were focused on the situation where a vehicle is following a lead vehicle under quasi-steady-state conditions",[29] that means without speed changes.

From 1964–1971, the OSU-program studied longitudinal control in an *asynchronous* environment – a network of cars without a system-wide synchronizing signal. In this scenario, a vehicle would be able to enter an automated highway system at any moment, only following other controlled vehicles. During field-tests the overtaking of a lead-car, emergency braking and car following were tested at speeds "up to 30 m/s", reports Fenton.[30]

In November 1967 Ohio State University tested a longitudinal control system with a 1965 Plymouth sedan.[31] Braking, acceleration and steering were actuated by an electrohydraulic closed-loop control system. Taking the control of the throttle valve as example, Fenton explains: "The input signal was compared with a signal from a potentiometer coupled to the system output – the position of the throttle valve. The resulting error signal was applied to an electrohydraulic control valve, which in turn controlled a hydraulic actuator connected to the throttle valve."[32] The same system was applied to the brake and steering mechanisms.

An analog computer with 9 solid-state operational amplifiers and 8 potentiometers for data collection was installed in the trunk of the car.[33] The data was recorded on a four-channel FM magnetic-tape recorder and later reduced on an analog computer in the laboratory.[34] A gasoline-driven motor-generator on top of the research vehicle was used to supply power to the on-board analog computer and the recording equipment.[35]

---

[27] Shladover, Steven E., Jane Lappin, Robert P. Denaro, and Bryant Walker Smith. 2018. Introduction: The Transportation Research Board's 2013 Workshop on Road Vehicle Automation. In *Road Vehicle Automation*, ed. Gereon Meyer and Sven Beiker. Cham, Heidelberg, New York, Dordrecht, London, p. 7 and Özgüner, Ümit, Tankut Acarman, and Keith Alan Redmill. 2011. *Autonomous Ground Vehicles*, p. 3. Boston, London.

[28] Fenton, Robert E. 1970 February. Automatic Vehicle Guidance and Control – A State of the Art Survey. *IEEE Transactions on Vehicular Technology VT* 19 (1), p. 155.

[29] Fenton, Mayhan 1991, p. 102.

[30] Fenton, Mayhan 1991, p. 103 and Fenton 1994, p. 18.

[31] Fenton, Robert E., Robert L. Cosgriff, Karl W. Olson, and Lyle M. Blackwell. 1968 April. One Approach to Highway Automation. *Proceedings of the IEEE* 56 (4), p. 559 and Bender, James G., Robert E. Fenton, and Karl W. Olson. 1971 November. An Experimental Study of Vehicle Automatic Longitudinal Control. *IEEE Transactions on Vehicular Technology* VT-20 (4), p. 118.

[32] Bender, James G., and Robert E. Fenton. 1969 November. A Study of Automatic Car Following. *IEEE Transactions on Vehicular Technology* VT-18 (3), p. 136.

[33] Fenton et al. 1968, p. 559.

[34] Bender et al. 1971, p.118.

[35] Fenton et al. 1968, p. 559.

A guide-wire was used for the lateral control of the vehicle, but the velocity and longitudinal distance between two vehicles also had to be controlled. Unfortunately, at the end of the 1960s, a sensor for a car-following control system was not available. A vehicle-mounted Doppler radar and an infrared system were in development,[36] but it was obvious that they "would not be available for many years".[37] This is a very important point because a technological substitute for the missing sensor had to be found.

In the first part of the testing program, *longitudinal* control was realized by the simulation of a "phantom" car to follow, represented by a signal that was fed into the on-board computer.[38] This virtual car provided the speed indications for the vehicle that followed the signal. In the second phase, headway and velocity between the controlled and the lead vehicle were obtained via a "mechanical take-up reel"[39] or "yo-yo",[40] also called "spring-wound tape".[41] A "take-up drum" with a cable was mounted on the front bumper of the test vehicle. The cable was connected to the rear of a lead car and was kept taut, giving the impression of a "towing arrangement", as a journalist observed.[42] With this physical connection, the distance between the lead vehicle and the following car was measured constantly, it was a way to simulate electronic feedback. The cable allowed the test vehicle to follow the lead car while maintaining a specified distance, called headway by the engineers. A similar system was also used at the Technische Hochschule Darmstadt in Germany – a cable between two vehicles was used to measure the distance to another vehicle in the front.[43]

It is obvious that this mechanical cable-system was not developed for general use on the road, it served experimental needs only. Today it is substituted by radar, lidar, vision systems and data from other vehicles and from the roadside.[44]

The OSU-engineers worked also on a radical modification of the human-machine interface of the car: The steering wheel and the accelerator pedal were removed,[45] only the brakes remained connected for safety reasons. If necessary, the vehicle could be manually controlled with a new "control stick", designed by Fenton: "Push

[36] Bender et al. 1971, p. 118.

[37] Fenton, Mayhan 1991, p. 103.

[38] Bender et al. 1971, p. 118, Fenton et al. 1968, p. 559.

[39] Fenton et al. 1968, p. 557.

[40] Bender et al. 1971, p. 118.

[41] Özgüner, Ümit, Tankut Acarman, and Keith Alan Redmill. 2011. *Autonomous Ground Vehicles*, p. 3. Boston, London.

[42] Hartford, Bill. 1967, April. You 'Fly' this Car with a Stick. *Popular Mechanics*, p. 207.

[43] Hartwich, Erwin, and Walter Zimdahl. 2012. Regelung von Straßenfahrzeugen: Ein Zeitzeugenbericht aus dem Institut von Prof. W. Oppelt, TH Darmstadt. *at-automatisierungstechnik* 60 (6), p. 341.

[44] Fenton, Robert E. 1994, December. IVHS/AHS: Driving into the Future. *IEEE Control Systems Magazine* 14 (6), p. 18.

[45] Bender et al. 1971, p. 118.

forward and you accelerate; pull back and you brake; lean it to the left or right and off you go in that direction".[46]

Unfortunately, the results with the control stick were not satisfactory because the complete vehicle control had to be performed with only two planes of motion. This required a good coordination to regulate the speed and steering commands simultaneously. As there was no "feel" to the brakes, excessive brake line pressure was the result. And steering was only satisfactory at speeds above 30 mph.[47]

Therefore, another control stick was developed, mounted between the driver and the passenger seat – like the Unicontrol in the Firebird III concept car that GM had presented in 1958. The engineers included a little finger "which recesses or protrudes from the front of the head an amount proportionate to deviations in headway."[48] This mechanism provided a "tactile assist for the driver". The laboratory investigated also "auditory and visual methods of informing the driver that he is drifting, but none was as effective as the tactile system."[49]

While the longitudinal control of vehicles was studied by many groups in the 1960s and 1970s, the OSU experiments and a bus-like vehicle study of GMC were the only two studies in the US that progressed to field tests of the controllers.[50] The experiments with the take-up reel and the control stick were conducted on a 4.5-mile straight township road[51] "near Ashville, Ohio and on an unopened 8.05 km section of Interstate 70 near Columbus, Ohio."[52] Later the tests were performed at the proving grounds of the Transportation Research Center (TRC) of Ohio that contained 70 miles of interconnected test tracks.[53]

According to Fenton, the system was tested on these roads with an average speed of 30 and 40 mph.[54] But in other publications, Fenton claims that the longitudinal control system was tested at speeds up to 65 mph (104 km/h)[55] and even "up to 70 mi/h".[56]

In 1969, Fenton and his colleague Karl Olson proposed to develop a centrally controlled system of automated highways with entry and exit points while all other roads should remain non-automated.[57] The main promise was that this *dual-mode system* would increase the lane capacity of highways and reduce traffic accidents. To realize this system, the evolutionary vision of Zworykin – detailed in the preced-

[46] Hartford 1967, p. 88.

[47] Fenton et al. 1968, p. 563.

[48] Hartford 1967, p. 89.

[49] Ibid., p. 207.

[50] Fenton 1994, p. 18.

[51] Fenton et al. 1968, p. 562.

[52] Fenton et al. 1968, p. 565 and Bender et al. 1971, p. 118.

[53] Fenton et al. 1968, p. 565.

[54] Ibid., p. 562.

[55] Fenton, Robert E., and Karl W. Olson. 1969, July. The Electronic Highway. *IEEE Spectrum*, p. 64.

[56] Fenton 1970, p. 155.

[57] Fenton, Olson, 1969, p. 60–66.

ing chapters – was adopted by the engineers. Future electronic highways should evolve from the one of today, following three development stages: Starting with driver aids, vehicles should be partially automated in the second stage and reaching the transition to complete automatic vehicle control in the final stage.[58]

The missing sensor "was the major reason for switching the OSU efforts to the point-following approach",[59] also called *synchronous* longitudinal control. Fenton and Mayhan explain that the car follows the synchronous signals of "an imaginary conveyor belt, which is divided into equally spaced slots, moving at a fixed speed through a highway network. No more than one vehicle would be assigned to each electronic slot, and that vehicle would be required to maintain itself at slot center." This system requires the synchronization of the whole network on a special lane of an automated highway system (AHS). However, all synchronous approaches "require active electronics in the roadbed or at roadside".[60] OSU worked on this sector control of vehicles in the mid-1970s.[61]

From the late 1960s to 1979[62] studies on automated lateral control with guide-wire cables producing a magnetic field were undertaken at OSU.[63]

As we have seen in the preceding chapter, this technology was first developed by RCA and GM[64] in the 1950s. In the 1960s, the British Road Research Laboratory worked also in this field.[65] In these early concepts, only a single cable was buried in the road. The "cable is excited by an alternating current which produces a magnetic field in the vicinity of the cable. This field induces a voltage in each of the tuned pickup coils which are mounted on either side of the vehicle centerline. The difference between these voltages is used to determine the location of the vehicle relative to the lane centerline and to actuate the steering control unit."[66]

A main problem of this system was the presence of ferrous materials on the road and the steel reinforcement of roads and bridges – both resulted in errors. Therefore in 1967 the OSU-researchers tested a system with two wires on both sides of the road and a phase-sensing system.[67] It had the advantage to enlarge the sensing range from 2 feet (space between the sensing coils in the single cable system) to 8 feet

58 Ibid., p. 61.

59 Fenton, Mayhan 1991, p. 103.

60 Fenton 1994, p. 18.

61 Fenton, Mayhan 1991, p. 111.

62 Ibid., p. 106.

63 Fenton, Olson 1969, p. 65.

64 Flory, Leslie E., et al. 1962 September. Electronic techniques in a system of highway vehicle control. *RCA Review* 23, pp. 293–310.

65 Giles, G.C., Martin, J.A. 1969 July. Cable installation for vehicle guidance investigations in the new research track at Crowthorne, Rept. RN/40 57/CGG, JAM, Road Research Laboratory, Crowthorne, England (cit n.: Fenton, Robert E., Olson, Karl W. The electronic highway. *IEEE spectrum*, p. 66).

66 Fenton et al. 1968, p. 564.

67 Ibid. and Fenton, Mayhan 1991, p. 107.

(space between the two wires).[68] Influenced by these US-American experiments, the Government Mechanical Laboratory in Japan also explored the two-wire system in May 1968 (see Sect. 3.2.2).

OSU developed an experimental sensing arrangement for the two-wire system: Two vertically and horizontally working sensing coils were mounted "in a plastic pod on the front of the vehicle".[69] The coil pod moved on a pivoted arm "which was mechanically linked to the steering mechanism."[70] The following image (Fig. 2.4) shows the front wheels steering left and the coil pod on the left side of the vehicle.

The system was tested with speeds up to 70 mph (112 km/h) but again it worked only on roads "without buried steel reinforcing rods".[71] As the noise inputs of ferrous materials caused large lateral accelerations with a great discomfort for the passengers, the system was "unusable" at speeds above 30 mph on those streets.[72]

Finally, the FHWA decided in 1977 to abandon the development of active wire-following reference systems at the side of the road or under its surface.[73] After restrictions in the DOT/FHWA budget the test facility was closed. According to Fenton and Mayhan, OSU then stopped all studies on guide-wire systems. Only passive devices like a guard rail reflector that could be detected by a side-looking

Fig. 2.4 The moving sensing coils of the 1965 Plymouth research vehicle of Ohio State University. (Credits: OSU. Image Source: Fenton et al. 1968, p. 564)

[68] Fenton et al. 1968, p. 564.
[69] Fenton et al. 1968, p. 564 and Fenton, Olson 1969, p. 66.
[70] Fenton et al. 1968, p. 564.
[71] Fenton, Olson 1969, p. 66.
[72] Fenton et al. 1968, p. 564.
[73] Fenton, Mayhan 1991, p. 107, 111.

radar were developed further. "Wire following requires an active source in the roadway, whereas the radar approach involves only a passive guardrail, with the active device in the vehicle."[74] With the advent of the Reagan-administration in 1981, BPR funding for traffic systems suffered severe budget cuts. Until the mid-1980s the research on automated transportation systems was stopped.[75]

2.1.3 Bendix: Guide-Wire Tests for a Personal Rapid Transit (PRT) System

Another system that used guide-wires should also be mentioned shortly. It was developed by the Aerospace division of the Bendix Corporation in Ann Arbor at the end of the 1960s. The research program was sponsored by the Applied Physics Laboratory of Johns Hopkins University and the Department of Transportation. Three Swedish DAF-based Kalmar small delivery vans (Tjorven) circulated on a 5500-foot roadway on a test site in Ann Arbor, under construction in 1971.[76] Guide wires kept the vehicles in contact with a computer control center that would decide a safe route and calculate the distance between the driverless vehicles, measured by microwave systems.[77]

The aim of this test facility was not to develop automated automobiles but to study headway sensing systems for a Personal Rapid Transit (PRT) System. These track-based systems consisted of fully automated small vehicles for 20 passengers, moving on computer controlled guideways. Another concept, called "People Mover", was developed to function on a scheduled basis, with much faster intervals than busses. Both systems were not conceived in opposition to the car system, but to revolutionize short distance public transport on airports or commercial centers.

2.2 The Advent of Mobile Robots

After this overview about the guidance-cable experiments, we will now come to a very different kind of technology development that arose around the same time in the robotics field. It is the history of autonomous mobile robots that started at the

[74] Fenton 1994, p. 16.

[75] Ibid., p. 15 and Ioannou, Petros A., 1997. Introduction. In *Automated Highway Systems*, ed. Ioannou, Petros A, p. 4. New York/London.

[76] Author unknown, Firm Devises Driverless Vehicle Controls. *Ann Arbor News*, June 30, 1972, online in: https://aadl.org/aa_news_1972060-firm_devises_driverless_vehicle_controls (accessed on June 18, 2019).

[77] Author unknown, Bendix Works On Automatic Vehicle Project. *Ann Arbor News*, June 19, 1970, online in: https://aadl.org/aa_news_19700619-bendix_works_on_automatic_vehicle_project (accessed on June, 13, 2019).

end of the Second World War. In a long-term perspective the developments in this field became much more important for all kinds of vehicle automation than the early concepts of cable-guidance.

One of the earliest applications were the turtels or "tortoise" robots, created in 1948 and 1949 by the neurophysiologist William Grey Walter (1910–1977).[78] Named Elmer and Elsie (Electromechanical Robot, Light-Sensitive), they were equipped with motors, relays and photo cells. They moved towards light sources, checked for obstacles and changed their direction to avoid collisions. This made them the first mobile robots that interacted meaningful through feedback with their environment.[79]

Another example of an early mobile robot controlled by on-board electronics was the *Beast* of Johns Hopkins University Applied Physics Laboratory (1964). Guided by sonar sensors and photocells, it could move around in "white-walled corridors looking for dark colored power plugs".[80] If its battery power was low, it plugged itself in and recharged.

These robots had in common that they were not stationary but able to move, using rudimentary perception techniques. The robots of Grey Walter even avoided obstacles. All these capabilities – perception, mobility, obstacle detection and avoidance – are crucial elements of autonomous driving. For that reason, the research efforts in the field of mobile robots deserve an important place in this book.

The following section begins with a brief historical introduction to the early steps of AI-development in the United States of America. We will also present the main actor in the field – DARPA, an agency of the Department of Defense that funded decades of AI- and robotics research. It had been decided to center the narration around two key dates: The Dartmouth Conference in 1956 and the Sputnik-Shock in 1957. The main part of the chapter then presents two of the most influential robots of the twentieth century, the Stanford Cart and Shakey. Much more mobile robots had been developed around the world, but we had to limit our presentation to the most important ones with regard to autonomous driving.[81]

2.2.1 AI-Research and the Creation of DARPA

Artificial Intelligence is a subfield of Computer Science, searching for computer intelligence. Knowledge management, perception, image analysis, speech recognition, robotics and machine learning are the main fields of AI. It is based on insights from other disciplines like psychology, economics, philosophy, pedagogics and neurosciences.

---

[78] Crevier, p. 31.

[79] Henderson, p. 15.

[80] Nilsson, Nils J. 2010. *The Quest for Artificial Intelligence, a History of Ideas and Achievements*, p. 162. New York: Cambridge University Press.

[81] For a presentation of further mobile robots see Meystel 1991.

Artificial Intelligence research started in 1956 with a 2-month workshop at the Dartmouth College in Hanover, New Hampshire. The conference had been organized by the computer scientist John McCarthy, the mathematician Claude Shannon, the cognitive scientist Marvin Minsky and the computer scientist Nathan Rochester. Today, this conference is considered as the "official birth date" of a new academic field that aims to build intelligent machines. It is common to start the history of AI with the Dartmouth Conference, but it should be mentioned that the idea of an electronic brain – a machine that can think – had not been possible without the contribution of three theoretical schools: computation (Alan Turing), theories of information (Claude Shannon) and Cybernetics (Norbert Wiener).

The conference program claimed that "every aspect of learning or any other feature of intelligence can in principle be so precisely described that a machine can be made to simulate it."[82] According to this assumption, the human brain and a computer have in common that they cannot access the world directly. Both depend on internal symbolic representations of the outside world. In this perspective intelligence is nothing but "the ability to process symbols" explains the AI-researcher Daniel Crevier. This ability is independent from the hardware that supports it – "arrays of electronic switches in a digital computer or firing neurons in a brain."[83]

It was John McCarthy who gave the project to make computers think the name artificial intelligence.[84] Marvin Lee Minsky, a student of McCarthy who later founded the AI-Lab at the Massachusetts Institute of Technology (MIT), defined AI as "the science of making machines do things that would require intelligence if done by men."[85] The ability to connect symbols following specific rules – called computation – was seen as the way to machine intelligence.[86]

Since it had been established as a research field, the development of AI followed regular cycles of enthusiasm and disappointment. The 10 years from 1965–1975 are often described as the "Golden Age" of the research on Artificial Intelligence. AI-research concentrated on experiments with big, stationary computers that explored the human brain and the ability to think.[87] At the end of the 1960s, computer programs were successful in solving problems in algebra and geometry, proving theorems, learning and communicating in simplified English and playing chess games.[88] These early advances seduced many AI-researchers to announce optimistic

[82] McCarthy, John; Minsky, Marvin; Rochester, Nathan; Shannon, Claude, A Proposal for the Dartmouth Summer Research Project on Artificial Intelligence (1955), cit. in: Crevier, Daniel. 1993. *The Tumultuous History of the Search for Artificial Intelligence*, p. 26. New York.

[83] Crevier, p. 48.

[84] Ibid., p. 49f., see also Moravec, Hans. 1988. *MIND Children, The Future of Robot and Human Intelligence*, p. 8. Cambridge, London.

[85] Crevier, p. 9.

[86] See Lenzen, Manuela. 2018. *Künstliche Intelligenz, Was sie kann & was uns erwartet,* p. 36. München.

[87] See Moravec 1988, p. 16.

[88] See Ibid., p. 8f.

timelines and to make exaggerated claims.[89] The AI-pioneer Herbert Simon predicted that "machines will be capable, within twenty years, of doing any work that a man can do." John McCarthy started the Stanford AI project in 1963 with the goal "of building a fully intelligent machine in a decade" and Marvin Minsky said in 1967 that only very "few compartments of intellect will remain outside the machine's realm".[90]

But at the beginning of the 1970s, the promised results failed to realize. The lack of processing speed and memory size of the computers was one factor that limited the advances. While computers could work on difficult geometrical problems, apparently simple tasks like facial reconnaissance or moving in a room without collision were very difficult. Paradoxically, "machines doing well things humans find hard, while doing poorly what is easy for us", emphasizes the computer scientist Hans Moravec.[91] Especially machine vision proved to be very complex and lacked progress in the 1970s. Crevier describes machine vision as "one of the toughest nuts to crack in AI"[92] because computers can only handle numbers and therefore each image must be represented as an array of numbers.[93] The Defense department was interested in this field and therefore they funded the research of Moravec with the Stanford Cart.

To understand the different concepts of autonomous driving that are detailed in the following chapters, two concepts of AI should be kept in mind:

The early approaches of Minsky and McCarthy "focused on developing artificial reasoning and problem-solving capabilities as well as finding ways to encode knowledge so it could be accessed and used automatically."[94] Crevier explains that these classical AI-approaches aimed at the creation of a symbol-processing machine that can reason about what it is doing and formulate a plan.[95] This focus on reasoning is also called the *top-down approach* of machine intelligence. As will be shown later, the robot Shakey followed this concept.

Hans Moravec advocates another approach: He emphasizes that intelligence is always linked with "the ability to perceive and explore" the environment.[96] According to him, AI-research should explore "perception and mobility" first. Sophisticated behavior should be generated from simple interactions.[97] This is called the *bottom-up approach*[98] represented for instance by the late Stanford Cart,

[89] See Crevier, p. 6

[90] Ibid., p. 109.

[91] Moravec 1988, p. 9, see also Crevier, p. 40.

[92] Crevier, p. 89, 91.

[93] The need to start with concepts for representing the spatiotemporal (four-dimensional: 4-D) world and motion processes of objects in it, has been recognized around 1980 as we will see in Sect. 4.3.1.

[94] Henderson, p. 90.

[95] Crevier, p. 36, 40.

[96] Moravec 1988, p. 16.

[97] Henderson, 90f.

[98] Moravec 1988, p. 17.

as we will see in the following sections. Moravec estimated in 1988 that fully intelligent machines should be possible in 40 years if the top-down and bottom-up approaches merge half on the way.[99]

During the 1970s most robotics researchers "viewed mobility as an unnecessary complication to an already overly difficult problem."[100] A great number of research laboratories worked with robot arms that were connected to computers, "but hardly any had robot vehicles", emphasizes Moravec. This explains the important role of the mobile robots and research vehicles that will be presented later.

Since the 1960s, most of the robotics research in the US had been funded by the *Advanced Research Projects Agency* (ARPA) of the US-military, later known as DARPA. For instance, the agency funded projects at the Massachusetts Institute of Technology (MIT), Carnegie Mellon University (CMU) and Stanford University with millions of dollars.[101] Therefore, a short portrait of this government agency should be given, based on two recent publications that retrace its history.[102]

ARPA had been created during the Cold War, following an event that occurred only 1 year after the Dartmouth conference: On October 4, 1957, the Soviet Union launched the first satellite into the Earth orbit. It was called Sputnik, meaning fellow traveler in Russian. The satellite weighed 184 pounds and "its sole function was to circle the earth, emitting a beep that could be tracked from the ground."[103] Its launch triggered the so called *Space Race* between the US and the Soviet Union during the Cold War, a fierce competition of both nations in space exploration.

In the first days following the launch of Sputnik, the Eisenhower administration downplayed its importance and there was no panic or hysteria that had later been addressed as "Sputnik choc", emphasizes Sharon Weinberger.[104] But Lyndon Johnson, Senate leader of the oppositional Democratic Party proclaimed that the Soviets may soon be able to throw bombs from space. In fact, the R-7 rocket used to launch Sputnik into space could also be used to send a nuclear warhead anywhere in minutes. This narrative was developed further in mass media when 1 month later, on November 3, 1957 the Soviets launched a second, bigger Sputnik that carried a dog named Laika. During the following weeks, panic was growing in the American public that the Soviets would soon be able to send a human into space.[105]

---

[99] Ibid., p. 20.

[100] Moravec 1988, p. 13.

[101] Crevier, pp. 52–107, 64–65.

[102] Jacobsen, Annie. 2015. *The Pentagon's Brain, An Uncensored History of DARPA, America's Top Secret Military Research Agency*. New York and Weinberger, Sharon. 2018. *The Imagineers of War: The Untold History of DARPA, the Pentagon Agency That Changed the World*. New York.

[103] Weinberger, p. 32.

[104] Ibid., p. 32f.

[105] On April 12, 1961 the Cosmonaut Juri Gagarin became the first human in space, orbiting the earth during 108 minutes.

The nuclear physicist Ernest Lawrence and the former Manhattan project scientist Charles Thomas proposed to the Secretary of Defense the creation of a central research agency for space research.[106] The military establishment was opposed to the idea but after the explosion of the Vanguard rocket on December 6, 1957 – built to bring a Navy satellite to space[107] –, Eisenhower approved the creation of the new agency that was called *Advanced Research Projects Agency* (ARPA).[108]

On February 7, 1958, ARPA was officially founded as independent agency that had to report to the Secretary of Defense. Its initial justification had been to catch-up with the Soviet Union in the Space Race and to prevent another technological surprise.[109] To reach this goal, ARPA had to identify needs before they existed, explained Neil McElroy, Secretary of Defense from 1957–1959.[110]

For a short time, ARPA was charged with "all civilian and military space programs" and became the successor of the National Advisory Committee for Aeronautics (NACA), founded in 1915.[111] The main mission for ARPA was to find "a way how to defend the United States from a nuclear attack."[112] One of the most significant early projects of ARPA was a missile shield in the magnetosphere that should destroy incoming Intercontinental Ballistic Missiles (ICBM). This military focus became even stronger when ARPA "was ordered in October 1958 to turn over its scientific satellite programs" to a newly formed civilian agency, the National Aeronautics and Space Administration (NASA).[113] In 1959 ARPA lost all responsibility for the Space Program.[114]

In the next decade, "the guiding principle" of ARPA had been to "look for promising researchers and just give them a budget, with which they could explore in their own unique and crazy directions. That was the spirit through the Shakey time, the late sixties" explains Crevier.[115] During this period, ARPA worked on controversial weapons projects, brain-driven computers, parapsychology and magic. Within the context of the American engagement in the Vietnam war (1955–1975) and growing public concerns about the power of a military-industrial complex, this policy changed in the last years of the war. As the spending of the Pentagon seemed to be out of control, a committee chaired by Senator Mike Mansfield, "an antiwar Democrat from Montana"[116] proposed to limit the funds of the Department of

[106] Weinberger, p. 36, Jacobsen, p. 51.

[107] Finally, it was the team of the former Nazi rocket scientist Wernher von Braun that successfully put the first American satellite into orbit (with its Explorer-1, based on the Jupiter-C, on January 31, 1958).

[108] Weinberger, p. 38.

[109] Weinberger, p. 204, see also Moravec 1988, p. 21, Jacobsen, p. 239.

[110] Jacobsen, p. 51.

[111] Weinberger, p. 44.

[112] Ibid., p. 46.

[113] Ibid., p. 51.

[114] Ibid., p. 89f.

[115] Crevier, p. 115.

[116] Jabobsen, p. 226.

Defense to military purposes only. On November 19, 1969, Congress passed the "Mansfield Amendment" to the Defense Procurement Authorization Act of 1970 (Public Law 91–121). It required that only research with "a direct and apparent relationship to a specific military function or operation" should be financed.[117] On March 23, 1972 this change in focus was also mirrored in the name of the agency: ARPA now became the Defense Advanced Research Projects Agency (DARPA).[118] Its new obligation was to fund only "mission-oriented direct research, rather than basic undirected research" as during the 1960's, explains a report of the NRC.[119] But according to Ceruzzi the amendment "did not intend to cripple basic research at universities, only to separate basic from applied research."[120]

The agency decided in 1973 to refocus on its original mandate – avoid technological surprise. With the rise of computers in the 1970s DARPA shifted its focus on command, control and communication technologies (C3).[121] ARPAs work was "speculative, looking ten to twenty-five years into the future" emphasizes Jacobsen.[122] ARPA-director Rechtin qualified the agency as a pre-requirement organization: Research had to be done before the need arises – when the need is there, it would be too late to start research. So even after the Mansfield Amendment, DARPA funded basic research as it often produces paradigm-shifting results in a long-term perspective. The military funding of scientific research did not aim at the immediate delivery of new weapons but more on the long-term accumulation of knowledge and know how. In the robotics domain for instance, "most of DARPA's programs were focusing on advancing different aspects of robotics, rather than building war robots."[123] Often new robotic systems should demonstrate a "specific ability", not destined to the battlefield, says Weinberger. This explains why just after the Mansfield Amendment the agency funded basic research on computer vision with the Stanford Cart.

To understand the dominance of DARPA-funding in the history of vehicle automation, it should be noted that the United States have no Ministry of Research and Technology on a federal level, even if the National Science Foundation must be mentioned, a government agency that funds basic research. For that reason many initiatives on future technologies are funded by the U.S. Department of Defense. DARPA does not operate own research laboratories but executes its program through

[117] Nilsson, p. 203, see also Jacobsen, p. 226, Dickson, David, 1984. The New Politics of Science, p. 122. New York (zit n. Ceruzzi, p. 258).

[118] Weinberger, p. 203.

[119] Fleck, James, 1982. Development and Establishment in Artificial intelligence. In *Scientific Establishments and Hierarchies, Sociology of the Sciences Yearbook*, eds. Elias N., Martins H., Whitley R. Vol. 6, Dordrecht, cit. without giving page numbers in: Committee on Innovations in Computing and Communications, et al. (ed.), Funding a Revolution: Government Support for Computing Research, 1999, p. 213. Washington.

[120] Ceruzzi, p. 258.

[121] Jacobsen, 243f.

[122] Ibid., p. 226.

[123] Weinberger, p. 329.

contracts "with industry, universities, nonprofit organizations, and federal R&D laboratories."[124] The agency is acting in between different academic research institutions, political and military stakeholders, investment interests, technological possibilities – forming a network that could be called *military-scientific complex.*

Finally it can be concluded that without the important financial resources of DARPA basic research on technologies that play a crucial role for autonomous driving had not been made available. Without this military funding, these technologies would probably not exist or at least in another form.

2.2.2 Research on Remote Control at Stanford Mechanical Engineering Laboratory (1960–1966)

It was in the context of the Space Race that the first mobile robot that contributed to the history of autonomous driving was created. It can be considered as the "real start of autonomous car technology".[125]

The first version of the research vehicle that later became famous under the name Stanford Cart had been developed in the context of the American efforts to realize a manned mission to the Moon: Financed under a NASA contract, its original purpose was to serve as a research platform to learn how to remotely control a lunar rover from Earth using video information.[126]

In 1960, James Adams, a graduate student of the Mechanical Engineering Department of Stanford University built the vehicle: "The Cart had four small bicycle wheels with electric motors powered by a car battery and carried a television camera with a fixed view in the forward direction. Tests were conducted using both 2-wheel steering, like a car, and 4-wheel steering, in which the wheels and television camera swivel together" explains the computer scientist Lester Earnest.[127]

From a historical view, the crucial point of this early version of the Stanford Cart was the idea to combine a vehicle with a camera. But the camera did not yet serve as eye of the cart – only as additional eye for a human: Earnest details that the images of the onboard TV camera had been sent with a "very long cable" to a human operator in a laboratory who manipulated the steering control with the help of a television display. To simulate the Earth-Moon communication delay, the

[124] Congressional Research Service, Defense Advanced Research Projects Agency: Overview and Issues for Congress, Updated August 19, 2021, online in: https://sgp.fas.org/crs/natsec/R45088.pdf (accessed June 27, 2022).

[125] Torchinsky, p. 63.

[126] Moravec, Hans, *Obstacle Avoidance and Navigation in the Real World by a Seeing Robot Rover*, Ph.D. thesis, Computer Science Department, Stanford University, March 1980, Chapter 2: History, in: http://www.frc.ri.cmu.edu/users/hpm/project.archive/robot.papers/1975.cart/1980.html.thesis/a02.html

[127] Earnest, Lester, Stanford Cart, 2018. https://web.stanford.edu/~learnest/les/cart.html (accessed on February 3, 2019).

steering commands were delayed for two and a half seconds before broadcasted to the cart. But with this delay, the Cart had the disadvantage to be very slow. It could not be controlled "if traveling faster than 0.2 mph", i.e. ~ 9 cm/sec.[128] In the year 1962 graduate student Paul W. Braisted added an analog computer that predicted future locations of the cart.[129] This increased its speed to some degrees: It could now be controlled at a speed of 5 mph, i.e. ~ 2.22 m/sec.[130]

But in the same year, the experiments at Stanford were interrupted by an acceleration of the political ambitions of the Space program: "We choose to go to the Moon" – with the intention to gather support for the Apollo-program, President John F. Kennedy announced on September 12, 1962 to a crowd of 40.000 people at Rice University (Rice Stadium) in Houston (Texas) that the United States would send a manned mission to the moon until the end of the decade. With this ambitious goal in mind, the immediate use of the remote-control technology that had been developed with the Cart was not seen as realistic and the vehicle became unused between 1963 and 1966. We will come back to the important advances of the Stanford Cart after a paragraph about another robot, called Shakey.

2.2.3 Shakey, the Robot – Stanford Research Institute (1966–1972)

Only eight miles away from Stanford University, another robot specialized in seeing and moving was developed at Stanford Research Institute (SRI), an industry- and public sponsored research facility, founded in 1946.[131]

In November 1963, the AI-pioneer Charles Rosen from SRI proposed to develop a mobile robot that could reason. It should be able to combine "the pattern-recognition and memory capabilities of neural networks with higher level AI-programs" explains Nils J. Nilsson, who became later one of the leaders of this

[128] Adams, James L. 1961. *Remote Control with Long Transmission Delays,* PhD in Mechanical Engineering. Stanford University.

[129] Braisted, Paul. 1963. *Study of a Predictor for Remote Control Systems Operating with Signal Transmission Delays, PhD in Mechanical Engineering*. Stanford University.

[130] Earnest 2018.

[131] SRI developed inkjet printing (1961), the computer mouse (1964) and created the company Siri (2007), acquired by Apple in 2010. ARPA was one of the SRI's major funders. Its greatest accomplishment was the ARPANET, precursor of the modern Internet. On October 29, 1969, a computer console at the SRI received the word Lo, the first transmission sent across the ARPANET (Weinberger, p. 220). In April 1969, Stanford's board of trustees decided to end the relationship with the Stanford Research Institute (Weinberger, p. 219). In 1970, SRI separated from Stanford University and became SRI International in 1977. This was a response to criticism of Vietnam War-protesters who feared that with funding from DARPA the university became part of the military-industrial complex (see also Julia, Luc. 2019. *L'intelligence artificielle n'existe pas*, p. 35. Paris).

project.[132] In 1964, the Advanced Research Projects Agency (ARPA) decided to finance the robot and two years later SRI obtained a $750.000 contract.[133]

One of the military goals of the project was to "develop automatons capable of gathering processing and transmitting information in a hostile environment", as an ARPA statement says.[134] The military hoped to get a device for information collection "behind enemy lines: in effect, a mechanical spy", explains Crevier.[135] Nilsson remembers that "someone had in mind dropping the robot on a road in China and have it count tanks".[136]

In the following years, the Artificial Intelligence Center (AIC) of SRI developed a mobile robot called *Shakey* (Fig. 2.5). Shakey got his name simply because it "shook when it came to an abrupt stop".[137] Nearly two meters tall, Shakey consisted of a three-wheeled platform with touch sensors on its bumpers, a rack with a camera control-unit and an on-board logic and on the top a television camera, a sonar range-finder and a radio antenna. His name and shape may have had a personifying effect on the robot. Scientists referred to Shakey as "he" and Life Magazine called him in 1970 the "first electronic person".[138]

Nilsson emphasizes that Shakey was the first mobile robot that had the ability to perceive its environment (locate walls, doors, objects), to reason about its own actions, to plan a course, to act autonomously (steer around objects) and to learn from the execution of its plans.[139] Shakey was primarily controlled "by programs that reasoned",[140] using software for perception, world-modeling and acting. With this configuration, the robot was an expression of the top-down approach.

Computer vision was a difficult task for Shakey "because the appearance of an object can change radically depending on the angle from which it is being viewed."[141] It is important to understand that Shakeys vision and motion software was only able to interpret camera images "of simple, uniformly colored, flat-faced objects".[142] Therefore Shakey lived in a blocks world, consisting of seven rooms connected by eight doorways.

---

[132] Nilsson, p. 162.

[133] Shakey the robot, in: https://history-computer.com/Dreamers/Shakey.html (accessed on December 5, 2019).

[134] ARPA, Automaton work statement, published in: Nilsson, p. 164.

[135] Crevier, p. 94.

[136] Cassel, David, 2017. Remembering Shakey, the First Intelligent Robot. *The New Stack*, March 5, 2017, online in: https://thenewstack.io/remembering-shakey-first-intelligent-robot/ (Accessed on December 5, 2019).

[137] Nilsson, p. 164.

[138] Darrach, Brad, 1970. Meet Shaky, the first electronic person. *LIFE Magazine*, November 20, 1970, p. 58 and 68.

[139] Nilsson, p. 175.

[140] Moravec 1988, p. 14.

[141] Henderson, p. 87.

[142] Moravec 1988, p. 14.

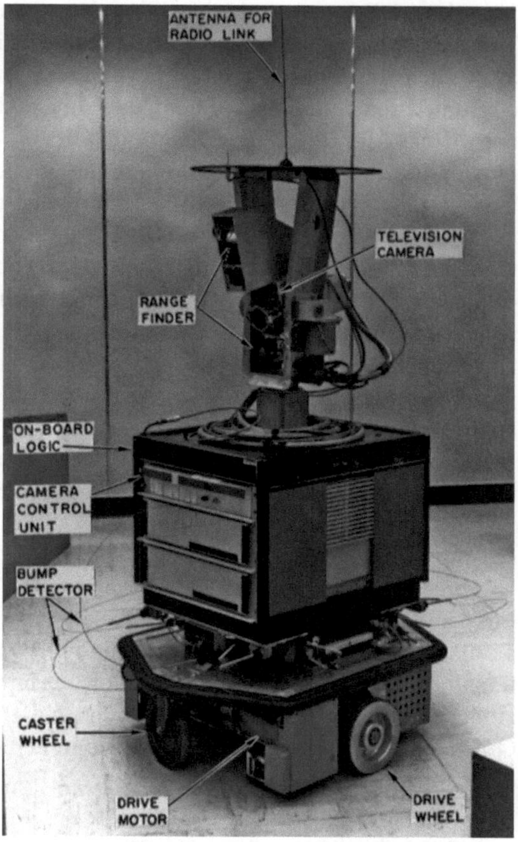

Fig. 2.5 Shakey, the robot, 1968. (Credits: SRI International. Image Source: https://fr.wikipedia.org/wiki/Shakey_le_robot/, CC BY-SA 3.0)

When a human operator gave Shakey a task in English – go through a door, move a block – the robot planned how to solve it. Shakey was connected via a two-way radio and video link to a DEC PDP-15 and a DEC PDP-10 computer in the laboratory. These off-board computers processed the transmitted video and sensor information, recognized patterns and sent motion commands back to the robot. Shakey could move big rectilinear wooden blocks on the floor of the lab environment.[143] After accomplishing the task, the action routine was memorized and added to the action models for future use. Shakey learned by analogy: If an image resembled an image that he saw already before, a decision could be made based on this resemblance.[144]

By counting its wheel revolutions, Shakey kept tracking of its own position – a method called dead reckoning navigation. This method plays an important role in all the research on autonomous driving. As a classic AI robot Shakey had "to create a

---

[143] Crevier, p. 95.

[144] Darrach, p. 61.

'world model' based on the incoming data and then plan and calculate actions."[145] To be able to navigate, Shakey had to update its internal map before he could move an obstacle in the room. The result was a very low speed of movement – each move consumed about "an hour of computer time and had a high likelihood of failure", notes Hans Moravec,[146] concluding that the robot "was impressive in concept but pitiable in action". According to him, the problem was not the planning of movements, it lay "in the programs that interpreted the raw data from the sensors and acted on the recommendations." But the problem of controlling its movement and the interpretation of sensory data were of "secondary importance to the designers" of Shakey, explains Moravec.[147] Without giving further evidence, Douglas W. Gage resumes even that Shakey was considered a "failure in its day" because the robot never functioned autonomously.[148] The project ended in 1972, when DARPA disengaged because basic research on robots did not fit the new emphasis on practical applications.[149]

Despite its limitations, Shakey became a legend in the robotics world. According to Nilsson, its search techniques, control mechanisms, learning and vision strategies are "widely used today",[150] especially "many of the visual processing techniques".[151]

Three examples for the "long term impacts"[152] of its algorithms can be given: First, a method was developed for Shakey that allowed the robot to identify when points in an image fall into a line. The video images from the camera were then transformed into a line representation. Originally developed in 1962 by Paul Hough (Hough transform) and adapted in 1972 by Peter E. Hart and Richard O. Duda, this feature extraction algorithm is today one of the "most widely-used algorithms in computer vision."[153] Since 2014 it is used in automobile lane keeping alert systems[154] and now it serves the development of several self-driving cars.[155]

Second, Shakeys control software was structured for the first time in three hierarchical layers: Programs in the lowest level captured the sensory information

---

[145] Henderson, p. 88.

[146] Moravec 1988, p. 15.

[147] Ibid, p. 14.

[148] Gage, Douglas W. 1995 Summer. UGV History 101: A Brief History of Unmanned Ground Vehicle (UGV) Development Efforts. *Unmanned Systems*, Special Issue on Unmanned Ground Vehicles 13 (3), p. 2.

[149] Nilsson, p. 175.

[150] Ibid., p. 165.

[151] Ibid., p. 174.

[152] Engineering and Technology History Wiki, Milestones: SHAKEY: The World's First Mobile Intelligent Robot, 1972, last modified on May 24, 2017, online in: http://ethw.org/Milestones:SHAKEY:_The_World's_First_Mobile_Intelligent_Robot,_1972 (accessed on December 5, 2019).

[153] Ibid.

[154] Ibid.

[155] Cassel 2017.

and drove the motors to control physical actions – roll and turn of the robot, pan and tilt of its camera. Programs in the intermediate level processed the images from the TV camera and planned actions like autonomous navigation to a goal position. Shakey could circumvent obstacles by defining subgoals and deal with unforeseen events: A 24-min film produced by the engineering team in 1969 shows how Charles Rosen – disguised as a Gremlin wearing a black cape – displaces the position of a box that Shakey had to push.[156] The robot then adapted its course to the new coordinates (error recovery). The highest level of programs planned the execution of more complex actions: They could process the incoming data and send commands to the circuits that controlled its electrical motors. This three-level architecture became a model for many future robots. For example, Stanley – the self-driving VW Touareg that won the DARPA Grand Challenge in 2005 – used like Shakey software with hierarchical layers, confirms project leader Sebastian Thrun.[157]

Third, the planning program Stanford Research Institute Problem Solver (STRIPS) combined intermediate level actions to perform new tasks, execute plans and recover from errors. It became "the basis of many subsequent planning systems". The A* path-finding algorithm is used in a variety of applications for the computing of driving directions,[158] e.g. plotting the path of the Mars rovers Spirit and Opportunity that reached Mars in 2004 and Curiosity, arriving on Mars in 2014.[159]

Named the fifth-best robot of all time by *Wired Magazine* in 2006,[160] Shakey is now exhibited at the Computer History Museum in Mountain View, California. In 2017, it had been declared a "milestone" of engineering history by the IEEE.[161]

---

[156] Stanford University Libraries (Youtube-account), Shakey: Experiments in Robot Planning and Learning (1972) (video title), Shakey, An experiment in robot planning and learning, produced by Peter Hart and Nils Nilsson, assisted by Michael Wilber, sponsored by the Advanced Research Projects Agency and the National Aeronautics and Space Administration, Research performed by the Artificial Intelligence Center at Stanford Research Institute, Menlo Park, California. Uploaded on September 23, 2016, online in: https://www.youtube.com/watch?v=GmU7SimFkpU (accessed on December 5, 2019).

[157] See Fig. 5 in Thrun, Sebastian, et al. 2006. Stanley: The Robot that Won the DARPA Grand Challenge. *Journal of Field Robotics* 23(9), pp. 661–692, online in: https://onlinelibrary.wiley.com/doi/epdf/10.1002/rob.20147 (Accessed Mai 7, 2018).

[158] Nilsson p. 168.

[159] Milestones: SHAKEY: The World's First Mobile Intelligent Robot, 1972. In Engineering and Technology History Wiki (last modified on May 24, 2017), online in: http://ethw.org/Milestones:SHAKEY:_The_World's_First_Mobile_Intelligent_Robot,_1972 (accessed on December 5, 2019).

[160] WIRED listed very different robots from engineering, fiction, literature, starting with Stanley, Astroboy, Spirit and Opportunity, Robby the robot, Shakey etc. (see Capps, Robert, 2006. The 50 best robots of all time. *WIRED*, January 1, 2006, online in: https://www.wired.com/2006/01/robots-3/ [Accessed October 1, 2019]).

[161] The milestone plaque in the lobby of the SRI in Menlo Park resumes that Shakey "could perceive its surroundings, infer implicit facts from explicit ones, create plans, recover from errors in plan execution, and communicate using ordinary English." (Engineering and Technology History Wiki, Milestones: SHAKEY: The World's First Mobile Intelligent Robot, 1972, February 16, 2017, in: http://ethw.org/Milestones:SHAKEY:_The_World's_First_Mobile_Intelligent_Robot,_1972).

2.2.4 *The Stanford Cart at the Stanford Artificial Intelligence Lab SAIL (1965–1980)*

The Stanford Artificial Intelligence Laboratory (SAIL) had been founded in 1963 by the Artificial Intelligence pioneer John McCarthy using funds of the Advanced Research Projects Agency (ARPA) of the US-Defense Department. Between 1963 and 1973, $10 million of DARPA-funds were spent by SAIL.[162] From 1965 to 1980, the laboratory was housed in the D.C. Power building on a remote location in the Santa Cruz Mountains.[163] Its architecture "resembled a flying saucer that had broken in two and crash-landed on the hilltop."[164] As we have already mentioned in Sect. 2.2.1, the goal of the laboratory was to build "a fully intelligent machine in a decade."[165] It became one of the leading centers for AI research worldwide.

The vision of autonomous driving was part of SAIL since the beginning. During this period, AI-enthusiasts "began dreaming of cars smart enough to navigate ordinary streets on their own" affirms Marc Weber from the Computer History Museum in California.[166] In 1965, the computer scientist Lester Earnest had the idea to create an experimental robot road vehicle using visual guidance.[167] He acquired the Stanford Cart – presented in Sect. 2.2.2 – from James Adams, who had become a Professor of the Mechanical Engineering Lab of Stanford University. As it had been stripped of its video and remote control equipment, Earnest recruited Rodney Schmidt, a graduate student and radio amateur. He had to build a new television transmitter and a radio control link – based on a model airplane – as part of his thesis on automatic cars.[168] The transmitter did send images to the SAIL computer and the radio link was used to send guidance commands to the Cart.

When Schmidt started in 1967 to work on the computer control of the cart, John McCarthy, Director of SAIL became interested in the project and took over supervision from Earnest. McCarthy discussed with Rodney Schmidt "the idea of replacing the emotional and sometimes irrational human automobile driver with a more

---

[162] Davies, Alex, 2021. *Driven, The Race to Create the Autonomous Car*, p. 80. New York, London, Toronto, Sydney, New Delhi.

[163] The building (1600 Arastradero Road, midway between Page Mill Road and Alpine Road) was named after Donald Clinton Power, president of the General Telephone & Electronics Corporation. After abandoning plans to create a research center, the company donated the building to Stanford University. It was demolished in 1986.

[164] Author unknown, Take me, I'm Yours, The autobiography of SAIL, 1991, online in: http://infolab.stanford.edu/pub/voy/museum/pictures/AIlab/SailFarewell.html

[165] Moravec 1988, p. 20.

[166] Weber, Marc. 2014. Where To? A History of Autonomous Vehicles, May 8, 2014, in: http://www.computerhistory.org/atchm/where-to-a-history-of-autonomous-vehicles/.

[167] Earnest 2018.

[168] Moravec, Hans. 1980, March. *Obstacle Avoidance and Navigation in the Real World by a Seeing Robot Rover*, Ph.D. thesis, Computer Science Department, Stanford University, Chapter 2: History, in: http://www.frc.ri.cmu.edu/users/hpm/project.archive/robot.papers/1975.cart/1980.html.thesis/a02.html

efficient, safer computer-controlled navigation system" reports Lisa Nocks.[169] During this time, McCarthy wrote a visionary paper about "Computer Controlled Cars",[170] envisioning a fully autonomous vehicle. He imagined a user who should not need to "be a driver" and even had not to "accompany the car". McCarthy outlined four use cases that are very similar to those mentioned in actual debates about the potential benefits of autonomous driving: The "computer driven car" could be used by "children, old people, and the blind", it could be send to a store to pick up purchases ordered by telephone, it could park automatically after discharging the passengers (valet-parking) and it could be shared "more easily than a conventional car."

The central technical element of McCarthys paper was the idea to equip the computer with a television camera that uses "the same visual input available to the human driver." As Jason Torchinsky has highlighted, this was a central step in the direction of machine vision – the camera had not the purpose to deliver images to a human operator, its mission was to give the cart the ability to see, to detect obstacles and to avoid them.[171] Mentioning the experiments with the Stanford Cart, McCarthy recognized also that the radio link to an offboard computer had to be replaced with an onboard computer. That would make the vehicle independent of a central computer. As it could run on dedicated freeway lanes, changes of the existing road infrastructure would not be necessary. He also stated that the Stanford Artificial Intelligence-Lab "is the only organization now working on computer control of a vehicle using vision". This is probably correct because the Japanese researchers only started at the beginning of the 1970s to orient their research to visual approaches.

For McCarthy, the key goal of autonomous vehicles was greater safety: He estimated that traffic fatalities could be "reduced to a fifth" of the present numbers, i.e. 10.000 per year. But he also admitted that "complete safety cannot be guaranteed". Computer control would even create new causes for accidents – but "only half" of the remaining fatalities would be allocated to the "unreliability of the electronics". He emphasized that "a computer program capable of reliably taking care of all the contingencies that can arise in driving a car will have to be more complex than any ever written". As the failure rates of present electronical systems were not acceptable, extensive development and testing efforts would be necessary. Finally, McCarthy estimated that the development of a computer controlled car would cost millions of dollars and its price would be up to $ 800.000. Even after 20 years it would "double the price of the car".

Supported by McCarthy, Rodney Schmidt "took on as his thesis project the problem of building a vehicle that could navigate autonomously in traffic along unmodified paths using onboard sensor information."[172] He wrote a program for the PDP-6

[169] Nocks, Lisa. 2007. *The Robot: The Life Story of a Technology*, p. 102. Westport.

[170] McCarthy, John, 2019. Computer Controlled Cars, (according to the author, the paper was written around 1968 and revised in the 1970s), online in: http://jmc.stanford.edu/commentary/progress/cars.pdf (accessed October 1, 2019).

[171] Torchinsky, p. 64.

[172] Nocks, p. 102.

(Programmed Data Processor-6) computer of the laboratory[173] that used a KA10 processor (0.65 MIPS). The system occupied "an area about the size of a basketball half-court", says Earnest.[174]

This program made the cart automatically follow a high contrast white line under controlled lighting conditions. But with a speed of about 0.8 mph (~ 0.36 m/sec), it was again extremely slow. And according to Moravec, the system worked only occasionally. Finishing his thesis in 1971,[175] Schmidt estimated that the technical challenges would make future experiments "increasingly difficult and unrealistic."[176] Nevertheless, this experiment made a step forward from remote control to autonomous driving by machine vision: As the computer vision system now controlled a moving machine, this was a big milestone.

From 1970–1973, graduate student Bruce Baumgart developed a system that permitted the cart "to deduce its own position by comparing" the TV images with images from a 3D-model of the laboratory and its surroundings. He replaced the analog link by a digital one but according to Moravec, the results again were not satisfying.

In 1973 Hans Moravec joined the AI Lab and used the cart with the aim to build an obstacle avoider. His research on robot navigation with the Cart had been funded from 1973–1980 by DARPA, the National Science Foundation, and the National Aeronautics and Space Administration (NASA).[177]

Moravec rebuilt the control link and developed the hardware and the motor subroutines of the cart further. His colleague Lynn Quam worked on a vision system that used stereo imaging to locate objects. During the International Joint Conference on Artificial Intelligence (IJCAI) in 1973, Moravec drove the cart by remote control around the AI lab for demonstration purposes. But an accident interrupted his research: The cart fell off a ramp and lying on its side, acid from its batteries spilled into the electronics.[178] While the TV camera was not damaged, the video transmitter took 9 months to repair (Fig. 2.6).

In October 1979, the cart successfully crossed a chair-filled room without human intervention, navigating "exclusively by vision."[179]

---

[173] The PDP-6 had been developed by the Digital Equipment Corporation (DEC) in 1963. DEC was a startup company from MIT (Crevier, p. 63). The PDP-6 used words of 36 bits like most of these early computers. Only 23 machines of this type had been sold by DEC worldwide.

[174] Earnest 2018.

[175] Schmidt, Rodney. 1971. *A Study of the Real-Time Control of a Computer-Driven Vehicle*. PhD in Electrical Engineering. Stanford University.

[176] Schmidt cit. in: Moravec 1980, Chapter 2: History, in: http://www.frc.ri.cmu.edu/users/hpm/project.archive/robot.papers/1975.cart/1980.html.thesis/a02.html

[177] Henderson, p. 133.

[178] Moravec 1980, Chapter 2: History, in: http://www.frc.ri.cmu.edu/users/hpm/project.archive/robot.papers/1975.cart/1980.html.thesis/a02.html

[179] Moravec 1980, Chapter 7: Stereo, online in: http://www.frc.ri.cmu.edu/users/hpm/project.archive/robot.papers/1975.cart/1980.html.thesis/p07.html

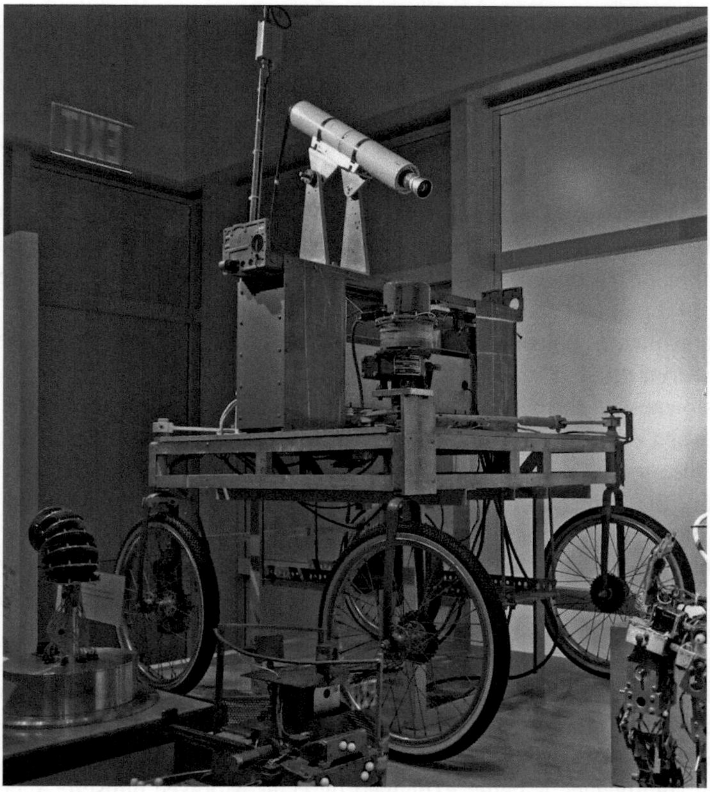

Fig. 2.6 The Stanford Cart, now in the Computer History Museum, Mountain View. (Credits: Don DeBold on FlickR, 2011, CC-BY-2.0. Image Source: https://flickr.com/photos/28156071@N00/5900601910)

The cart proceeded in different steps: After the start of the *obstacle avoiding program*, the cart asked for a destination. After being told to move a specific distance forward, the camera on the cart moved on a slider mechanism from one side to the other and took nine pictures at 6.5 cm intervals.[180] These images had been compared by the *interest operator* according to their contrast, searching for distinctive features. Another program, the *camera solver* then determined their three-dimensional spatial position and mapped the obstacles around the cart. This map allowed the moving cart to plan a path that was adjusted according to the incoming new images from the camera. The *correlator* then attempted to find all features in the other eight images. In the next step, the computer *navigator* planned a path that avoids these obstacles with a large safety margin. After receiving steering and drive

[180] Moravec 1980, Chapter 3: Overview, online in: http://www.frc.ri.cmu.edu/users/hpm/project.archive/robot.papers/1975.cart/1980.html.thesis/p03.html

commands from the off board-DEC KL10[181] computer (2.5 MIPS)[182] in the laboratory, connected by a TV link, the cart moved one meter. In the following step, the camera again took nine images and the correlator tried to find all the features that had been identified in the previous images. This reflection took always 10 to 15 min. The process continued "until the cart arrives at its destination or until some disaster terminates the program", Moravec notes sarcastically.[183]

With a velocity of "3 to 5 meters an hour",[184] the indoor course of 20 meters took about 5 h, reports Moravec: "The cart moved 1 m every 10–15 min, in lurches. After rolling a meter it stopped, took some pictures, and thought about them for a long time. Then it planned a new path, executed a little of it, and paused again." Even if this indoor run was considered a success, Moravec admits that "about one time in four, however, it failed".[185]

The slowness of the cart became difficult in outdoor runs because in the 15 minutes between each picture the sun moved and the shadows also and the system mistook the moving shadows with its edges for obstacles, ignoring the real obstacles like chairs that were overrun.[186] The "program's idea of the cart's own position became increasingly wrong".

Despite these technical difficulties, the Stanford Cart occupies an important place in the history of technology: It represents a "gigantic leap in robotics"[187] and can be considered as "the most influential robot" of the 1970s.[188] "After the years of slow progress following Shakey", it redirected attention to the robotics field, emphasizes computer scientist Robin Murphy.[189] While most of the literature about mobile robots, coming from Computer Science,[190] Computer engineering[191] and Mechanical Engineering[192] focus their narratives on the technical specifications of the Cart, its historical significance is better explained by Hans Moravec himself. According to

[181] The KL10 was the first microprogrammed PDP-10 processor, and had been commercialized by the Digital Equipment Corporation in 1975.

[182] Earnest 2018.

[183] Moravec 1980, Chapter 3.

[184] Moravec 1980, Chapter 9: Evaluation, online in: http://www.frc.ri.cmu.edu/users/hpm/project.archive/robot.papers/1975.cart/1980.html.thesis/p09.html

[185] Moravec 1988, p. 32.

[186] Moravec 1980, Preface, online in: http://www.frc.ri.cmu.edu/users/hpm/project.archive/robot.papers/1975.cart/1980.html.thesis/index.html

[187] Torchinsky, p. 64.

[188] Murphy, Robin R. 2000. *Introduction to AI Robotics*, p. 67. Cambridge/London: The MIT Press.

[189] Ibid. p. 254.

[190] Murphy 2000, p. 67, Dudek, Gregory, and Michael Jenkin. 2010. *Computational Principles of Mobile Robotics*. Cambridge, New York, Melbourne, Madrid: Cambridge University Press. (first edition: 2000). Nilsson, Nils J. 2010. *The Quest for Artificial Intelligence, a History of Ideas and Achievements*. New York: Cambridge University Press, Chapter 12.2.

[191] Meystel gives a good overview about all types of mobile robots (Meystel, pp. 36–39).

[192] Stone, Wesley L. 2005. The History of Robotics. In *Robotics and Automation Handbook*, ed. Thomas R. Kurfess, p. 6. Boca Raton, London, New York, Washington: CRC Pess.

him, it was "the first mobile robot controlled by a large computer that did *not* reason, and the first testbed for computer vision."[193] The most important point is that its approach diverged from the philosophy of Shakey by "getting its initial understanding of the world directly from its sensors" – from the bottom up – and not from a preprogrammed internal world model, emphasizes Lisa Nocks from the IEEE History Center.[194]

In 1980, the Stanford Artificial Intelligence Laboratory was shut down and its activities were merged into the Computer Science Department at the Stanford campus.[195] The cart went into storage until 1987 when it started a second career as an exhibition object of the Computer Museum in Boston. It became part of the Smart Machines Theater or Robot Theater, a "collection of artifacts on stage, lit up in sequence with some moving in their moments of glory, synched to a video, proving that even old robots can have a second career in show business."[196] After the Boston museum shut down, the robots and other artifacts were sent to its successor, the Computer History Museum in Mountain View, California. Beginning in 2011 the Stanford Cart was put on static display at the Museum, where it can be seen today.

2.3 The Mobile robots of Carnegie Mellon University

After being awarded his PhD by Stanford University in 1979, Hans Moravec moved to the Robotics Institute of Carnegie Mellon University (CMU) in Pittsburgh that had been founded in the same year by Raj Reddy, another SAIL graduate. The institute dedicated its mission explicitly to mobile robots. While the CMU Rover (1981) and the Terregator (1984) were focused on indoor purposes, the Navigation Laboratory (NavLab) of CMU began in 1986 to develop a series of outdoor robot vehicles. In more than 30 years, eleven different experimental vehicles evolved from the research. Most of the funding had been provided by DARPA.[197]

---

[193] Moravec 1988, p. 20.

[194] Nocks, p. 102.

[195] The SAIL reopened in 2004, with Sebastian Thrun becoming its new director.

[196] Earnest 2018.

[197] Thorpe, Charles E. 1990. Chapter 1: Introduction. In *Vision and Navigation: The Carnegie Mellon Navlab*, ed. Charles E. Thorpe, p. 6f. Boston.

2.3.1 *The CMU Rover (1981–1984)*

With the aim to continue the work on visual navigation that had been started with the Stanford Cart, Moravec developed a new mobile robot, the CMU Rover, later called Pluto.[198] It was one meter tall, weighed 198 pounds (90 kg) and had a cylindrical shape with a diameter of 30 cm.

The Rover can be characterized by two innovations, its steering and its sensing equipment: Its first aim was to overcome the limitations of "the kinematic structure" of the Stanford Cart that was based on a car-like steering system as we have seen in the preceding chapter.[199] The CMU Rover had "three individually steerable wheel assemblies",[200] assuring a high steering flexibility that was necessary for the use of an arm on the rover: Its mobility could be used to substitute the shoulder joint of the arm. Therefore "it could open a door with its arm, travel a straight line through the doorway, rotating about its vertical axis to maintain the arm contact holding the door open."[201]

The second aim was to improve the sensing equipment of the robot: The CMU Rover was equipped with a modified camera mount that permitted the TV camera to pan, tilt and slide, movements that were not possible with the Stanford Cart.[202] The rover also used short range infrared and long range sonar proximity sensors and "a dozen on-board processors"[203] for "high speed local decision-making, servo control and communication."[204]

But again, the most important computer power – necessary for image processing – had been located *off-board* in the laboratory: a VAX 11/780 host computer, a FPS-100 array processor and a special analog data device. These new image processing arrays, "that made picture analysis about a hundred times faster than with the old Cart",[205] were linked with the rover via a remote-control link.

But according to Chuck Thorpe, Pluto never did work very well. The advantage of the robot – "that you could control the steering and velocity of each of the three wheel sets independently" – was at the same time its disadvantage: If the wheels were not exactly synchronized, "it was fighting itself and mostly came to a stop."[206]

---

[198] Moravec, Hans. 1982. The CMU rover. *AAAI Proceedings*, p. 377.

[199] Dudek, Gregory, and Michael Jenkin. 2010. *Computational Principles of Mobile Robotics*. Cambridge, New York, Melbourne, Madrid: Cambridge University Press (first edition: 2000), without page numbers, chapter Early Autonomous Robots.

[200] Moravec 1982, p. 377.

[201] Stone, p. 7.

[202] Dudek, Gregory, and Michael Jenkin. 2010. *Computational Principles of Mobile Robotics*. Cambridge, New York, Melbourne, Madrid: Cambridge University Press (first edition: 2000), without page numbers, Chapter Early Autonomous Robots.

[203] About half of the processors were 16 bit MC 68000 s (Moravec 1982, p. 377).

[204] Moravec 1982, p. 377.

[205] Henderson, p. 135.

[206] Mail of Chuck Thorpe to the author, February 3, 2020.

Fig. 2.7 The second CMU Rover (Neptune). Image provided by CMU

For that reason, Pluto was followed by Neptune, a much simpler robot (Fig. 2.7). With a fork-mounted wheel and two parallel wheels, it resembled a child's tricycle.[207] "It would sit still while it steered its front wheel, then move at a fixed speed and fixed steering angle (executing an arc of a circle), then stop and change steering position."[208] According to Thorpe, this minimalist design worked very well and allowed experiments with vision and sonar.

The third robot was called Uranus and its complexity and ability can be situated in between Pluto and Neptune. It was an omni-directional mobile robot that used four Mecanum wheels from Sweden. Each of these special wheels had twelve free-spinning rubber rollers around its circumference.[209] This gave Uranus the capability to move in all directions.

[207] Podnar, Gregg W. 1985a, February. The NEPTUNE Mobile Robot. In *Annual Report, Technical Report CMU-RI-TR-86-04*, Robotics Institute, Carnegie Mellon University, pp. 123–125.

[208] Mail of Chuck Thorpe to the author, February 3, 2020.

[209] Podnar, Gregg W. 1985b, February. The URANUS Mobile Robot. In *Annual Report, Technical Report CMU-RI-TR-86-04*. Robotics Institute, Carnegie Mellon University, pp. 127–129.

2.3.2 *The Terregator (1984)*

While the CMU Rover was limited to indoor tests, the first outdoor tests had been made with a vehicle called Terregator (Terrestrial Navigator), launched in May 1984.[210] Use cases were defined as hazardous environments like "construction, mining, nuclear facility maintenance, material handling, and the military".[211]

The mission of the six-wheeled skid-steered vehicle (Fig. 2.8) was to follow the road autonomously, using video cameras, a distance measuring sonar and a power generator. An on-board computer received high-level commands from an off-board

Fig. 2.8 The Terregator research vehicle, 1984. (Credits & Image Source: CMU)

[210] Champeny-Bares, Lee, Coppersmith, Syd, Dowling, Kevin, 1991, May. *The Terregator Mobile Robot, Field Robotics Center*. Pittsburgh: The Robotics Institute, Carnegie Mellon University. http://citeseerx.ist.psu.edu/viewdoc/download?doi=10.1.1.68.6410&rep=rep1&type=pdf, p. 14 (Accessed on December 13, 2019).

[211] Ibid., p. 9.

supervisor and executed low-level control commands to the motors.[212] In May 1985 a laser range finder was bolted on the vehicle. It also used "acoustic sensing for obstacle avoidance".[213] Davies reports that in the late 1980s Terregator was the first autonomous vehicle that had been equipped with a LIDAR.[214]

In October 1984, the Terregator "had made its first runs under the control of its vision system, albeit slowly and with frequent halts" reports the historian Alex Roland.[215] But a CMU-video states that while the movements of the Stanford Cart and the CMU Rover had been interrupted by regular stops, the vehicle "kept moving continuously". Instead of processing single images, the Terregator processed image streams.[216] Thorpe explains this contradiction by the fact that the vehicle "was used by multiple groups, using different techniques".[217] While the early work on stereo processing needed to stop and start, a more advanced version of Terregator with sonar processing allowed it to move continuously.

The Terregator mapped a portion of a mine,[218] "was capable of climbing stairs and bouncing across railroad tracks, though it did not run well on grass."[219] The outdoor runs were performed in Schenley Park near the campus, where the engineers tested techniques to detect edges or color contrasts of a road. But it was very difficult for the black-and-white cameras to distinguish the edges of the grey asphalt in Schenley Park. The robot sometimes mistook tree trunks with their sharp, firm edges for the road "and tried to climb it", reports Moravec.[220] This is also confirmed by William Whittaker.[221] Finally the researchers had even "to use masking tape to mark the roads".[222] As a lesson learned from these experiments, the Autonomous Land Vehicle (ALV), developed by Martin Marietta (see Sect. 2.4) did not rely on straight lines only.

The main problem was that the Terregator remained radio-controlled by either an off-board human supervisor or a remote stationary 1-MIPS-computer.[223] It moved

---

212 Ibid., p. 28.

213 Cmurobotics (Youtube-account), Terregator – CMU – Field Robotics Circa 1983 (video title). In Youtube, uploaded on October 13, 2016, online in: https://www.youtube.com/watch?v=beGRM9ZmWlI (Accessed on December 13, 2019).

214 Davies, p. 100.

215 Roland, Alex, and Philip Shiman. 2002. *Strategic Computing, DARPA and the Quest for Machine Intelligence*, p. 229. Cambridge, London.

216 Todd Jochem (Youtube-account), Navlab 84–94 (video title), Carnegie Mellon University, Navlab: Vision & Navigation 1984–1994, Promotional Video, 1994. In Youtube, uploaded on January 19, 2018, online in: https://www.youtube.com/watch?v=yfVxt-eBVLo, 2:44 (accessed on December 3, 2019).

217 Mail of Chuck Thorpe to the author, February 3, 2020.

218 Champeny-Bares et al. 1991, p. 7.

219 Roland, p. 236.

220 Moravec 1999, p. 43.

221 Schefter, Jim. 1985, October. Look, Ma! No Driver. *Popular Science*, p. 114.

222 Roland, p. 230.

223 Champeny-Bares et al. 1991, p. 9, 28.

with a speed of only a few centimeters per second because the "communications bandwidth between the vehicle and its off-board computers was far too low to transmit all of the data with sufficient speed," explains Alex Roland.[224] A secondary source reports a speed of 1 km/h and a cycle time of 3 seconds.[225] The limited processing power, the use of black and white cameras and early algorithms were other reasons for the slow movement, adds Thorpe.[226]

It now was evident that a robot car should function independently from off-board equipment. Therefore, the CMU-team replaced the Terregator in 1986 with a new vehicle, the Navlab 1, that avoided the transmitting problem by an onboard solution. The new configuration also allowed outdoor tests far away from the laboratory. "Mobile robots are where Artificial Intelligence meets the real physical world"[227] emphasizes Chuck Thorpe: "Only by building real robots, running tests in realistic environments, then going back to the lab to start anew, are we able to claim confidently that our theories are appropriate and that our systems work."[228]

2.3.3 The NavLab (1986–1990), Later Called NavLab 1

As part of the big national program "On Strategic Computing" – that will be presented in more detail in Sect. 2.5 – DARPA funded in 1984 the research on the Autonomous Land Vehicle.[229] At the same time, Chuck Thorpe pursued his doctoral degree in the Computer Science department at CMU. He "wrote a proposal to DARPA describing techniques for driving under hazardous conditions. As a result, the agency funded the first NavLab."[230]

In 1986, the team of William "Red" Whittaker equipped a commercially purchased blue Chevy panel van (Fig. 2.9) with forward and backward facing color TV-cameras and a big ERIM laser-range finder. It scanned 40 meters of the road ahead, providing a 3-D image of obstacles.[231] Sonar sensors were also used for finding obstacles.

Called NavLab, later NavLab 1, the Chevy was used by Takeo Kanade and Chuck Thorpe to build a new generation vision system. It was the first self-contained vehicle of CMU. As a lesson learned from the Terregator project, the van carried its

[224] Roland, p. 236.

[225] Meissner, p. 1.6

[226] Mail of Chuck Thorpe to the author, February 3, 2020.

[227] Thorpe, Charles E. 1990. Chapter 1: Introduction. In *Vision and Navigation: The Carnegie Mellon Navlab*, ed. Charles E. Thorpe, p. 1. Boston.

[228] Ibid., p. 2.

[229] Kanade, Takeo. Foreword. In *Vision and Navigation: The Carnegie Mellon Navlab*, ed. Charles E. Thorpe, p. xiii.

[230] Nocks, p. 106.

[231] Thorpe, Charles, Martial Hebert, Takeo Kanade, and Steven Shafer. 1991a, August. Toward Autonomous Driving: The CMU Navlab, Part I, Perception. *IEEE Expert* 6 (4), p. 37.

Fig. 2.9 The Navlab 1 Chevy van during experiments in snow. (Credits & Image Source: CMU)

own power supplies, was equipped with on-board computer equipment and provided even enough space for five graduate students.[232] A WARP supercomputer (10*10 MFLOPS) and four Sun-3 Workstations were charged with image feature extraction, route planning and analysis. The image evaluation times were 10 seconds (1 km/h) with the SUN-computers and 4 seconds (3 km/h) with the WARP-computer.[233] An inertial navigation system helped tracking the position of the vehicle. It also had a "68,020-based processor" for controlling the steering wheel and longitudinal accelerations.[234] The computer equipment had a worth of $1/2 million.[235]

Unlike the Autonomous Land Vehicle, the NavLab could also be driven by a human driver to reach a test site. It was even tested – without permission – on public streets.[236] But a safety driver was always behind the steering wheel and a big red Stop-button was available.

[232] Roland, p. 243.

[233] Thorpe, Thorpe, Charles, Martial H. Hebert, Takeo Kanade, and Steven A. Shafer. 1988, May. Vision and Navigation for the Carnegie-Mellon Navlab. *IEEE Transactions on Pattern Analysis and Machine Intelligence* 10 (3): 362–373. See also: Mysliwetz, Birger, 1990. Parallelrechner-basierte Bildfolgen-Interpretation zur autonomen Fahrzeugsteuerung, Diss. UniBwM/LRT/ISF, p. 14.

[234] Pomerleau, Dean A. 1993. *Neural Network Perception for Mobile Robot Guidance*, p. 4. New York.

[235] Nocks, p. 106.

[236] Thorpe, Chuck, An oral history conducted in 2010 by Peter Asaro and Selma Šabanović, Indiana University, Bloomington Indiana, for Indiana University and the IEEE, in: http://ethw.org/Oral-History:Chuck_Thorpe#Interview (accessed on Mai 7, 2018).

During early tests, the Navlab 1 ran at "two centimeters per second along clean sidewalks",[237] later "at ten centimeters per second"[238] and with the vision and range algorithms of the Warp machine at "50 cm/s".[239] The engineering team explains that the acquisition rate of the laser range finder was "too slow for high-speed highway driving".[240] In 1987, the research vehicle advanced with a speed of 2.7 mph (4,3 km/h), following a preprogrammed path. This is about walking speed, as Thorpe confirms in a TV-emission from History Channel.[241]

To control the van at these very low speeds, the vehicle was equipped with a computer controlled hydraulic drive train: Members of the NavLab team indicate in an article from 1990 that "low gear supports low-speed experimentation (0-20 km/h); high gear (0 – 40 km/h) transports the vehicle along public roads in manual mode".[242] That means that at this time the experiments on automated driving were mechanically limited to 20 km/h. But a report from 1991 gives a slightly higher speed indication, affirming that "on more typical structured roads, the Navlab drives up to its mechanical limit of 28 kilometers per hour." (17,3 mph).[243] A report for the 1995 Intelligent Vehicles Conference states that at the end of the 1980s, Navlab 1 reached its top speed of 20 mph[244] (32,1 km/h) "on a variety of real roads".[245] Hans Moravec reports that the van reached in 1990 – using faster computers (10 MIPS) – a top speed of 40 km/h (24,8 mph). But he also adds that "almost every time a new road type or condition was encountered, the programs need tinkering."[246] A statement that Todd Jochem confirmed as "100 % true".[247]

---

[237] Thorpe, Charles, Martial Hebert, Takeo Kanade, and Steven Shafer. 1991b, August. Toward Autonomous Driving: The CMU Navlab, Part II, System and Architecture. *IEEE Expert* 6 (4), p. 51.

[238] Roland, p. 243.

[239] Thorpe, Charles, Martial H. Hebert, Takeo Kanade, and Steven A. Shafer. 1988, May. Vision and Navigation for the Carnegie-Mellon Navlab. *IEEE Transactions on Pattern Analysis and Machine Intelligence* 10 (3), p. 372.

[240] Thorpe, Charles, Martial Hebert, Takeo Kanade, and Steven Shafer. 1991a, August. Toward Autonomous Driving: The CMU Navlab, Part I, Perception. *IEEE Expert* 6 (4), p. 37.

[241] Cmurobotics (Youtube-account), History Channel 1998: Driverless Car Technology Overview at Carnegie Mellon University (video title). In Youtube, uploaded on September 29, 2017, online in: https://www.youtube.com/watch?v=2KMAAmkz9go, min. 4:06 (accessed on December 3, 2019).

[242] Dowling, Kevin, Robert Guzikowski, Jim Ladd, Henning Pangels, Sanjiv Singh, and William Whittaker. 1990. Navlab. An Autonomous Navigation Testbed. In *Vision and Navigation. The Carnegie Mellon Navlab*, ed. Charles E. Thorpe, pp. 273–275. Boston/Dordrecht/London.

[243] Thorpe, Charles, Martial Hebert, Takeo Kanade, and Steven Shafer. 1991a, August. Toward Autonomous Driving: The CMU Navlab, Part I, Perception. *IEEE Expert* 6 (4), p. 31.

[244] Jochem, Todd, Pomerleau, Dean, Kumar, Bala, Armstrong, Jeremy, PANS: A Portable Navigation Platform. In *Proceedings of the Intelligent Vehicles '95. Symposium*, p. 107, also in: https://www.cs.cmu.edu/~tjochem/nhaa/navlab5_details.html (accessed on August 3, 2018).

[245] Thorpe, Charles, Martial Hebert, Takeo Kanade, and Steven Shafer. 1991b, August. Toward Autonomous Driving: The CMU Navlab, Part II, System and Architecture. *IEEE Expert* 6 (4), p. 51.

[246] Moravec 1999, p. 43.

[247] Mail of Todd Jochem to the author, May 10, 2020.

The sources cited do not permit to obtain a more precise description of the circumstances during which these speeds were obtained. It can only be resumed that around 1990 NavLab 1 reached a speed between 17,3 and 24,8 mph (28–40 km/h). Todd Jochem confirms to the author of this book that "about 20 mph is about right". As the exact road and obstacle type is not mentioned in the sources, the value of these indications is nevertheless limited.

2.3.4 The NavLab 2 Army Vehicle (1990)

From 1990–1996, CMU focused on the Demo II program of DARPA that aimed to demonstrate Unmanned Ground Vehicles (see Sect. 2.5). In the 1990s, the research focus shifted "from on-road operation to operation in cross-country environment, including rough terrain, multiple vehicles, complex missions."[248] For that reason, a modified High Mobility Multipurpose Wheeled Vehicle (HMMWV) – pronounced Humvee – was used as new test vehicle. It was provided by the United States Army Tank-automotive and Armaments Command (TACOM) and called NavLab 2. The NavLab1 Chevy Van was not replaced, the laboratory just doubled its testing capability with a second vehicle, explains Jochem.[249]

The military jeep was seen by the engineering team as "a good platform for off-road navigation, where extra ruggedness is necessary and short (less than 10 miles) missions are the norm. It is not well suited for on-road driving research because of its size, complexity, and temperamental operational nature."[250] For instance, its excessive weight of 2361 kg was responsible for poor acceleration performance.

The vehicle was equipped with computer controlled actuators on the throttle, brake and steering mechanisms. Stereo cameras and a laser range finder were used for obstacle avoidance, a video camera for road following and teleoperation, a Doppler radar for ground velocity and sensing, as well as a GPS and an inertial navigation system for attitude and position sensing.[251] Data processing had been assured by three Sparc 10 workstations (20-MIPS). With these computers, obstacle detection was achieved "in 600 milliseconds" and the acquisition time averaged 500 milliseconds according to the engineering team.[252] But the speed of obstacle detection does not say anything about its quality: Some sources indicate that Navlab 2 was

[248] Hebert, Martial, Charles E. Thorpe, and Anthony Stentz. 1997. Introduction. In *Intelligent Unmanned Ground Vehicles, Autonomous Navigation Research at Carnegie Mellon*, ed. Martial H. Hebert, Charles E. Thorpe, and Anthony Stentz, p. 1. New York.

[249] Mail of Todd Jochem to the author, May 10, 2020.

[250] Jochem et al. 1995, p. 107.

[251] Hebert et al., p. 3.

[252] Thorpe, Charles, Martial Hebert, Takeo Kanade, and Steven Shafer. 1991a, August. Toward Autonomous Driving: The CMU Navlab, Part I, Perception. *IEEE Expert* 6 (4), p. 38.

"unable to distinguish between rocks and tall grass, trees and hillsides, or even mud and hard ground."[253]

The goal of the NavLab 2 testbed was "to develop techniques that enable artificial neural networks to guide mobile robots using visual input" explains Dean Pomerleau, at this time a PhD student in the Navlab group.[254] He developed a trainable road tracker named *Autonomous Land Vehicle In a Neural Network (ALVINN)* that served as navigation system. In contrast to earlier rule-based systems that used if-then instructions – *if* this condition is met, *then* the program will execute this action – neural networks can choose the best solution from several options and most importantly, they are able to learn: An artificial neural network has to be trained by an operator how to drive.

Two Motorola 68,000-computers were used for low level control mechanisms[255] and two other computers for perception processing.[256]

They used a machine learning approach: While driving his car, Pomerleau filmed the road and fed images of road markings and data about the steering direction to the computer system that learned when it had to turn the steering wheel.[257] Moravec recalls that early versions of the neural network "required hours of training, provided by simulator scenes or long videotapes of human road trips."[258] Later, the learning time was reduced to 5 min. "For long trips, the system was provided with the neural interconnection weights for many road types. It tested against the current road, using consistency of the steering output as measure of suitability."[259] It received as input visual road information as well as the current position of the steering wheel and produced as output the steering commands.

Again, different indications of the image evaluation time can be found in the literature: A journal article mentions that ALVINN needed "at about a fifteenth of a second per image"[260] in the road following mode. Another source indicates that it digitized 10 images per second.[261] The difference between 10 Hz and 15 Hz might be due to "what computer hardware ALVINN was running on; it evolved over sev-

---

[253] De Vries, Marten Johannes, Kachroo, Pushkin, and Ozbay, Kaan, eds. 1998. Intelligent transportation systems. In *Proceedings of the Society of Photo-Optical Instrumentation Engineers (SPIE)*, vol. 3207, 15–17 October 1997, p. 98. Pittsburgh.

[254] Pomerleau 1993, p. 5

[255] Jochem et al. 1995, p. 107.

[256] Todd Jochem (Youtube-account), Navlab 2 on Flagstaff Hill Pittsburgh People, Places, and Things (video title), TV-emission from 1994, Reporter: Lisa Beatty, published on January 7, 2018, in: https://www.youtube.com/watch?v=btJDzWt82wU (accessed on August 3, 2018).

[257] Davies, p. 83.

[258] Moravec 1999, p. 43.

[259] Ibid., p. 43f.

[260] Thorpe, Charles, Martial Hebert, Takeo Kanade, and Steven Shafer. 1991a, August. Toward Autonomous Driving: The CMU Navlab, Part I, Perception. *IEEE Expert* 6 (4), p. 36.

[261] Todd Jochem (Youtube-account), Navlab 84–94 (video title), Carnegie Mellon University, Navlab: Vision & Navigation 1984–1994, Promotional Video, 1994. In Youtube, uploaded on January 19, 2018, online in: https://www.youtube.com/watch?v=yfVxt-eBVLo, min. 2:44 (Accessed on December 3, 2019).

eral years and several computer architectures, including the WARP", explains Thorpe.[262]

These technical modifications may also explain why the speed indications for the Navlab 2 differ from year to year: Moravec reports that in 1991 ALVINN "traversed 30 km of busy highway at 70 km/h" (43 mph).[263] A 1992 promotional video from CMU indicates that Navlab 2 did "run as fast as 55 mph and as far as 22 miles".[264] 55 mph is also confirmed in a History Channel film.[265] An article in the Proceedings of the Intelligent Vehicle Conference 1993 precises that the Navlab 2 "safely followed a lead vehicle on a winding road in light traffic at 32 km/h." (19 mph).[266] In 1994, Todd Jochem indicated that the Navlab 2 drove up to 25 mph (40 km/h) on city roads.[267] A CMU-Video from 1994 indicates that ALVINN has driven as far as 100 kilometers "and at speeds as over 110 km/h" (68 mph),[268] an information that was confirmed by Todd Jochem.[269] But according to Thorpe, this result was not reached with the Navlab 2 army jeep but with Navlab 5, a Pontiac Minivan (see Sect. 2.3.6).

To understand these exceptional performances for an automated vehicle, it must be highlighted that ALVINN was only doing lane following and did not detect and avoid obstacles, explains Chuck Thorpe.[270] A report of the engineering-team for the Intelligent Vehicle-conference 1995 precises that the software steered the vehicle "over rough terrain, avoiding obstacles, at speeds up to 6 mph and on-road at 70 mph."[271] But Jochem comments that full obstacle detection was only achieved "in 1996 and 1997 on the same roads" using NavLab 6 and 7.[272]

---

[262] Mail of Chuck Thorpe to the author, February 3, 2020.

[263] Moravec 1999, p. 44.

[264] Todd Jochem (Youtube-account), Navlab 2 Cross Country Navigation 1992 (video title), Official Video from CMU, 1992. In Youtube, uploaded on January 8, 2018, in: https://www.youtube.com/watch?v=HmLafSKdCRs, min 1:43 (Accessed on December 3, 2019).

[265] Cmurobotics (Youtube-account), History Channel 1998: Driverless Car Technology Overview at Carnegie Mellon University (video title). In Youtube, uploaded on September 29, 2017, online in: https://www.youtube.com/watch?v=2KMAAmkz9go, min. 4:06 (Accessed on December 3, 2019).

[266] Sukthankar, R., RACCOON: A Real-time Autonomous Car Chaser Operating Optimally At Night. In Proceedings of the Intelligent Vehicles '93 Symposium, p. 37.

[267] Todd Jochem (Youtube-account), Navlab 2 on Flagstaff Hill Pittsburgh People, Places, and Things, TV-emission from 1994, Reporter: Lisa Beatty, published on January 7, 2018, in: https://www.youtube.com/watch?v=btJDzWt82wU, min. 6:18 (Accessed on December 3, 2019).

[268] Todd Jochem (Youtube-account), Navlab 84–94 (video title), Carnegie Mellon University, Navlab: Vision & Navigation 1984–1994, Promotional Video, 1994. In Youtube, uploaded on January 19, 2018. In Youtube, online in: https://www.youtube.com/watch?v=yfVxt-eBVLo, min. 2:59 (Accessed on December 3, 2019).

[269] Mail of Todd Jochem to the author, May 10, 2020.

[270] Mail of Chuck Thorpe to the author, February 3, 2020.

[271] Jochem et al. 1995, p. 107 (Note: For better readability we decided to substitute m.p.h. in the original source with mph).

[272] Mail of Todd Jochem to the author, May 10, 2020.

2.3.5 The NavLab 3 and 4 Vehicles (1994)

The NavLab 3 and 4 vehicles should also be mentioned, even if Moravec classifies them as "unimportant".[273] Thorpe reports that in 1994, Dean Pomerleau preferred to use another research vehicle that is easier to drive than the Hummer. He had the idea to take his own Honda Accord, modifying it with a "robot surplus motor driving a belt to turn the steering wheel". He also "suction-cupped a camera at the rearview mirror, he put a laptop computer in there and plugged it into the cigarette lighter, and he had a robot car."[274] The vehicle, called Navlab 3, was used to test the algorithms with less expensive computer equipment. It drove 1000 miles on the highway, in traffic and attained a top speed of 60 mph "but only doing lane following".[275] As the operator controlled speed and braking, it was only a semi-automated vehicle.[276]

Like the Navlab 2, the Navlab 4 was also a Humvee but it was equipped with a "new scanning laser rangefinder jointly developed by K2T and CMU."[277]

2.3.6 The NavLab 5 Test Drive No Hands Across America (1995)

On July 23, 1995, Dean Pomerleau and Ph.D. student Todd Jochem from the Robotics Institute of Carnegie Mellon University used a semi-automated car for a journey from Pittsburgh, Pennsylvania to San Diego. The test drive had been promoted as "No Hands Across America" and it aimed at steering by vision while the human operators handled throttle and brake. The research vehicle – a 1990 Pontiac Trans Sport donated by Delco Electronics – was called NavLab 5 (Fig. 2.10).[278]

The main idea was to make as few modifications as possible to the vehicle, and take advantage of improved efficiency of electronics to power everything from the existing vehicle generator. The test drive was also made to see how the system will react to different types of roads across the country. It was part of a research project

[273] Moravec 1999, p. 44.

[274] Thorpe 2010.

[275] Mail of Chuck Thorpe to the author, February 3, 2020.

[276] Mail of Todd Jochem to the author, May 10, 2020.

[277] Thorpe, Chuck. 1998. Mobile Robots and Smart Cars. In *Field and Service Robotics*, ed. Alexander Zelinsky, p. 3. Berlin, New York.

[278] The research program had been financed by Delco Electronics, AssistWare Technology and Carnegie Mellon University, with additional support from Andrew Corporation and RDI Computer Corporation.

Fig. 2.10 The NavLab 5, a Pontiac Trans Sport. (Credits & Image Source: Todd Jochem)

funded by NHTSA and the National Automated Highway System Consortium (NAHSC)[279] that will be presented in Sect. 2.6 of this book.

The size of the equipment was much smaller than that of the earlier Navlabs. It consisted of a portable laptop computer (50 MIPS), a windshield camera over the rearview mirror and a GPS receiver, all powered from the cigarette lighter of the car.[280] The vision system was very simple: The computer used the images from the camera to determine the position of the vehicle on the road. It used lane markings, road edges, ruts, lane discolorations and the position of the car ahead to identify the center of the lane. "Autonomous lane keeping, lateral roadway departure warning" and a map based curve warning mechanism ran on the *Portable Advanced Navigation Support (PANS)* that used input from GPS and a fiber optic rate gyro. The system had the advantage to be based on commercially available components that could be used in a passenger vehicle. The position estimation and vehicle control had been processed on a Sparc LX class portable workstation. As no detailed road maps existed in 1995, the GPS was only used to calculate the speed, not to navigate. At this time, GPS had a selective ability, meaning that a high-accuracy positioning was very expensive.

[279] U. S. Department of Transportation, Request for Applications to Establish a National Automated Highway System Consortium, Contract number DTFH61-94-X-0000, December 15, 1993.

[280] Jochem, Todd, Pomerleau, Dean, No Hands Across America General Information, (without date) https://www.cs.cmu.edu/~tjochem/nhaa/general_info.html (Accessed on August 3, 2018).

During this time, CMU abandoned the neural network approach: Around 1994 the general learning mechanism of ALVINN was replaced by a more powerful program with the name *RALPH (Rapidly Adapting Lateral Position Handler)* that could handle lane changing in heavier traffic. As it was almost impossible to teach domain knowledge to a neural network, "the known constraints were built in, and only the changeable parts were learned" explains Moravec.[281] It could also learn more quickly, "almost instantly", adds Jochem.[282] This was a big key to its success.

But it must be emphasized that only the steering commands (lateral control) had been given by RALPH. The human drivers had to handle throttle and brake mechanisms (longitudinal control) of the Pontiac – even if in limited traffic the cruise control did much of the longitudinal control.[283]

During the *No Hands Across America* test-run, the Navlab 5 drove successfully "in the dark, in heavy snow, and in driving rain."[284] Only newly paved roads without lane markings, low sun angles and shadows forced the security drivers to retake the steering wheel.

When they arrived at their destination on July 31, 1995, RALPH had successfully steered the vehicle during 98.2 percent of the 2800-mile trip without intervention of a human driver. The participants claimed that the average speed was above 60 mph (100 km/h).[285]

After the successful test drive, the NavLab team had been invited to take part in a demonstration of an Automated Highway System in San Diego 1997. At this presentation of automated vehicles, the NavLab buses and cars drove with speeds around 60 mph in a closed section of new high occupancy vehicle (HOV) lanes of Interstate 5. This event is further detailed in Sect. 2.6.3.

2.4 The Strategic Computing Initiative and the ALV-Program of DARPA (1984–1988)

While AI research suffered of funding cuts in the 1970s, the technological and cultural contexts of the 1980s were more favorable. The success of personal computers like the IBM-PC 5150 (1981), the first flight of the Space shuttle (also 1981) and the success of Science Fiction Films like Star Wars V and VI (1980, 1983) are only three examples for a decade that embraced technology. The field of computer vision

[281] Moravec 1999, p. 44.

[282] Mail of Todd Jochem to the author, May 10, 2020.

[283] Mail of Chuck Thorpe to the author, February 3, 2020.

[284] Jochem, Todd, Pomerleau, Dean, No Hands Across America Journal. http://www.cs.cmu.edu/~tjochem/nhaa/Journal.html (Accessed on August 3, 2018).

[285] On a close test track the NavLab 5 reached even 90 mph according to the engineers (Jochem et al. 1995, p. 107).

developed considerably and the demand of the private sector boomed, focusing on knowledge based systems.[286]

In this historical context, DARPA made a push into computer science and important funding became available. In October 1983, the agency prepared a ten-year plan with a budget of $1 billion on AI-research, the *Strategic Computing Initiative (SCI)*.[287] To justify the massive spending, DARPA pointed to the Fifth Generation Computer System (FGCS) in Japan, started by MITI in 1981. This research program aimed at the development of multi-processor computers for knowledge processing, based on massive data-bases. It was an attempt to make Japan the leader in computer technology. Armed with a budget of $850-million, Japan had the ambition to create an artificial intelligence that could "carry on casual conversations, translate languages, interpret pictures, and reason like human beings."[288] For the American Government, this project was seen as a menace of its technological dominance in the market of computer technology. Another Sputnik-shock had to be avoided.

But finally, the main goal of the FGCS-project – develop a software that could think like human intelligence can – was never realized. The failure of the Fifth Generation Computer System-project showed once again that the bottleneck of AI is not the hardware but the software development. While the hardware has always "super-human capabilities" in speed and sensing, software remains the main problem, confirms Jochem.[289]

DARPA's Strategic Computing Initiative had the goal of developing the world's most advanced computer system for real-time applications. Full machine intelligence – a machine that can see, hear, speak and think like a human – was the goal.[290] Three areas have been considered, corresponding to the aims of the three main branches of the military – Air Force, Navy and Army:

The Pilot's Associate program had the goal to develop "an interactive computer system that would aid a combat aircraft commander".[291] Besides this "R2D2"-like electronic copilot,[292] a "ship borne strategic knowledge base" was to be developed: The Battle Management System should assist the commander-in-chief of the U.S. Pacific fleet to supervise the operation of nearly 300 ships in the Pacific and Indian Ocean regions.[293] The third project funded was the Autonomous Land Vehicle (ALV), developed by the aerospace company Martin Marietta that later merged with Lockheed. It started in August 1984 and had been funded with $10.6 million over a period of 42 months.[294] The idea of using a vehicle to develop AI technologies can be seen as "a key element separating the U.S. and Japanese fifth-generation com-

[286] Crevier, p. 201f.

[287] Weinberger, p. 279f.

[288] Crevier, p. 211f.

[289] Mail of Todd Jochem to the author, May 10, 2020.

[290] Roland, p. 4.

[291] Nilsson, p. 289.

[292] Crevier, p. 214.

[293] Nilsson, p. 290.

[294] Ibid., p. 292, Roland, p. 222.

puter programs" because it forced the engineering teams to test their algorithms quickly on hardware.[295] This advantage was also recognized in Japan where the ALV-project inspired MITI together with Nissan and Fujitsu to start the "Personal Vehicle System" (see Sect. 3.5.1).

More than the other projects, the ALV was "the showcase program" of the SCI-initiative, emphasizes the historian Alex Roland.[296] The New York Times called the ALV one of the "most publicized goals" of the Strategic Computing Initiative.[297] Public demonstrations of the vehicle were held and the press was invited.

Already in 1981 DARPA decided that it was time to develop autonomous vehicles able to travel large distances "into war zones or other hazardous areas" without a human operator, writes Hans Moravec.[298] Daniel Crevier specifies that the ALV-project aimed at a robot vehicle for "missions as weapons delivery, reconnaissance, ammunition handling, and rear area supply."[299] It should "rove over rugged terrain as fast as a running soldier, skirt around obstacles" and "travel insectlike". Nilsson emphasizes that the ALV "was to be the forerunner of military vehicles that could move, unguided, on roads and over rough terrain".[300] Computer vision programs were seen as crucial to inform these vehicles about their environments. These mobile robotic vehicles "would have to avoid such hazards as other vehicles, rocks, trees, ditches, water obstacles, and steep or muddy terrain. They would also have to be able to identify landmarks and other significant objects in their immediate surroundings. They were to be entirely self-contained with all computing to be done on board."

In 1984, officials from DARPA and the Engineering Topographical Laboratory at Fort Belvoir selected Martin Marietta as prime contractor. "Martin had experience in robotics and the construction of remotely operated vehicles, and it had a research site outside of Denver, ideal for testing the vehicle."[301] Azriel Rosenfeld from the University of Maryland who developed an image understanding program with DARPA, assisted Martin Marietta. According to Roland, he was also the first who suggested an ALV. Chuck Thorpe and Takeo Kanade from Carnegie Mellon University were charged with the road following and parallel algorithms. Their experience with Terregator and Pluto/Neptune, two earlier testbeds of CMU for on-board sensors (see Sects. 2.3.1 and 2.3.2), made them the "obvious choice", writes Roland.[302]

Other contractors from academia and industry participated also in the project: The Massachusetts Institute of Technology worked on parallel algorithms, the University of Rochester developed the parallel programming environment, SRI

[295] Schefter, Jim. 1985, October. Look, Ma! No Driver. *Popular Science*, p. 116.

[296] Roland, p. 233.

[297] Weinberger, p. 280.

[298] Moravec 1988, p. 21, see also Nocks, p. 105.

[299] Crevier, p. 210f.

[300] Nilsson, p. 292.

[301] Roland, p. 222.

[302] Ibid., p. 224.

Fig. 2.11 The Autonomous Land Vehicle (ALV) weighed eight tons and was 2.7 meters wide, 3 meters high and 5.2 meters long. (Credits: Martin Marietta. Image Source: https://paleofuture.com/blog/2013/4/30/darpa-spent-1-billion-trying-to-build-a-real-life-skynet-in-the-1980s)

worked on spatial representation, Hughes Research Labs on obstacle avoidance, ADS on object modeling, Columbia University on vision algorithms, the University of Southern California in San Diego and the University of Massachusetts in Amherst on target motion detection and tracking, Honeywell on the tracking of moving targets and General Electric developed three-dimensional models based on two dimensional visual images.

The ALV was an eight-ton vehicle based on an all-terrain chassis with four axles and eight not steerable wheels mounted on each side (Fig. 2.11). Driven by a diesel engine, the wheels were linked by chains so the vehicle could navigate by skid steering "like a tank, stopping one set of wheels while the other continued to move".[303] It had also a very sophisticated active suspension system – "when it wanted to turn it would unweight the corner wheels, putting more of the weight on the central wheels, which let it turn much more easily."[304]

The one-piece body shell, "nine feet wide, ten feet high, and fourteen feet long",[305] was made of fiberglass and painted in white. Its design made the historian Alex Roland describe the ALV as "large, ungainly, box-shaped monster" with a

[303] Roland, p. 226.

[304] Mail of Chuck Thorpe to the author, February 3, 2020.

[305] Roland, p. 226.

"cyclopean eye on the front". It "looked more 1950s camp science fiction than *Terminator*",[306] comments Sharon Weinberger.

An officer of the army complained that the ALV would be useless for the military – huge, slow and painted in white, it would become easily a target on the battlefield.[307] He misunderstood the aim of the vehicle: The vision system was the "primary capability that DARPA wanted to demonstrate on the ALV", emphasizes Roland.[308] DARPA funded vision research already in the 1960s, and image understanding since the 1970s. But "high-level, real-time, three-dimensional" image understanding had not been developed yet.

Roland points out that computer vision is a difficult task: The image of a video camera must first be transformed in a binary form that a computer can read. Second, the computer had to create an internal representation of this image – at least in the ALV. Third, it should extract information from the image – and "this is the hardest part". All these functions are computationally intensive, they take a lot of time – but a moving vehicle must do this in real-time. This would only be possible with parallel processing, the simultaneous execution of calculating processes.

To assure perception, two "eyes" were mounted at the front of the vehicle: A color video camera – not analog, but with several CCDs – on the top of the ALV tracked road edges.[309] In 1985, the scanning of a single video frame took about 1/30 s, but the processing of the image in a computer took about 2 s. For that reason, real-time decisions were not possible. But according to Azriel Rosenfeld this was already a progress – when the program was started 1 year before, image processing took 2 min.[310]

The second "eye" was hidden behind a big rectangular window at the front of the ALV. It contained a two-axis laser range finder from the Environmental Research Institute in Michigan – like the one that was used for the NavLab 1.[311] The laser scanned the road for ½ second, and during the next 3 s the Vicom computer analyzed the image.[312] The system detected obstacles with at least 0.5 m in height.[313] Both methods, video and lidar, had specific advantages: Video gave better angular resolution and laser better range accuracy, explains David Morgenthaler, head of Martin Marietta's computer-vision unit.[314]

The images were analyzed by onboard computers powered by a 12-kw diesel auxiliary electrical power unit. The ALV contained three racks of several Intel computers – the main computer based on the Intel 80286 processor – for vehicle

[306] Weinberger, p. 328.

[307] Roland, p. 246.

[308] Ibid., p. 220.

[309] According to Thorpe, the output of the CCDs was still analog, and had to be converted into digital inputs later (mail of Chuck Thorpe to the author, February 3, 2020).

[310] Schefter, Jim. 1985, October. Look, Ma! No Driver. *Popular Science*, p. 114.

[311] Roland, p. 229.

[312] Schefter, p. 114.

[313] Strachan, p. 53.

[314] Schefter, p. 94.

navigation, a VICOM image processor and an AC system. The computer equipment worked at a speed of 50 MIPS.[315] The data that had been obtained by the vision system was used "to generate a model of the scene ahead. Higher level reasoning was performed by goal seeker and navigator modules, which then passed the desired path to the pilot module that actually steered the vehicle", explains Douglas W. Gage.[316]

In May 1985, the ALV successfully managed its first demonstration at Martin's test site near Denver.[317] It followed a straight paved road of 1 km with an average speed of 3 km/h.[318] According to a paper of the engineering team, the ALV traversed in June 1986 a 4.2-km test track in Denver at speeds of up to 10 km/h, negotiating sharp curves and varying pavement types.[319] In October 1986 the ALV "steered around obstacles while remaining on the road, and speeds up to 20 km/hr were achieved on a straight, obstacle-free road."[320] But Mysliwetz[321] and Meissner[322] object that with an image evaluation time of 2–3 s – as mentioned above -, the ALV then advanced at this speed completely blind during 10–15 meters.

In 1987, cross-country experiments started. The ALV had been improved with "two VICOM computers for stereo vision, a Butterfly for navigation-reasoning, and a Warp parallel machine with a SUN front end for image processing."[323] In August 1987, the ALV "performed the first autonomous cross-country traverse based on sensor data"[324] along a 0.6 km off-road track at speeds up to 3 km/h "while avoiding ditches, rocks, trees, and other small obstacles", claims Gage.[325] According to Roland, this was "the first time an autonomous vehicle drove itself cross country according to a route it selected itself, based only on digital map data and sensory input. The vehicle successfully drove around gullies, bushes, rock outcrops, and steep slopes."[326]

Other reports claim that the ALV finally achieved in November 1987 a top speed of 20 km/h "with an improved color segmentation algorithm using a Bayes'

315 Moravec 1999, p. 41.

316 Gage, p. 3, see also Nilsson, p. 292.

317 Roland, p. 226.

318 Ibid., p. 228. Confirmed in Turk, p. 357f.

319 Turk, Mathew A., David G. Morgenthaler, Keith D. Gremban, and Martin Marra. 1988, May. VITS–A Vision System for Autonomous Land Vehicle Navigation. *IEEE Transactions on Pattern Analysis and Machine Intelligence* 10 (2), p. 358, 342.

320 Turk et al., p. 342.

321 Mysliwetz, p. 13.

322 Meissner, p. 1.8.

323 Roland, p. 244.

324 National Research Council. 2002. *Technology Development for Army Unmanned Ground Vehicles*, p. 153. Washington.

325 Gage, Ibid. Other sources indicate a speed of 3,5 km/h, and a longest autonomous navigation path of 735 m (Strachan, p. 53).

326 Roland, p. 245.

classifier".[327] Its average speed was 14 km/h – four more than required[328] – "through varying pavement types, road widths, and shadows, while avoiding obstacles".[329] Curiously, these results were never presented on international conferences and video sources are missing that could document these successes.

Despite considerable progresses, some observers consider the ALV or Smart truck – discontinued in the fall of 1987 -,[330] as one "of the largest unfulfilled promises" of AI technology, as Daniel Crevier said.[331] Like the Japanese Fifth Generation Project, the Strategic Computing Initiative did not reach its goal of high-level machine intelligence – we should remember that initially, the vehicles top speed was expected to be about 60 km/h.[332]

Several authors confirm that the most challenging task of the ALV was to assure correct computer vision with limited hardware: Moravec calls the ALV project "overambitious"[333] and "premature" because the computing power was not sufficient. "Sensor processing time was cited as the limiting factor for the top speed of the vehicle", confirms Roland.[334] Crevier also emphasizes that "more powerful vision hardware might have made all the difference."[335] The "size, cost and power consumption" of the laser range-finder and the computers was also a major hindrance for progress, adds Larry Matthies from the Jet Propulsion Laboratory.[336]

According to other observers, it was not the hardware but the software that constituted the main obstacle. Moravec reports that the simple road identification techniques of the ALV – and also the early NavLab – "were often fooled, and both were unlikely to stay on a road for a whole kilometer."[337] Roland reveals that the vision system of the ALV "worked differently from month to month, day to day, and even test to test."[338] It could reliably distinguish the pavement of the road from dirt on its shoulders but it had difficulties to understand dirt on the road, to act under different lighting conditions and to distinguish asphalt patches from road repairs and shad-

---

[327] Olin, K.E., Vilnrotter, F.M., Daily, M.J., Reiser, K. 1987. Developments in knowledge-based vision for obstacle detection and avoidance. In *Proceedings of the Image Understanding Workshop*, Los Angeles, p. 78–86, cit. in: National Research Council. 2002. *Technology Development for Army Unmanned Ground Vehicles*, p. 153. Washington.

[328] Roland, p. 245.

[329] Gage, p. 3.

[330] Roland, p. 246. Note: The undercarriage ended up at CMU and was used as the transport vehicle for one of the robotic attempts to descent Mt. Erebus (see Wettergreen, David, Chuck Thorpe, and William Whittaker. 1993. Exploring Mount Erebus by Walking Robot. *International Conference of Intelligent Autonomous Systems*, pp. 72–81).

[331] Crevier, p. 210.

[332] Adam, John A. 1986, January. Aerospace and Military. In *IEEE Spectrum*, p. 79.

[333] Moravec 1999, p. 41.

[334] Roland, p. 245.

[335] Crevier, p. 311.

[336] Matthies, Larry H. 1998. Robotic Perception for Autonomous Navigation of Mars Rover. *1998 Symposium on Frontiers in Engineering*, September 1998, p. 1.

[337] Moravec 1999, p. 41.

[338] Roland, p. 234, see also Nilsson, p. 294.

ows from trees[339] – like the Terregator. It confused "inanimate objects, like a tree trunk, for the edge of the pavement."[340] The "smart truck" had also problems to differentiate between rocks and shadows, writes Weinberger.[341] Finally it can be said that the developers "spent too much time getting fancy hardware and too little developing better algorithms" emphasizes Jochem.[342] Weinberger also concludes that the algorithms did not work very well.[343] One algorithm worked only on flat ground, the other did not work in curves.[344] According to Roland, one of the most important lessons learned was that "no single perception system could address all possible road configurations". It "had to be tailored to specific circumstances, conditions, and tasks".[345]

Kanade criticized the program as "too much demo-driven".[346] According to Roland "ad-hoc, off-the-shelf solutions" were used to meet the straight demonstration deadlines.[347] He agrees that the "tyranny of the demonstration" and its risk of "public failure" frightened Martin Marietta's engineers to develop its innovation potential. Therefore, the pull effect did not take place.

Finally, it is also seen as a problem that the ALV "was conceived, funded and managed with little interest or support from the army, the expected end user of the system."[348] The army did not yet have "any requirement for robotic combat vehicles" during these years, explains Roland.[349] In consequence, the following UGV program, described in the next chapter, was conducted jointly by DARPA and the army.

Other engineers do not share these mostly critical opinions. Thorpe sees the ALV as "major milestone" of the Strategic Computing Initiative. According to him, it pointed out to improvements – "better ladars, better computing, and better algorithms" – but it also prepared the ground for "road following, obstacle avoidance, planning and traversing cross-country routes" and "led directly to both Demo II/III and to Automated Highways."[350] It should also be mentioned that laser scanners, video cameras and inertial navigation units – as used during the SCI ALV-program – are the basic technology of most driverless car projects today.

The ALV was discontinued in 1987 but the algorithms, sensors and computer systems tested with the ALV had been integrated around 1987 into two military

339 Ibid., p. 234.

340 Weinberger, p. 328.

341 Ibid., p. 280.

342 Mail of Todd Jochem to the author, May 10, 2020.

343 Weinberger, p. 328.

344 Roland, p. 234.

345 Ibid., p. 247.

346 Ibid., p. 245.

347 Ibid., p. 243.

348 Ibid., p. 248.

349 Ibid., p. 246.

350 Mail of Chuck Thorpe to the author, February 3, 2020.

vehicles, developed during the first intelligent vehicle project of the army – the Advanced Ground Vehicle Technology (AGVT)-program: First, TARDEC cooperated with Lockheed Martin to integrate the robot technology into a M113 armored personnel carrier tank of the Food Machinery Corporation (FMC), Santa Clara. It was reported that the tank attained 24 km/h in autonomous mode and 8 km/h with obstacle detection.[351] Second, General Dynamics tested the technology with a four-wheeled armored Rapid Reaction vehicle (RRV).[352]

2.5 The Demo I-III Programs of the US-Army

Interest in robotic warfare grew after 1990, especially in the wake of the Persian Gulf war. With the use of Unmanned Aerial Vehicles – drones – for reconnaissance missions during the operations Desert Shield and Desert Storm (1990/1991) the US-army started to recognize the potential of Unmanned Ground Vehicles (UGV).

An UGV is a remotely controlled, semi- or fully autonomous vehicle that is operating on the ground without human presence on-board. According to a definition of DARPA, an autonomous UGV receives its mission by a human, but "having been learned from and respond to its environment", it is "self-sufficient" and "operates without further human intervention".[353] As most robotic systems, UGVs can be used for dull (patrolling), dirty (nuclear, biological and chemical detection) and dangerous tasks (bomb disposal).

In 1990, the US-Congress united all ground vehicle robotics into a Tactical Systems Office within the Department of Defense and created the first Unmanned Ground Vehicle Master Plan.[354] The army Laboratory Command (LABCOM) that united a number of army laboratories,[355] merged its research efforts with those of the Tank Automotive Research And Development and Engineering Command

[351] Kuan, D., and U.K. Sharma. 1987. Model Based Geometric Reasoning for Autonomous Road Following. *Proceedings of the IEEE Int. Conf on Robotics and Automation*, Raleigh, NC, USA, March 31 – April 3, 1987, pp. 416–423.

[352] Ibid., Kuan, D., and G. Phipps. 1987. A-Shuan Hsueeh, Autonomous Land Vehicle Road Following. *Proceedings of the First International Conference on Computer Vision*, London England, June 1987, pp. 557–566, Tamaney, L.S., 1987. Real Time Intelligent Control of a Mobile Robot, NATO Advanced Research Workshop on Mobil Robot Implementation, Porto, Portugal, 28. Sept. 2. Oct.

[353] Defense Advanced Research Projects Agency, Next Generation Explosive Ordnance Disposal Remote Controlled Vehicle (NGEODRCV), Common Systems Architecture, Draft version 5, Arlington 2001, cit. in: National Research Council 2002, p. 24.

[354] Roland, p. 246, Department of Defense, Unmanned Ground Vehicle Master Plan, July 1992. https://apps.dtic.mil/dtic/tr/fulltext/u2/a277757.pdf (accessed June 25, 2020).

[355] The LABCOM united eight laboratories: the Army Research Office (ARO), the Atmospheric Sciences Laboratory (ASL), the Ballistic Research Laboratory (BRL), the Electronics Technology and Devices Laboratory (ETDL), the Harry Diamond Laboratories (HDL), the Human Engineering Laboratory (HEL), the Materials Technology Laboratory (MTL) and the Vulnerability Assessment Laboratory (VAL) (Moye, William T., The Genealogy of ARL, May 1997, p. 5).

(TARDEC) and started three Unmanned Ground Vehicle programs managed by DARPA.[356]

Named Demo I-III, these research programs were "explicitly directed toward military application",[357] and had a special military use case for future UGVs in mind – the Reconnaissance, Surveillance and Target Acquisition (RSTA) mission, also called scout-mission.[358] Its aim is to provide the military with detailed intelligence about the capabilities and locations of enemy forces in a hostile area. Therefore in the Anglo-Saxon context they are also called intelligence, surveillance, target acquisition and reconnaissance (ISTAR)-missions. Obtained by surveillance technologies, the intelligence gathered can lead to the detection and acquisition of potential targets. On the basis of this increased "situational awareness", informed decisions can be made. RSTA missions have also the aim to provide combat assessment during and after military operations – for instance they are used to confirm if a target was effectively destroyed or not.

Today, satellites and Unmanned Aerial Vehicles are regularly used for these purposes. But RSTA missions on the ground continue to play an important role in many war zones. As they are among the most dangerous missions in military operations, they are obviously one of the most important scenarios for the use of ground robots.

As the ALV-program of the 1980s, the Demo programs were not a direct answer to military demands, emphasize Stuart I. McMillan and Jason G. McPhee from the Naval Postgraduate School in California.[359] According to the authors, new solutions are often pushed to the military. Only afterwards, the military – precising its own demands – could create a pull dynamic.

2.5.1 The Demo I Program (1990–1992)

The Demo I program was started in 1990 and "focused primarily on teleoperation",[360] as this technology had been proven successful in the control of Unmanned Aerial Vehicles. The technology had been developed by several research laboratories of the LABCOM that merged with other Army research elements and became the U.S. Army Research Laboratory (ARL), founded in October 1992. The ARL had the mission "to form a centralized laboratory concentrating on scientific research, technology development, and analysis."[361]

---

356 McMillan, Stuart I., and Jason G. McPhee. 2014. *An Analysis of the First Fifteen Years of the Department of Defense Framework for Unmanned Ground Systems*, Thesis, Naval Postgraduate School, Monterey, p. 28. California.

357 Roland, p. 246.

358 Chun, Wendell H., and Todd M. Jochem. 1994. Unmanned Ground Vehicle Demo II: Demonstration A. *Unmanned Systems* 12 (1), p. 14.

359 McMillan, McPhee, p. 283.

360 Chun, Jochem, p. 14.

361 ARL History in: https://www.arl.army.mil/www/default.cfm?page=516 (accessed June 20, 2020).

Demo I used six different High Mobility Multi-purpose Wheeled Vehicles (HMMWV) to demonstrate three types of mobility – computer aided teleoperation, semi-automated control and highly autonomous modes like *retrotraverse* and road following. All Humvees were controlled from a remote vehicle that served as control center. Equipped with steering yokes and brake pedals, the control vehicle allowed steering two tele-operating HMMWVs simultaneously.

In April 1992, a concluding demonstration of tele-operated Unmanned Ground Vehicles developed during the Demo I program was held at Aberdeen Proving Ground, Maryland.[362]

Charles Shoemaker from the Army Research Laboratory describes the differences between the six vehicles (Fig. 2.12)[363]:

Fig. 2.12 The six HMMWVs of the Demo I program from left: Harry Diamond Laboratories (HDL), Ballistics Research Laboratory (BRL), two from the Tank Automotive Research And Develoment and Engineering Command (TARDEC), National Institute of Standards and Technology (NIST), Jet Propulsion Laboratory (JPL). (Credits & Image Source: Shoemaker (2006), p. 290)

[362] Gage, p. 5, McMillan, McPhee, p. 53.

Aberdeen Proving Ground is the oldest proving ground of the US-army. It was created during World War I (October 20, 1917) and served to test weapons, especially chemical agents. It is also part of the history of computing: The earliest electronic computers – ENIAC and EDVAC – were installed on the proving ground at the end of the 1940s (Ballistics Research Laboratory).

[363] Shoemaker, Charles. 2006. Development of Semi-Autonomous Robotic Ground Vehicles, DoD's Ground Robotics Research Programs: Demo I through Demo III. In *Intelligent Vehicle Systems: 4 D/RCS Approach*, ed. Raj Madhavan, Elena R. Messina, and James Albus, 283–315. New York.

The first vehicle from Harry Diamond Laboratories (HDL) demonstrated the electronic control of steering, throttle, brake and transmission. With a low video data transmission rate, the operator could tele-operate the vehicle at slow speed (16 km/h). In contrast to all civil research vehicles mentioned in this book, not only the driving functions but also specific military functions had been automated: The Humvee was equipped with an automatic target acquisition module based on 4 CCD cameras that could detect a target located at a range of 1 km. The second HMMWV from the Ballistics Research Laboratory (BRL) had the same configuration as the first but its target acquisition module was laser based: mounted on a turret, the laser could be pointed on a distant target that triggered a smoke charge to confirm accurate target acquisition. It should be mentioned that the Demo-programs did not develop automatic target *engagements*, meaning that a call for fire remained in the hands of human operators. For that reason, they were not part of the development of Lethal Autonomous Weapon Systems. We will come back to this point at the end of the chapter.

The third vehicle from TARDEC employed a teleoperation system from the Jet Propulsion Laboratory (JPL). JPL had developed near real-time stereo algorithms "producing range data from 64 x 60 pixel images at 0.5-Hz frame rate" that showed automatic obstacle detection and safe stopping, but no obstacle avoidance.[364] According to a report of the National Research Council, "this stereo system used area correlation and produced range data at every pixel in the image, analogous to a LADAR." The fourth vehicle, also from TARDEC, used an autonomous road following technology from the former ALV-program that could identify edges.

The fifth HMMWV from the National Institute of Standards and Technology (NIST) had the same technology as the first from HDL, but featured a "retrotraverse" mode that allowed the vehicle to follow a road, retrace a path and let it come back on the same path autonomously.[365] The system was based "on the output of an inertial dead reckoning system" only, reports Strachan in his PhD-thesis.[366]

The sixth vehicle from the Jet Propulsion Laboratory (JPL) was – like the third jeep from TARDEC – equipped with a teleoperation technology from JPL. But it supported a longer path and the car was also equipped with a tactical radio system.

All tele-operating systems had the disadvantage to be very labor intensive because they always needed a human operator.[367] They also relied on a large communications bandwidth.[368] Therefore, the path retrace mode was "seen as one of the key breakthroughs" of Demo I because it allowed the operating soldier to conduct

[364] National Research Council 2002, p. 155.

[365] Strachan, Jamie Robert, 2009. *NOVA: Nottingham Offroad Vehicle Architecture*, PhD thesis, University of Nottingham. http://eprints.nottingham.ac.uk/10842/1/Microsoft_Word_Complete_Thesis_issue_M.pdf, p. 54.

[366] Ibid., p. 54.

[367] McMillan, McPhee, p. 21.

[368] Spofford, John R., Raymond D. Rimey, and Steve H. Munkeby. 1997. Overview of the UGV / Demo II program, online in: https://pdfs.semanticscholar.org/72fa/74930bb02e33e2c7d887f561dd78d4bf4271.pdf, Accessed on April 8, 2020, p. 4.

other operations while the UGV steered home autonomously.[369] Unfortunately, obstacle detection functioned only with rather large objects (2 to 3 feet) and was not ready for different types of terrain.[370] The detection of negative obstacles – ditches and large holes in the surface – was also a major challenge. The road following technologies of the early 1990s were restricted "to daylight, dry conditions, no abrupt changes in curvature, and good edge definition."[371]

2.5.2 *The Demo II Program (1992–1996)*

The Demo II program started in 1992 and focused on scout missions with "a team of autonomous vehicles that could perform reconnaissance missions with minimum human supervision."[372] This scenario is illustrated in Fig. 2.13. Scout units "generally sustain high loss rates during a conflict, as they are lightly armed and contact the enemy first" emphasizes Spofford.[373] Therefore these missions are strong

Fig. 2.13 Scenario of a scout mission with unmanned ground vehicles and a drone, operator in the foreground. (Credits: Unknown artist. Image Source: Spofford et al., p. 2)

---

[369] Ibid., p. 39.

[370] McMillan, McPhee, p. 39.

[371] National Research Council 2002, p. 155.

[372] Strachan, p. 54.

[373] Spofford, et al., p. 2.

candidates for the use of robotic systems as already mentioned in the introduction to this chapter.

Two experimental scenarios were developed. In an offensive operation, a group of cooperating UGVs would be used to overwatch hostile terrain, to direct artillery fire at the enemy and perform a reconnaissance mission of a village. In a defensive operation, three UGVs would monitor reconnaissance efforts of enemy forces and provide intelligence. The test vehicles demonstrated "supervised autonomy"[374] and were called Surrogate Semi-autonomous vehicles.

Demo II was cosponsored by DARPA and the Office of the Secretary of Defense. In difference to the first demo, Demo II was not integrated under the group of Army Labs but under Martin Marietta. Most of the co-contractors had participated in the ALV-project (see Sect. 2.4).[375] Shoemaker emphasizes that "much of the perception, and intelligent control technology used through Demo II was developed by a team from CMU's Robotics Institute."[376] The cross-country navigation technology of NavLab 2 (presented in Sect. 2.3.4) became "the core software" used on each of the Demo II vehicles – autonomous road following, waypoint navigation and obstacle avoidance with range points.[377]

Three interim demonstrations, Alpha (1993), Bravo (1994) and Charlie (1995) were conducted at the old test track of the Autonomous Land Vehicle at Martin Marietta's Waterton test facility.[378]

Demo A, held on July 7–8, 1993 showed the teleoperation and autonomous navigation of a single M998-Humvee from AM General. Its throttle, brakes and transmission were remotely controlled by push-pull cables, steering was actuated through a power belt. Five racks with control electronics, radio communication, GPS, computers (two SUN Sparc workstations), video and image recording and processing equipment were located in the payload area. Two pairs of stereo cameras were located over the cab with a wide field of view for navigation and with a narrower field of view for target acquisition.[379]

Developed for "area surveillance and overwatch",[380] the vehicle autonomously followed a preplanned path with the help of three technologies: First, an Inertial Navigation System (INS) that combines gyroscopes and accelerometers to measure the rotation and acceleration of the vehicle in 3D was used.[381] Together with GPS, a highly accurate determination of the UGV position was the result.

---

[374] Chun, Jochem, p. 14.

[375] Carnegie Mellon University, Hughes Research Laboratories, Advanced Decision Systems, SRI, Teleos, Jet Propulsion Laboratory, University of Massachusetts, University of Michigan (Chun, Jochem, p. 15).

[376] Shoemaker, p. 296.

[377] Strachan, p. 54.

[378] Chun, Jochem, p. 15.

[379] Ibid., p. 17, Strachan p. 54.

[380] Chun, Jochem, p. 15.

[381] Strachan, p. 50.

Second, two different algorithms were used, STRIPE for off-road terrain[382] and ALVINN for paved ground: The *Supervised TeleRobotics using Incremental Polygonal Earth geometry* (STRIPE) algorithm allowed the tele-operation of the Humvee in rugged terrain where it followed a path of 3D-points that the remote human operator had chosen with the help of the camera images. When the vehicle arrived at a hill and a target vehicle was detected, a simulated call for fire was initiated. When the target was destroyed, the UGV departed again.

Arriving at a paved road, it then switched to the road following algorithm *Autonomous Land Vehicle in A Neural Network (ALVINN)* (see Sect. 2.3.4) that had also been developed by Carnegie Mellon University.[383] It permitted the Humvee to drive "autonomously down the road" using images from a single camera.[384] The system architecture combined hierarchical, top-down and behaviorist's bottom-up approaches based on feedback loops. As for Shakey (see Sect. 2.2.3), three hierarchical layers were combined: the low-level control layer covered basic functions like safety checks. The middle-level local action layer contained behavioral elements. The highest level was the global action layer with cognitive and reasoning functions like object and target recognition, route and task planning and mission monitoring. This approach is called *Subsumption Architecture*, developed in the 1980s by Rodney Brooks.[385]

Demo B, held during 2 days in June 1994 on a route of 3 kilometers, aimed at "semiautonomous navigation".[386] Three types of ground surfaces were selected: To drive on a paved or dirt road, the vehicle created a map and planned a path. The perception system then detected objects annotated in the map and followed the road avoiding obstacles using a laser range finder and a stereo vision system – a black and white Charged Coupled Device (CCD) – that identified the edges of the road. According to Shoemaker, this allowed the HMMWV "to move along the road" at a maximum speed "of 25 km/h".[387]

Third, the vehicle managed an off-road course following GPS-waypoints. This method did not include obstacle detection so the Humvee could "easily collide with an object such as a fallen tree or another vehicle", emphasizes Shoemaker.[388] On roads, computer vision was possible at 3 frames per second only, so the vehicle velocity was "less than 10 km/h".

Demo C was held during 3 days in July 1995 and showed vehicle cooperation, controlled by GPS. A shared map allowed to communicate obstacles from the first to the second vehicle that was then able to avoid them. This was the first

---

[382] See Kay, Jennifer S. 1997, January. *STRIPE: Remote Driving Using Limited Image Data*, PhD thesis. Carnegie Mellon University, Computer Science Department.

[383] Chun, Jochem, p. 16.

[384] Shoemaker, p. 296.

[385] Brooks, Rodney. 1986. A Robust Layered Control System for a Mobile Robot. *IEEE Journal of Robotics and Automation* 2 (1): 14–23, cit. in: Strachan, p. 24. See also Lenzen, p. 84f.

[386] McMillan, McPhee, p. 66.

[387] Shoemaker, p. 298.

[388] Ibid., p. 299

time that a "tactical behavior" was demonstrated in a military robotics program, claims Shoemaker.[389] Like in Demo A, the vehicle had the objective to move upon a hill and to "pulse its laser rangefinder in the direction of the expected target." Targets were represented by three tanks: a Bradley Fighting Vehicle, an M60 tank and an M113 armored personnel carrier.[390] The Humvees had a security driver onboard, but he could not take the steering wheel, only hit a "kill button" to restart the electronics. The speed of each vehicle cross-country was up to 5 mph.

The final demo had been held over 3 weeks in May and June 1996 at Ft. Hood in Texas. Four HMMWVs "operating cooperatively under supervised autonomy"[391] and engineered by Lockheed-Martin,[392] participated in the final event of Demo II. They were equipped with stereo vision for passive navigation, laser radar (ladar) and autonomous navigation systems integrated by Martin Marietta. It should be remembered that stereo vision is attractive for military purposes because electromagnetic emissions could reveal the position of an UGV that would become an easy target. With these technologies, the vehicles could create maps, determine their position, detect obstacles, plan a path and communicate. The three operational modes of the vehicles consisted of teleoperation by a human, of technology supervised remote driving and of semi-autonomous mobility.[393]

Three offensive and defensive scenarios were explored: In the *Urban Assault* mission three Humvees were deployed simultaneously to detect enemy soldiers in a town. In the *Forward Observer* mission a single Humvee had been sent into hostile terrain with the objective to "call in live fires on a target selected by the operator".[394] The mounted *Scout operation* was demonstrated with vehicles driving in obstacle detection mode avoiding scrub trees at speeds of max. 8 km/h, reports Shoemaker.[395] The Demo II vehicles detected obstacles of 0.5 m height from a range of 15 meters.[396] But like in Demo I, the stereo vision system had problems with negative obstacles like holes and ditches, emphasizes Shoemaker.[397]

---

389 Ibid, p. 300.

390 Spofford et al. p. 17.

391 Gage, p. 5.

392 Strachan, p. 54.

393 Chun, Jochem, p. 1

394 Shoemaker, p. 301.

395 Ibid., p. 272.

396 Strachan, p. 55.

397 Shoemaker, p. 302.

2.5.3 The Demo III Program (1998–2002)

Funded by the Office of the Secretary of Defense through the Joint Robotics program, Demo III was started in 1998 and managed by the Army Research Laboratory.[398]

It was the first time that relevant technologies for unmanned ground vehicles were developed in a transnational cooperation between the United States and Germany. A Memorandum of Understanding (MoU) between the US-American Department of Defense and the German Federal Ministry of Defense (BMVg), signed in March 1995, led to the Autonomous Vehicle Navigation Control System (AUTONAV) agreement. It codified a technology transfer between the Demo III program and the ground vehicle program Intelligente Fahrzeug-Funktionen (IFF) funded by BMVg at UniBwM in Germany.[399]

A report from the US-Army highlights the interest of the American side in this agreement: "AUTONAV will save the United States the cost of developing comparable technology. These funds can then be used more efficiently to develop new advanced core technologies required for more intelligent machine control systems. Although the cooperative research project is primarily directed toward military application, many of the advanced technologies under development through the AUTONAV program will have dual-use applications in civilian and commercial sectors such as transportation."[400]

The agreement was made between several industrial, academic and military partners in both countries: The German side was represented by UniBwM and Dornier GmbH Friedrichshafen. UniBwM delivered the theoretical background for the development of the 4D/RCS approach of the Demo III program. On the American side, the National Institute of Standards and Technology (NIST) and the David Sarnoff Labs[401] participated in the AUTONAV agreement. According to Albus, the Jet Propulsion Lab and the SAIC Center for Information Systems also participated in the technology support.[402]

The Demo III program aimed at "night mobility, increased speeds on improved surfaces and over rough terrain, increased obstacle detection and avoidance, with the ultimate goal of achieving supervised autonomous operations", report McMillan and McPhee.[403] The Department of Defense outlined the three main goals of the

[398] Shoemaker, p. 302f.

[399] Albus, p. 297.

[400] US Army Material Command, 1998. *The U.S. Army's Report on International Cooperative Programs*, Urbana-Champaign 1998, p. 1.

[401] The history of the Sarnoff labs goes back to the Radio Corporation of America (RCA), an electronics company founded in 1919 who worked in the 1950s with GM on automated driving. In 1951 RCA gave its laboratory the name David Sarnoff Research Center. General Electric bought RCA in 1986. In 1988 the David Sarnoff Research Center was transferred to SRI International. Ten years later, it became Sarnoff Corporation and in 2011 it became part of SRI.

[402] Albus, p. 297.

[403] McMillan, McPhee, p. 123.

program: "Cross-country maneuver over vegetated terrain at speeds of up to 20 mph during daylight, 10 mph at night or in moderately inclement weather, and at speeds of up to 40 mph when driving on road. (2) Weight under 2500 pounds and be V-22 transportable. (3) Perform RSTA operations commensurate with the scout mission".[404]

To achieve these goals, the Jet Propulsion Laboratory developed the stereo vision system of Demo II further and worked on thermal infrared stereo cameras for night vision, named Forward Looking Infra-Red (FLIR) – by the way a Vietnam-era invention.[405] The National Institute of Standards and Technology was charged with the obstacle detection system using a two-axis scanning Schwartz Electro-Optics LADAR. LADAR is the most powerful perception technology because "it works day and night in any lighting conditions", representing an advantage over camera based obstacle detection, explains Strachan.[406] A "differential GPS backed up by an inertial navigation system and odometry" was used for the localization of the vehicle.[407]

General Dynamics built the new test vehicle that was much smaller than a HMMWV: The eXperimental Unmanned Vehicle (XUV) was the first truly unmanned vehicle because it had no space to carry a passenger. It utilized a four-wheel drive and four-wheel steering system. The wheels were hydrostatically driven by a hydraulic pump powered by a 78-hp turbocharged diesel engine (VW NATO spec). The XUV was "designed to weigh less than 1500 kg, be transportable by helicopter" and to be useful for missions up to 24 h.[408] The XUV platform was based on the Modular Detection and Response – External (MDARS-E) platform.[409] The following image (Fig. 2.14) shows the vehicle with its two axis Schwartz Electro Optics LADAR laser range imaging camera (on the left) and a pan/tilt head with two FLIR cameras and a stereo pair of color cameras (on the right).

Demo III Alpha was held in October 1999 at Aberdeen Proving Ground and showed the prototypes of two semi-autonomous XUVs, guided by human operators that planned their missions. Independently from each other, the vehicles drove off-road during daylight conditions with a speed around 10 mph in stereo obstacle detection mode.[410] The XUVs used both daylight CCD cameras and thermal imagers "to meet day and night requirements", reports Shoemaker.[411] Negative obstacles were again difficult to identify.

---

[404] Department of Defense (DOD), Office of the Secretary of Defense, Joint robotics program master plan, Washington 2000, p. 45, cit. in: McMillan, McPhee, p. 126.

[405] Jacobsen, p. 248. Night vision by infrared had been researched by DARPA since 1961 and was developed during the Vietnam war (Jacobsen, p. 276).

[406] Strachan, p. 63.

[407] Ibid., p. 56.

[408] Albus, p. 296f.

[409] Shoemaker, p. 304.

[410] McMillan, McPhee, p. 126, Shoemaker, p. 309.

[411] Shoemaker, p. 305.

Fig. 2.14 The Demo III eXperimental Unmanned Vehicle. (Credits: General Dynamics. Image Source: Strachan, p. 56)

Demo III Bravo, conducted 1 year later at Fort Knox showed again two XUVs. They now had been equipped with a LADAR scanner that improved the quality of obstacle mapping. With improved software, the vehicles drove now at a speed of 15–20 km/h, according to Shoemaker.[412] Other sources precise that the Demo III XUV drove with a speed of up to 60 km/h on the road, 35 km/h off-road during daylight conditions and with a speed of 15 km/h off-road at night or during bad weather.[413] Jochem comments that the 35 km/h off road "with any kind of obstacles is unlikely" unless they drove in "very simple environments".[414]

Demo III Charlie, the final event, held during 6 days in October 2001 in the forest area of Ft. Indianatown, showed four XUVs crossing rugged terrain, followed by manned Humvees that surveilled its movements.[415] Shoemaker resumes that the vehicles now demonstrated "fully autonomous RSTA capabilities".[416] The XUVs used a 4D/RCS software architecture developed by James Albus from the National Institute of Standards and Technology (NIST) in Gaithersburg, Maryland.[417] It combined a priori information like maps and operator inputs with real time sensor data.

[412] Ibid., p. 310.

[413] Madhavan, R., Hong, T., Messina, E., 2006. Temporal Registration of Sensed Range Images for Autonomous Navigation. In *Intelligent Vehicle Systems: 4 D/RCS Approach*, eds. Madhavan, Raj, Messina, Elena R., Albus, James, p. 162. New York. See also Albus, p. 297.

[414] Mail of Todd Jochem to the author, May 10, 2020.

[415] McMillan, McPhee, p. 124, Shoemaker, p. 278.

[416] Shoemaker, p. 311.

[417] Albus, James, and Anthony Barbera. 2006. 4D/RCS Reference Model Architecture for Unmanned Ground Vehicles. In *Intelligent Vehicle Systems: 4 D/RCS Approach*, ed. Raj Madhavan, Elena R. Messina, and James Albus, 1–26. New York.

The 4D/RCS architecture integrated the Real-Time Control System of NIST with the 4-D approach to dynamic machine vision developed by the team of Dickmanns at UniBwM (see Sect. 4.3).[418]

This cooperation led finally to a solution for the resurgent problem to recognize ditches: At the end of the AUTONAV-project in 2003, the first stereo system for full-size video images and full frame rate has been demonstrated successfully with the VaMoRs-van of UniBwM. This system early and precisely recognized ditches while driving cross-country.

It can be resumed that Demo I developed the teleoperation and the path-retrace capability of UGVs and Demo II advanced with autonomous road following. Obstacle detection remained a challenge until Demo III when the problem of negative obstacles (ditches) had been resolved. Nevertheless it should be mentioned that the efforts of DARPA, CMU and the other partners finally did not lead "to a fielded robot beyond the teleoperation capability."[419] For that reason McMillan and McPhee conclude that the Demo-programs to create tactical UGVs were not successful.[420] Even if in some cases, semi-autonomous robots performed path retrace to their original deployment location,[421] only remote-controlled systems are used by warfighters in the field until today as will be detailed in the next chapter. Especially RSTA missions remain "extremely complex and difficult for UGVs".[422]

Despite these difficulties the political pressure for the introduction of UGVs increased even when the Demo III-program was not terminated: In 2001, Congress mandated that by 2015 one third of all ground vehicles shall be unmanned.[423] The decision to push the UGV-development further might have been motivated by the rapid development of drones, a technology that is "at least 10 years ahead of UGVs"[424] – the environment in the air is very much easier to handle than environment at the ground. The first successful test of an armed drone took place in Nevada in February 2001. But in order to realize fully autonomous UGVs a "major technological breakthrough" was missing and the Department of Defense stopped the creation of UGV master plans in 2004. Another approach taking advantage of the newly realized GPS-system seemed to be more promising since it reduced the task for a vision system considerably by pre-specifying the path via GPS-waypoints: In 2004 DARPA started its first Grand Challenge.[425] A prize of 1 million Dollar was offered for building a robot vehicle that could drive autonomously 142-miles through a pre-specified off-road course in the Californian desert, resembling those

[418] Albus, James S., and Alexander M. Meystel. 2001. *Engineering of Mind, An Introduction to the Science of Intelligent Systems*, p. 297. New York.

[419] McMillan, McPhee, p. 274.

[420] Ibid., p. 69.

[421] Ibid., p. 274, 281.

[422] Ibid., p. 254.

[423] US Congress 2001, S. 2549, Sec. 217.

[424] Krishnan, Armin, and Killer Robots. 2009. *Legality and Ethicality of Autonomous Weapons*, p. 11. Surrey/Burlington.

[425] McMillan, McPhee, p. 274.

in Iraq and Afghanistan. "If a vehicle could navigate the Mojave, it could handle Afghanistan and Iraq."[426] The main task now was avoiding obstacles in this course. It was hoped that an open competition between different academic and industrial players would lead faster to results than military research programs only.

2.5.4 Outlook: UGV-Development and the Relation Between Civil and Military Research

For several reasons, the Demo-programs are important for the perspective of this book. On the one hand, they direct our attention to the different application fields of technologies that promise autonomous decisions. When technologies for the automatization of car traffic are combined with military technologies like automatic target *acquisition*, a completely different system with different use cases and different consequences for society emerges.

To understand better what is at stake here, our presentation of the historical development steps of UGVs will be followed by a short look at the state-of-the-art of UGVs, especially their present and possible future use cases. It can be shown that the development of armed UGVs does not always follow a linear path. To conclude the chapter, the dual-use problem of autonomous driving technologies will be addressed.

In 2008, more than 6.000 Unmanned Ground Vehicles were in use by the Pentagon, reports Armin Krishnan.[427] They are among a total number of 11.000 robots enlisted to the US military.[428] The National Research Council in Washington published in 2002 a report that differentiates four types of Unmanned Ground Vehicles – Searcher, Mule, Wingman and Hunter-Killer – that are associated with different use cases.[429] They can also be seen as successors in a chronological timeline.

Interestingly, the only UGV that is fielded today is teleoperated, not armed and has no autonomous capabilities. The *Searcher* is a small teleoperated robotic ground vehicle primarily used to protect the life of soldiers in extremely hazardous environments. Its most dominant use case remains the removal of improvised explosive devices (IEDs). In the years 2003–2007, IEDs were responsible for 63% of the US soldiers killed and for 69% of the wounded in Iraq, as reported by the Washington Post.[430] According to Carafano and Gudgel the first UGVs for demining arrived in

[426] Davies, p. 24.

[427] Krishnan, p. 10.

[428] Ibid., p. 2.

[429] National Research Council 2002, p. 10, 20, Strachan, p. 19.

[430] Author unknown, More attacks, Mounting casualties. *The Washington Post*, September 30, 2007, online in: https://www.washingtonpost.com/wp-dyn/content/graphic/2007/09/28/GR2007092802161.html (accessed July 7, 2020).

Iraq in April 2004.[431] Their number increased from 150 in 2004 to 5.000 in 2005. At the end of 2005, this type of UGV had been used to neutralize more than 1.000 IEDs.[432] Until 2013 the U.S. Army had purchased 7000 of these robots and 750 had been destroyed in demining operations.[433]

The most famous model of the Searcher is the Packbot from iRobot. In 2002, it was used for the first time by US forces in Afghanistan "to clear caves and bunkers, search buildings and cross live, anti-personnel mine fields."[434] Packbots were also used to search the debris of the World Trade Center after the terrorist attack in 2001. In 2007 they were used to detect the acoustic signatures of sniper fire in the Middle East.[435] After the 2011 earthquake in Japan they examined the damaged nuclear plant in Fukushima. With a weight of 18 kg the Packbot is man-portable and can be dropped from a height of nearly 2 meters. To detect IEDs, it can be steered with a fiber optic cable to avoid radio signals that could trigger an explosion of the device. By 2008, Packbots have disarmed over 10.000 IEDs.[436]

The next three types of robots described in the NRC-report are in the development stage and – according to the information that is publicly available – not yet fielded:

The *mule* or *donkey* is a semiautonomous follower that can be used for logistics (supply convoys, assure medical evacuation, carry casualties), following a predefined path. It should reach a speed of 40 km/h.[437] Its promise: If military convoys could be driven autonomously, human casualties among US-soldiers could be avoided.[438]

But the RAND-Corporation concluded recently in a report that the scenario of automated trucks following the path of manned trucks in a convoy operation – called partially manned concept – would be too risky. Only the employment of "automated trucks, each containing a single soldier" and following a manned truck – called minimally manned concept -, is seen by RAND as "ready for Army adaption" but only "in urban and highway environments."[439] The report emphasizes that a human

---

[431] Carafano, James Jay, and Andrew Gudgel. 2008. The Pentagon's Robots: Arming the Future. In *Competitive Technologies for National Security*, eds. James Jay Carafano, Andrew Gudgel, and Alane Kochems, 27–32. Washington: The Heritage Foundation. online in: https://www.heritage.org/node/15537/print-display. Accessed on June, 24, 2020, p. 29.

[432] Soon, it became a famous icon in popular films – a HD-1 from Remotec was shown in the 2010 Oscar winning film "The Hurt locker" from Kathryn Bigelow.

[433] Main, Douglas. 2013. Robots to the Rescue. *Popular Science*, April 8, 2013, online in: https://www.popsci.com/trp-sponsored-article-slideshow/. Accessed July 7, 2020.

[434] Merrin, William. 2019. *Digital War, A Critical Introduction*. Oxon/New York.

[435] Morris, Jefferson. 2006. iRobot's Dyer Sees Missions Multiplying for PackBot. *Aerospace Daily & Defense Report 217* (36), February 24.

[436] Merrin, William. 2019. *Digital War, A Critical Introduction*. Oxon/New York.

[437] National Research Council 2002, p. 48.

[438] Weinberger, p. 334.

[439] McKay, Shawn, Matthew E. Boyer, Nahom M. Beyene, Michael Lerario, Matthew W. Lewis, Karlyn D. Stanley, Randall Steeb, Bradley Wilson, and Katheryn Giglio. 2020. Automating Army

operator is needed to "compensate for the current shortcomings in automated technology". As this scenario reduces the number of soldiers in the truck from two to one, the primary motivation – "reducing soldier casualties" – would be maintained.

The *Wingman* could be used to assure reconnaissance and scout missions (RSTA), to detect biological or chemical agents but also to counter sniper fire. It should be "the eyes and ears" of the manned section leader and reach a speed of 100 km/h.[440] This type corresponds mostly to the UGV-scenarios developed during the Demo-programs.

Finally, the NRC-report mentions the *Hunter-Killer*, a "network-centric autonomous ground vehicle", that could be deployed in teams of several robots. It should be able to move at least with a speed of 120 km/h.[441] A main task could be their deployment for RSTA missions but according to the report they are also seen as useful to carry out ambushes.[442] As its name already reveals, the Hunter-Killer carries weapons for "lethal direct fire"[443] and "could be programmed with rules of engagement for enemy encounters"[444] – meaning that "the team would kill enemy forces in the kill zone with overwhelming lethal force" as the NRC-report details.[445] Therefore it must be able to identify non-combatants.

The Hunter-Killer will fall under the category of a Lethal Autonomous Weapon System (LAWS), defined as a machine that uses sensors and algorithms to select a target and engage an onboard weapon system to destroy the target without human control of the system.[446]

In 2015, a campaign to stop these Killer Robots – "the third revolution in warfare, after gunpowder and nuclear arms" – was started by over sixty nongovernmental organizations (NGOs), leading AI-researchers and entrepreneurs.[447] This alert was pronounced because US policy does not prohibit the development or employment of these *human out of the loop* systems. In 2012, a policy directive (3000.09) of the Department of Defense stated only that autonomous weapons systems should allow "appropriate levels of human judgment over the use of force".[448] This means that a direct manual human control of the weapon system is

---

Convoys, Technical and Tactical Risks and Opportunities. Santa Monica: RAND Corporation. in: https://www.rand.org/content/dam/rand/pubs/research_reports/RR2400/RR2406/RAND_RR2406.pdf. Accessed on June 24, 2020.

440 National Research Council 2002, p. 27, 48.

441 Ibid., p. 48.

442 Ibid., p. 26ff.

443 Ibid., p. 3.

444 Strachan, p. 19.

445 National Research Council 2002, p. 27.

446 Congressional Research Service, Defense Primer, U.S. Policy on Lethal Autonomous Weapons Systems, December 19, 2019.

447 Future of Life Institute, AUTONOMOUS WEAPONS: AN OPEN LETTER FROM AI & ROBOTICS RESEARCHERS, online in: https://futureoflife.org/open-letter-autonomous-weapons/?cn-reloaded=1. Accessed March 5, 2020.

448 Department of Defense, Directive, N° 3000.09, Autonomy in Weapon Systems, November 21, 2012, p. 2, online in: https://www.esd.whs.mil/Portals/54/Documents/DD/issuances/dodd/300009p.pdf

not required. It is only recommended that humans should not be completely out of the loop. The directive can be understood as support of a *human on the loop* system, meaning that the robot would be able to "autonomously select and engage targets but a nearby human operator can intervene to turn off the system" explains Alexander Velez-Green from the Center for a New American Security.[449] In contrast, a *human in the loop* system would wait for a decision of the human operator before engaging the target.

This differentiation helps to understand that the Demo-programs of the US-military did not aim at the development of a weapon system that functions autonomously. But in contrast to civilian institutional research frameworks they worked not only on automated navigation technologies for UGVs but also on "automatic target recognition technologies"[450] and simulated tele-operated target engagements. While automatic target *recognition* was a goal of the Demo-programs, the *engagement* of targets remained in the hands of human operators. This means that the technologies developed during the Demo-programs were *human in the loop systems* as explained above. They did not aim at the development of Lethal Autonomous Weapon Systems.

The classification of the NRC-report is also revealing because it suggests successive steps in the way to tactical autonomous UGVs that in the actual development may not be followed by all actors that are involved.

First, the defense industry did not wait to attain the Wingman-stage to arm a robot. Instead, they armed the Searcher: The US robotics company Qinetiq-NA developed a weaponized version of the Foster-Miller TALON bomb disposal robot. Called Special Weapons Observation Reconnaissance Detection System (SWORDS), it can be armed with M249 machine-guns, an M16 assault rifle or a grenade launcher. In 2007, three of these weaponized tracked SWORDS-robots were deployed to Iraq.[451] This was the first time that a weaponized UGV was sent into a war zone. According to the political scientist Armin Krishnan, these robots patrolled the streets of Baghdad.[452] But Krishnan emphasizes also that the robots remained tele-operated, incapable of firing their weapon without human control. He also reveals that their weapons remained unused and the army stopped funding the system.[453]

Second, according to the political scientist Peter W. Singer of the public policy think tank New America Foundation the first time a robot was used to kill a human was in the early 2000's when a soldier taped a Claymore mine on a remote

---

[449] Velez-Green, Alexander. 2015. *The Foreign Policy Essay: The South Korean Sentry—A "Killer Robot" to Prevent War*, March 1, 2015. https://www.lawfareblog.com/foreign-policy-essay-south-korean-sentry—killer-robot-prevent-war. Accessed March 3, 2020.

[450] Spofford et al., p. 1.

[451] The Inside Story of the SWORDS Armed Robot "Pullout" in Iraq, in: Popular Mechanics, October 1, 2009, https://www.popularmechanics.com/technology/gadgets/a2804/4258963/

[452] Krishnan, p. 29.

[453] Ibid.

controlled Marcbot surveillance robot to kill an insurgent in Iraq.[454] In July 7, 2016, the same procedure was employed for the first time in the US homeland when Police forces used a remote controlled ANDROS Mark V-A1 military robot to kill the heavily armed gunman Micah Xavier Johnson with a 1 pound C-4 explosive. Johnson had murdered five Dallas police officers.[455]

Both examples show that a normalization of a new technology – teleoperated robots – can lead to improvised adaptions of their function. Long before a dedicated weaponized robot will be officially sold, they are adapted for tactical needs. This gives some lessons on technology development in general. A first lesson is that the military use of armed robots does not follow a linear deployment path. Armed robots may be developed earlier than expected (SWORDS) but that does not mean that they are immediately used. A second lesson is that a new technology (PackBot) can be adapted very fast to become a deadly weapon, not only in warzones but also in other areas like the homeland that are beyond its initial development goals.

To conclude the chapter, the problem of dual-use technologies should be addressed shortly. As we have seen, autonomous steering, navigation and obstacle detection technologies can be used for both, civil and military purposes. Our investigation of the research history has even shown that a strict separation between civil and military development never existed in the US. It is obvious that all underlying technological devices for autonomous driving – radars, sonars, infrared cameras, head-up displays, etc. – were first developed as expensive military devices as highlighted for instance by Broggi et al.[456] But it had been overlooked until now that the common development goals go beyond the hardware components: Three engineers from Lockheed Martin Astronautics argue in a paper that "a good portion" of the technology developed for Demo II had "direct relevance" for the AHS program of the US Department of Transportation. They emphasize that "there are several candidate technologies that have dual-use potential for AHS such as road following."[457] Especially "the target detection work (RSTA) has some application to detecting and verifying obstacles."[458] Paraphrasing the example of the Lockheed Martin-engineers, we could resume that target identification via computer vision can be applied to a pedestrian on the road, triggering a brake reaction of the vehicle – or identifying an

[454] Thielman, Sam. 2016. Use of Police Robot to Kill Dallas Shooting Suspect Believed to be First in US History. *The Guardian*, July 8, 2016, online in: https://www.theguardian.com/technology/2016/jul/08/police-bomb-robot-explosive-killed-suspect-dallas. Accessed June 24, 2020.

[455] Gershgorn, Dave. 2016. Police Used Bomb Disposal Robot to Kill a Dallas Shooting Suspect. *Popular Science*, July 8, 2016, online in: https://www.popsci.com/police-used-bomb-disposal-robot-to-kill-dallas-shooting-suspect/ and Farivar, Cyrus. 2016. Dallas deployment of robot bomb to kill suspect is "without precedent". *Ars technica*, July 8, 2016, online in: https://arstechnica.com/tech-policy/2016/07/is-it-ok-to-send-a-police-robot-to-deliver-a-bomb-to-kill-an-active-shooter/. Accessed June 24, 2020.

[456] Bertozzi, Massimo, Alberto Broggi, and Alessandra Fascioli. 2000. Vision-Based Intelligent Vehicles: State of the Art and Perspectives. *Robotics and Autonomous Systems* 32 (1), p. 2.

[457] Chun, Wendell, Faulconer, J. Walt, Munkeby, Steven, UGV Demo II Reuse Technology for AHS. In *Proceedings of the Intelligent Vehicles '95 Symposium*, p. 252.

[458] Ibid.

enemy position in the field, triggering a command to shoot. This overlap of civil and military research interests can be proven by empirical findings. For example, the research on pedestrian detection of the University of Parma was funded by US-Army, TACOM (contract number N68171–01-M-5857).[459]

As we have seen, military research contributes to civil applications. But it is also productive to turn this perspective the other way around, as it is common in discussions about dual-use technologies: Civil research contributes to military applications. The NRC-report emphasizes that the research on road vehicle automation funded by the Department of Transportation since the 1960s is highly relevant for the development of UGVs. The authors highlight the economic advantage of dual-use technology development because it could "lower the costs" of unmanned army vehicles. The report identifies the "key technologies" from civil road transportation research – sensing, navigation, target identification, control and communication technologies – that "could be applicable to UGVs".[460] Giving some examples of overlapping technologies, the report mentions for instance the automotive night vision sensors from Raytheon, commercialized for the first time by GM in the Cadillac DeVille in 2000. Other examples are laser or millimeter wave radar sensors that can detect other vehicles in a distance of 150 meters and lane-tracking vision sensors that are crucial for the orientation of both, civilian and military vehicles in an unknown environment. Wireless communication standards like dedicated short-range communications (DSRC) "could be applied to any off-road battlefield environment for communications among UGVs and even between UGVs and UAVs."[461] And finally automatic vehicle following technologies are mentioned that would be useful for army supply convoys.

It can be concluded that automated driving and automated warfare systems use the same technology base. Both domains contribute to each other, and that in both directions. So it is possible that the big investments in autonomous driving during the last decades will have an accelerating impact on the UGV development in the near future. This observation should be integrated in the public discourse about the benefits and risks of autonomous driving.

---

[459] Bertozzi, Massimo, Alberto Broggi, Massimo Cellario, Alessandra Fascioli, Paolo Lombardi, and Marco Porta. 2002. Artificial Vision in Road Vehicles. *Proceedings of the IEEE* 90 (7), p. 1269.

[460] "Sensing of vehicle position relative to road vehicle lanes, sensing of absolute vehicle position (global positioning system (GPS) and inertial navigation system (INS) in combination with others, sensing of proximity to other vehicles (ranges up to 150 m), identification and hazard assessment of targets within sensor range, vehicle-vehicle data communication, vehicle-roadside data communication, vehicle dynamics control, road condition sensing and estimation, automatic steering control and vehicle-following speed and spacing control." [National Research Council 2002, p. 37]).

[461] Ibid., p. 40.

2.6 The Research on Automated Highway Systems (1980–2000)

After having presented 40 years of robotics research we have to go back in our chronology and refocus our attention to the vision of an automated highway system that had been started with the experiments of GM, RCA and Ohio State University (Sect. 2.1).

The 1970s and most of the 1980s were a difficult period for automated transport systems in the United States. In analogy to the AI-winters, Steven Shladover called this period "hibernation".[462] As the Reagan Administration (1981–1989) believed that "further safety standards were simply too onerous for industry",[463] the Department of Transportation (DOT) was ordered "to cut transportation R&D to the bone and not take any new initiatives that would incur financial commitments" reports Shladover.[464]

In the second half of the 1980s, the situation changed, first on the federal level. Important congestion problems in the Los Angeles and San Francisco Bay Area motivated California to begin research in a field that was later called Intelligent Transportation Systems (ITS) – meaning the application of advanced information and communication technologies to improve the transportation infrastructure and the vehicles.[465] Long before a national ITS-program was started in the US, Caltrans – the Californian Department of Transportation – and the Institute of Transportation Studies of the University of California in Berkeley initiated in 1986 the research program Partners for Advanced Transit and Highways (PATH). Its most active participants came from the fields of Mechanical Engineering, Electrical Engineering and Computer Science.[466]

---

[462] Shladover, Steven E. 2008. AHS Research at the California PATH Program and Future AHS Research Needs. *Proceedings of the 2008 IEEE International Conference on Vehicular Electronics and Safety*, Columbus, September 22–24, p. 4.

[463] The Diebold Institute for Public Policy Studies, Transportation Infostructures: The Development of Intelligent Transportation Systems, Westport 1995, p. 103.

[464] Shladover, Steven. 2007, December. PATH at 20 – History and Major Milestones. *IEEE Transactions on Intelligent Transportation Systems* 8 (4): 584–592, cited version (with other page numbers) online in: https://path.berkeley.edu/sites/default/files/path-at-20.pdf, p. 3 (accessed September 11, 2020).

[465] The term Intelligent Transport Systems had been approved by the US and Japan in 1991 and in Europe in 1994 during the first World ITS Congress in France. According to a directive of the European Community (Directive 2010/40/EC), ITS is defined as a system that connects users, infrastructure and vehicles through an information and communication technology network (Nowacki, G. 2012, September. Development and Standardization of Intelligent Transport Systems. International Journal on Marine Navigation and Safety of Sea Transportation 6 (3), p. 404). Since 1978, the research efforts were summarized under the term telematics, introduced by the French experts Simon Nora and Alain Minc, merging the words *télécommunication* and *informatique* (Nora, Simon, and Alain Minc. 1978. L'Informatisation de la Société: rapport à M. le Président de la République. Paris: La Documentation Française).

[466] Shladover 2007, p. 5.

The pioneering role of California influenced the creation of several institutions and laws on the national level: In June 1988, the expert panel Mobility 2000 was created, stimulated by Caltrans and PATH. This informal group of industry, academia and government representatives tried to influence the US administration to support Intelligent Vehicle-Highway Systems (IVHS), known later under the term ITS.[467] The work of this advocacy group led in August 1990 to the creation of a formal coordinating organization – the Intelligent Vehicle and Highway Society of America (IVHS America).[468] In 1991 IVHS America became a federal advisory committee of the US Department of Transportation. Its task was to develop a strategic plan for the implementation of ITS as a whole. The introduction of an Automated Highway System (AHS) in the US was only one element in that plan – the most futuristic and controversial part.[469]

The acronym AHS is a narrower term than ITS and IVHS and signifies a set of designated lanes on a limited access roadway where specially equipped vehicles are operated under completely automatic control.[470] It does not include a generalized form of automated driving on all roads. But it would also be an oversimplification to summarize AHS as exclusively infrastructure based – both, roadway and vehicles have to be modified.

Parallel to the founding of IVHS America the US-Congress signed in 1991 the Intermodal Surface Transportation Efficiency Act (ISTEA). This public law (102–240) can be qualified as "the first transportation bill of the post-Interstate era".[471] According to a report of the Department of Transportation it represented a "shift in focus from the building of a surface transportation system to the operational management and maintenance of that system". We can summarize the legislation as an ITS-driven rejuvenation of the highway system infrastructure, especially the main highway "corridors" between big urban agglomerations.

---

[467] The acronym IVHS includes five fields: (1) traffic management systems that can predict congestion and provide alternative routing instructions, (2) traveler information systems that assist people to reach their destination, (3) vehicle control systems enhancing vehicle safety (as collision warning, autonomous emergency braking, automated lines), (4) automated systems for commercial vehicles and (5) systems for public transportation.

[468] Strategic Plan for Intelligent Vehicle Highway Systems in the United States, IVHS America, Report No: IVHS-AMER-92-3, May 20, 1992, Page II-1.

[469] In autumn 1994 IVHS America was renamed Intelligent Transportation Society of America (ITS America) to reflect a broader multi-modal approach not limited to automobiles. ITS America consists of about 50% private sector companies and 50% academia, government, and associations (Cheon, Sanghyun. 2003. *An Overview of Automated Highway Systems (AHS) and the Social and Institutional Challenges They Face*, UC. Berkeley: University of California Transportation Center. online in: https://escholarship.org/uc/item/8j86h0cj, p. 23. Accessed on March, 28, 2020.

[470] Rillings, Jim. 1997, February. *The National Automated Highway System: Cooperative Program*. National Automated Highway System Consortium, p. 4.

[471] Auer, Ashley, Shelley Feese, and Stephen Lockwood. 2016. History of Intelligent Transportation Systems, Performed by U.S. Department of Transportation, Intelligent Transportation Systems Joint Program Office, Report No. FHWA-JPO-16-329, Produced by Booz Allen Hamilton, May 2016, p. 14.

In strong contrast to the political priorities of the 1980s, the ISTEA legislation set free important financial resources. According to a report from 1997, the total ITS funding following the ISTEA legislation was $991.1 Million.[472] This amount was invested in several hundred small research programs, covering seven categories: traffic management, traveler information, public transportation, electronic payment, commercial vehicle operations, emergency management and advanced vehicle control and safety systems. The domain last mentioned covered assistance systems like collision avoidance warnings, automatic braking controls – and also "automated highways systems, on which vehicles could move without drivers."[473]

The long list of research topics shows that the project of an automated highway system was only a small part of the whole program. According to the aforementioned report, the AHS program was funded from 1991 to 1996 with $49.8 million, only 5% of the total ITS funding of $991.1 Million.[474] Nevertheless the AHS program was one of the most visible parts of this initiative. One passage of the ISTEA law that is cited very often is Section 2.6, requiring the Secretary of Transportation to "develop an automated highway and vehicle prototype from which future fully automated intelligent vehicle-highway systems can be developed. (…) The goal of this program is to have the first fully automated roadway or an automated test track in operation by 1997".[475]

In response to this request, the Department of Transportation created in 1992 the National Automated Highway System Research Program (NAHSRP). The mission to conduct a federal AHS program was given to the National Automated Highway System Consortium (NAHSC), founded in December 1994 after a successful proposal in response to a request of the Federal Highway Administration (FHWA) – a division of the Department of Transportation in charge of highways.[476]

Headed by General Motors, NAHSC was a public-private partnership of nine core organizations from government, industry and academia, including General Motors, the California Department of Transportation (Caltrans), Carnegie Mellon University, the aerospace company Lockheed Martin, the engineering companies

472 US Department of Transportation. 1997 March. The National Intelligent Transportation Systems Program: Where We've Been & Where We're Going, Report No. FHWA-JPO-97-0027, p. 3.

473 The Congress of the United States, Congressional Budget Office, High-Tech Highways: Intelligent Transportation Systems and Policy, October 1995, online in: https://www.cbo.gov/sites/default/files/104th-congress-1995-1996/reports/High-techHighways.pdf, p. 2 (accessed March, 23, 2020).

474 US Department of Transportation 1997, p. 3.

475 Section 6054(b) of the Intermodal Surface Transportation Efficiency Act of 1991, 23 U.S.C. 307, 105 Stat. 291.

476 While the FHWA is in charge of highways, the National Highway Traffic Safety Administration (NHTSA) is in charge of vehicles. Both cooperate in events like Demo '97.

Bechtel and Parsons Brinckerhoff, the electronics companies Delco and Hughes, and the University of California-Berkeley.

PATH "had the lead role in the NAHSC for the development of modeling and simulation tools, for the second stage of operational concept development and evaluation, and for the demonstration of an automated platoon of eight cars" reports Shladover.[477] Nearly 100 stakeholders from "vehicle industry, government agencies, the highway design industry, vehicle electronics, environmental interests, trucking operators, transit operators, transportation users and the insurance industry" also participated in the consortium.[478]

NAHSC had the mission to conduct the federal AHS program under a cost-shared cooperative agreement – 20% of the investments had to be funded by its members. NAHSC proposed to invest $200 Million in the research program from 1994–2002,[479] but according to Shladover finally only $50 Million were spent when the program ended in 1998.[480]

2.6.1 The Demonstration of an Automated Highway System in San Diego (1997)

The cooperation of the NAHSC-partners culminated in a public demonstration called Demo '97, known today as the first demonstration of an Automated Highway System (AHS) in North America. Two years earlier, the first Japanese AHS demo was held by the Ministry of Construction (see Sect. 3.5.3).

Demo '97 consisted of a series of live demonstrations shown in August 7–10, 1997 on a new 7.6-mile section of car-pool lanes, separated by concrete barriers from the lanes of the Interstate-15 freeway north of downtown San Diego, California. During the event, 21 vehicles traversed 4468 miles in various modes of automated driving. The event had been attained by 1000 members of the public, 100 journalists, 1400 VIPs and 1000 members of the industry.

While testing and development work prior to the demo were done in evenings and weekends when the car-pool lanes were closed to the general public, the demo in August included weekdays. But the lanes were used during the mid-day off-peak period when they were closed to the public. This is important because other

[477] Shladover 2007, p. 7.

[478] Cheon, p. 10.

[479] Bishop, Richard, Kevin Dopart, and Steven Shladover. 2020. Commemorating the 20th Anniversary of the National Automated Highway Systems Consortium Demo '97 in San Diego, Powerpoint-Presentation, online in: https://path.berkeley.edu/sites/default/files/demo97foravs17v6.pdf. Accessed March 30, 2020, p. 3.

[480] Mail from Steven Shladover to the author, March 11, 2020.

demonstrations took place in real highway traffic (see Sect. 4.4 about Prometheus in Europe).

Demo '97 had the aim to show the feasibility of an AHS system, defined by NAHSC as a "specially equipped roadway on which vehicles can be operated automatically. (…) Drivers who choose to use the AHS would steer their specially equipped vehicle onto designated lanes of the highway, then release control of the vehicle to the system. (…) When the vehicle reaches the exit selected by the driver, it would be steered into a transition area where the driver would resume manual driving."[481]

Two main reasons were given to promote automation.[482] First, driver errors – caused by fatigue, inattention, the influence of alcohol etc. – had been identified as leading cause in 90% of the 40.000 fatalities and 5 million injuries that occurred on the US-highways each year. Automation promised to reduce the number of fatalities and to increase road safety.

Second, it was hoped that automation could increase the efficiency of the existing highways and "reduce congestion on highways by achieving significantly higher traffic flow through closer packing of automatically controlled vehicles into platoons."[483] At the time of Demo 97, traffic was doubling every decade but it was not possible to double the lane-miles of the highways. Adding a new lane could cost $100 million per lane-mile.[484] With manually driven vehicles, the capacity of the roads is limited to about 2000 vehicles per lane per hour. New technologies promised to double or triple the flow rate of traffic without a construction of new highways. Environmental factors like the reduction of fuel consumption and exhaust emissions were another topic that was mentioned.[485] But it was also discussed if automation would incite people to live further from work and to do more driving. Today, all these questions are again on the agenda.

As it was the case during decades of research on road vehicle automation, some participants of Demo '97 promoted overly optimistic and ambitious timelines. They estimated that automated highway systems "could be integrated into existing highways within 20 years",[486] a timespan that accompanies the research history since the 1930s as outlined in the introduction to this book.

---

[481] NAHSC, What is An Automated Highway System?, Demo '97-Flyer, in: http://www.path.berkeley.edu/sites/default/files/my_folder_76/NAHSC_Brochure-Demo%26Groups-971.pdf, p.1 (accessed June 17, 2018a).

[482] Thorpe, Chuck, Todd Jochem, and Dean Pomerleau. 1998. The 1997 Automated Highway Free Agent Demonstration. In *Proceedings of Conference on Intelligent Transportation Systems 1997*, p. 496.

[483] Rajamani, Rajesh, Han-Shue Tan, Boon Kait Law, and Wei-Bin Zhang. 2000. Demonstration of Integrated Longitudinal and Lateral Control for the Operation of Automated Vehicles in Platoons. *IEEE Transactions on Control Systems Technology* 8 (4), p. 695.

[484] Thorpe, Jochem, Pomerleau 1997, p. 496.

[485] NAHSC, Where the research meets the road, p. 11 (accessed June 17, 2018b).

[486] Ohio State University, Alternative technologies, Demo '97-Flyer, in: http://www.path.berkeley.edu/sites/default/files/my_folder_76/NAHSC_Brochure-Demo%26Groups-971.pdf, p. 48.

But the event also recalled that in more than 50 years of research, the idea of automatic guidance systems never left the sphere of basic research: The Firebird II (XP-43) concept car of General Motors was shown at the '97 Demo with the remark that it had been "restored to running order".[487]

Presented in 1956, the Firebird II was a show car that promoted the idea of an automobile that should follow a guide wire buried in an express lane of the highway. A GM-brochure from 1956 precises that "none of these features is operative in the Firebird II. They serve merely to complete the demonstration of how an electronically controlled car should be equipped to function on the electronic Safety Autoway."[488] It should be kept in mind that sufficiently small-sized computing power for highway automation did only become available 40 years later.

In fact, GM built two versions of the Firebird II, a functional car with a fiberglass body, serving as turbine research vehicle. It was tested at the 5-mile Mesa Proving Grounds in the desert of Arizona that had been opened in 1952. The other non-functional Firebird II with a titanium body served as showcar only for the Motorama.[489] While the functional car had a steering wheel and instruments that provided data about the Whirlfire gas turbine, the showcar was equipped with simulated systems for the Highway of Tomorrow: A double-handed aircraft-style steering wheel and two TV-screens with cathode-ray tubes for information readouts. This showcar was used to make the movies *Key to the Future* and *Design for Dreaming* about an automatic highway called "Auto-Way". Both films were shown to the visitors of the Motorama.

David W. Temple emphasizes that all the automated driving-systems of the Firebird II show car "were only simulated".[490] But he also reports that "about 2 years after the 1956 GM Motorama, functional systems were installed on the road-test car and were successfully demonstrated to the press by GM engineers." According to him, the concept was successfully tested at the Mesa Proving Grounds in Arizona. As we have indeed seen in Sect. 2.1.1, the underlying system had also been successfully demonstrated at GM's Technical Center in Michigan in 1958.

Forty years later, the situation had changed completely: Demo '97 showed seven scenarios for highway automation. The following sections will focus on the two largest demonstrations, the platooning demo of PATH and the free agent demo of Carnegie Mellon University. Both of these demonstrations were based on use of

---

487 Thorpe, Chuck, Jochem, Todd, Pomerleau, 2020. Dean, Automated Highways and the Free Agent Demonstration, online in: https://www.ri.cmu.edu/pub_files/pub2/thorpe_charles_1997_1/thorpe_charles_1997_1.pdf, p. 2 (accessed September 11, 2020).

488 General Motors, 2020. Firebird II, "Three-Zero-Four", The Gas Turbine Family Car, General Motors latest "Laboratory on Wheels", p. 13, in: The Old Car Manual Project, Brochures, GM Corporate and Concepts, http://www.oldcarbrochures.com/static/NA/GM%20Corporate%20and%20Concepts/1956_GM_Firebird_II/1956%20Firebird%20II-12-13.html (accessed on April 23, 2020).

489 Temple, David W. 2015. *Motorama, GM's Legendary Show and Concept Cars*, p. 170f. North Branch.

490 Ibid., p. 174.

technologies provided by GM, Delco Electronics (later Delphi Automotive) and Hughes Aircraft, which were integrated into the demonstrations by the two universities. The PATH demo was based on a series of prior experimental work on other vehicles, analogous to the CMU work that was detailed in earlier chapters. The other scenarios presented by organizations outside the core consortium are mentioned in a third section.

2.6.2 The Platooning Project of PATH, University of California

The Partners for Advanced Transit and Highways (PATH) program of the University of California in Berkeley showed in San Diego a completely automated platooning scenario. Platooning, the coupling of several cars in a line with reduced gaps, was mainly promoted as solution for traffic congestion. Equipped with this technology, each lane would accommodate "at least twice as much traffic as it can carry today", promised PATH.[491] As already reported, the scenario was also associated with an increase in safety and the reduction of energy consumption.

According to Steven Shladover, former technical director of PATH, the idea to form automated vehicles in a platoon dates back to the 1970s, when research on an Automated Guideway Transit (AGT) was undertaken in several countries.[492] Following the model of a railway train, these research efforts aimed at the mechanical coupling of automated vehicles. Platoons, realizing their coupling by control systems only, were first explored by the French engineering company Engins MATRA in the context of a project on the automated people mover Aramis. A pioneering effort was also made by the German automotive company Volkswagen that worked in 1989 on a multiple-vehicle platoon named "convoy driving" (see Sect. 4.4.4).

PATH started experiments with vehicle platoons guided by magnetic markers in the road in 1988.[493] In 1991, a test track of 300 m had been built at the Richmond Field Station, a satellite campus of the University of California in Berkeley to conduct automatic steering control experiments. Aiming at accurate measurements of the position within the lane, usable under all weather conditions – in contrast to camera-based systems – the PATH-team decided to use references in the road surface for lateral control – ceramic magnets were placed with a distance of 1 meter in the middle of the test track. Automatic steering was tested with a Toyota Celica,

---

[491] California PATH, Platoon Demonstration, Demo '97-Flyer, in: http://www.path.berkeley.edu/sites/default/files/my_folder_76/NAHSC_Brochure-Demo%26Groups-971.pdf, p. 15. Accessed June 17, 2018.

[492] Shladover, Steven E. 1997. Reasons for Operating AHS Vehicles in Platoons. In *Automated Highway Systems*, ed. Petros A. Ioannou, p. 15. New York, London.

[493] Shladover 2007, p. 6.

equipped with a hydraulic steering actuator from IMRA and line scan cameras. Other lateral control tests were made with an AMC Hornet.

A first platooning experiment of PATH was performed in 1992.[494] Lateral control experiments with speeds from 20–70 km/h were staged in different types of curves and longitudinal control was tested with a manually driven vehicle followed by another car „under automatic longitudinal control but manual steering control".[495] Speeds between 55 and 75 mph (88–120 km/h) were achieved. "Low speed tests were conducted at the Richmond Field Station, while high-speed tests have been performed at the California Highway Patrol Academy track in West Sacramento, CA and the I-15 commuter lanes in San Diego"reports Shladover.[496] Four vehicles donated by Ford were equipped with throttle and brake actuators, forward ranging radars and wireless LAN communication systems.[497] An IBM-compatible 80,386 computer was used for control functions and a VORAD Doppler radar for obstacle detection. While the experiment with one car following another had been completed in 1992, experiments with a four-car platoon were "in progress" during this time according to Shladover.[498]

This four-vehicle platoon capability was then demonstrated for visitors "on the I-15 HOV lanes in San Diego" in 1994.[499] As a video shows,[500] the four Lincoln Town Car (model 1990–1994) vehicles donated by Ford were used for the experiments (see Fig. 2.15). This 1994 PATH-demo can be considered as the first public demonstration of an AHS-system in history, if we exclude the VW-experiments from 1989 as they were not public and the MATRA-demo 1972 as it did not use passenger cars.

After this historical overview we will now present the configuration of PATH for the 1997 demo: PATH used eight Buick LeSabre, modified by Delco Electronics, General Motors, Hughes and the University of California (Fig. 2.16).[501] By choosing nearly identical vehicles, development costs were reduced and complications in redesigning the interface between the control system and different vehicle platforms were avoided. It was also easier to achieve consistent acceleration and braking performance when the vehicles were closely coupled with each other, explains

[494] E-mail from Steven Shladover to the author, March 11, 2020.

[495] Shladover, Steven E. 1992. The California PATH Program of IVHS Research and Its Approach to Vehicle-Highway Automation. In *Proceedings of the Intelligent Vehicles '92 Symposium*, p. 350f. Detroit.

[496] Ibid, p. 351.

[497] Shladover 2007, p. 7.

[498] Shladover 1992, p. 351.

[499] Shladover 2007, p. 7.

[500] Automatedhighway (Youtube-account), NAHSC Demo 97 Platoon Scenario (video title), uploaded on April 23, 2011, online in: https://www.youtube.com/watch?v=aaLPBtS8b9E. Accessed on January 25, 2020.

[501] NAHSC, Where the research meets the road, Demo '97-Flyer, in: http://www.path.berkeley.edu/sites/default/files/my_folder_76/NAHSC_Brochure-Demo%26Groups-971.pdf, p. 9. Accessed June 17, 2018b.

Fig. 2.15 The first platooning-demo in history, demonstrated by PATH 1994 on the I-15 in San Diego, using three Lincoln Town Cars following a Ford Crown Victoria. (Credits: PATH. Image Source: Screenshot in: Automatedhighway (Youtube-account), NAHSC Demo 97 Platoon Scenario (video title), uploaded on April 23, 2011, online in: https://www.youtube.com/watch?v=aaLPBtS8b9E (accessed on January 25, 2020)

Fig. 2.16 Demo '97: The platoon demonstration with eight Buick LeSabre from the PATH-program, on the I-15, San Diego. (Credits: PATH. Image Source: http://www.path.berkeley.edu/sites/default/files/my_folder_76/NAHSC_Brochure-Demo%26Groups-971.pdf, p. 14)

Shladover.[502] Doing a platoon with diverse vehicles is much more difficult because they accelerate and shift in different ways. For that reason, the Japanese AHS-experiments with 19 vehicles in September 1996 made also use of similar car models – four Toyota Crown, three Nissan Cima, two Honda Accord and two Mitsubishi Diamante (see Sect. 3.5.4).

Lateral control of the Buicks was realized by measuring their position relative to permanent magnet markers buried “every 1.2 m along the path that the vehicle follows.”[503] Shladover explains that the magnetic marker reference system was chosen “because of its very high accuracy (consistently measuring location to an accuracy of better than 1 cm), ability to encode location reference information in a binary code based on the sequence of north and south poles of the magnets, and robustness under all weather conditions (unaffected by rain, snow or ice on the road).”[504] In total, 93,000 magnetic guidance markers were installed in the centers of both lanes.

Three flux-gate magnetometers under the front bumper and three under the rear bumper of the Buicks tracked the magnetic fields of the markers in the center of the road. A Pentium computer in the trunk (Fig. 2.17) used the measurements to determine the lateral position and the yaw angle of the car. This information was then “used to generate commands to an electric motor mounted in the steering column.”[505]

Longitudinal control was realized by range information from a modified Delco-radar in the front of the cars. The radar was only used to detect the vehicle ahead. It was coupled with a vehicle-to-vehicle communication system that obtained speed and acceleration information from the other vehicles 50 times per second.[506] In addition, digital radios from Hughes communicated the movement of the lead vehicle to the platoon at 50 Hz.[507] Its speed was predefined – a computer issued commands to the throttle and brake actuators. All of the eight vehicles maintained an inter-vehicle space of 5.5 meters which made them look if they were tailgating. During the development process they were even tested with a gap of only 4 meters.[508] For the passengers that had been invited to participate in the experiment, this felt like a mechanical coupling because the control system maintained the gap with such high accuracy.

[502] E-mail of Steven Shladover to the author, March 11, 2020.

[503] California PATH, Automated Driving Mini-Demonstration. http://www.path.berkeley.edu/sites/default/files/my_folder_76/NAHSC_Brochure-Demo%26Groups-971.pdf, p. 34. Accessed June 17, 2018.

[504] E-mail of Steven Shladover to the author, March 11, 2020.

[505] California PATH, Automated driving mini demonstration. http://www.path.berkeley.edu/sites/default/files/my_folder_76/NAHSC_Brochure-Demo%26Groups-971.pdf, p. 34.

[506] California PATH, Platoon Demonstration, Demo ‘97-Flyer. http://www.path.berkeley.edu/sites/default/files/my_folder_76/NAHSC_Brochure-Demo%26Groups-971.pdf, p. 15, 18. Accessed June 17, 2018.

[507] Thorpe, Jochem, Pomerleau 1997, p. 4.

[508] E-mail of Steven Shladover to the author, March 11, 2020.

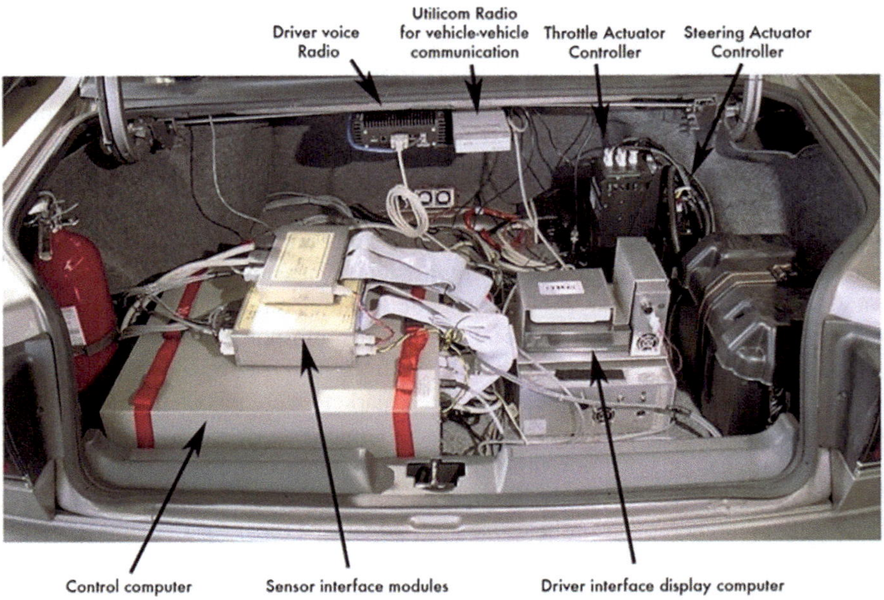

Fig. 2.17 Computer equipment in the trunk of the Buick LeSabre. (Credits: PATH. Image Source: Bishop, Richard, Dopart, Kevin, Shladover, Steven, Commemorating the 20th Anniversary of the National Automated Highway Systems Consortium Demo '97 in San Diego, Powerpoint-Presentation, online in: https://path.berkeley.edu/sites/default/files/demo97foravs17v6.pdf (accessed March 30, 2020), p. 8)

The eight-car-platoon demonstrated acceleration, deceleration, lane changing and split movements to allow other vehicles to join and leave the platoon.[509] They operated for 6–8 hours, each day, for 3 weeks – at highway speed.[510] The engineering team confirms 60 mph in a journal paper, mentioning that experiments demonstrated the feasibility of the system up to speeds of 80 mph.[511] A video documentation about the platoon scenario gives evidence of speeds of 58–61 mph.[512] Shladover explains that the speed of 58 mph "was chosen to remain below the last shift point on the automatic transmissions of the cars. If we went a bit faster than that, we would have been at the boundary between the two highest gears, and the automatic transmissions would have been shifting between those gears unpredictably because we could not control them electronically. We also tested at significantly higher

[509] NAHSC, Where the research meets the road, p. 9 (accessed June 17, 2018).

[510] Rajamani et al., p. 707.

[511] Ibid., p. 699, 705.

[512] Automatedhighway (Youtube-account), NAHSC Demo 97 Platoon Scenario (video title), uploaded on April 23, 2011, online in: https://www.youtube.com/watch?v=aaLPBtS8b9E. Accessed on January 25, 2020.

speeds, and indeed tested the automatic steering of an individual vehicle at over 100 mph to make sure that it would work well at a much higher speed".[513]

The most important argument against this guidance concept was the tremendous financial effort that would have to be mobilized to equip every highway with magnetic nails. Today vision systems onboard the vehicle (cameras and/or lasers) are preferred, needing no additional infrastructure in the road. But one could also argue – as Steven Shladover – that a magnetic system remains more accurate and more robust to weather conditions than any camera-based system. According to him, "its accuracy is only matched by the highest-performance (and most expensive) lidars today". He also precises that "the idea was never going to be to apply this to every mile of roadway in the country, but only to the major highways that carry high volumes of traffic." He estimates that "the cost of installing the magnets was less than 1% of the cost of building the highway lane".[514]

2.6.3 The Free Agent Demonstration of CMU (NavLabs 6–10)

The scenario of Carnegie Mellon University included five vehicles (Fig. 2.18): Two fully automated Pontiac Bonneville (NavLab 6 and 7), one partially automated Oldsmobile Silhouette Minivan (Navlab 8) and two fully-automated New Flyer city buses (Navlab 9 and 10) participated in the demo. The buses were provided by the Houston Metro who was interested to put buses closely together along the HOV lanes, increasing their capacity.[515]

The Navlabs followed another approach of automation than PATH: While the platoon vehicles actively coordinated their motions, the Navlabs operated independently from vehicle-vehicle communications.[516] For that reason CMU called its presentation Free Agent Demonstration. In contrast to the PATH-scenario, they were also independent from modifications of the road infrastructure (magnetic markers). The engineering-team explains: "The philosophy behind the Free Agent scenario is to surround the vehicles with sensors, putting all the sensing and decision-making on board the vehicles. When the vehicles see other automated vehicles, they can communicate with them and drive close to each other. But they have enough perception and reasoning capability so that they can drive by themselves, mixed in with

[513] E-Mail of Steven Shladover to the author, March 9, 2020.

[514] E-Mail of Steven Shladover to the author, December, 29, 2022.

[515] Houston Metro, Demo '97-Flyer. http://www.path.berkeley.edu/sites/default/files/my_folder_76/NAHSC_Brochure-Demo%26Groups-971.pdf, p. 45. Accessed June 17, 2018.

[516] See also: Kanaris, Alexander, Petros Ioannou, and Fu-Sheng Ho. 1997. Spacing and Capacity Evaluations for Different AHS Concepts. In *Automated Highway Systems*, ed. Petros A. Ioannou, p. 136. New York, London.

Fig. 2.18 Two New Flyer city buses (Navlab 9 and 10), an Oldsmobile Silhouette Minivan (Navlab 8) and two Pontiac Bonneville (Navlab 6 and 7). (Credits: CMU. Image Source: https://www.cs.cmu.edu/Groups/ahs/images/navlab_6_11_images/jpeg/nl_family1.jpg)

conventional manually driven vehicles. This ability to run by themselves is why they are called "free agents"".[517]

All five Navlabs were equipped with the same systems – video cameras, radar and ladar: To maintain a gap to the lead vehicle, a 77 GHz headway radar from Delco tracked targets "within a 12-degree field of view, out to a range of 150 meters".[518] The vehicles were also equipped with side-looking short-range radars from Eaton Vorad and rear-looking ladars from Riegel with a field of view of 20 degrees. Each vehicle had a human safety driver who could override the system.[519] Computer vision was realized with the RALPH system, developed by Dean Pomerleau. The overhead image of a forward-looking video camera located in front of the rear-view mirror was "processed to find the road curvature" that most closely

[517] Thorpe, Chuck, Todd Jochem, and Dean Pomerleau. 1998. The 1997 Automated Highway Free Agent Demonstration. In *Proceedings of Conference on Intelligent Transportation Systems 1997*, p. 497.

[518] Thorpe, Chuck, Jochem, Todd, Pomerleau, Dean, Automated Highways and the Free Agent Demonstration, online in: https://www.ri.cmu.edu/pub_files/pub2/thorpe_charles_1997_1/thorpe_charles_1997_1.pdf, p. 6.

[519] For the first time, the Navlab 7 was equipped with a realistic driver interface. It consisted of "customized steering wheel controls, a color flat panel display integrated into the dashboard, a headup display unit, and audio alert capability." (Team Navlab, Carnegie Mellon University, Demo '97-Flyer, in: http://www.path.berkeley.edu/sites/default/files/my_folder_76/NAHSC_Brochure-Demo%26Groups-971.pdf, p. 46. Accessed June 17, 2018.

followed all road segments.[520] This system was especially important in California where highways are equipped with raised dots to demarcate lane boundaries – instead of painted lines for which most systems are trained.

Using these techniques, the CMU scenario "demonstrated speed and headway control, lane following, lane changing, obstacle detection, and cooperative obstacle avoidance maneuvers."[521] According to Thorpe, during the 5 weeks before the demo and the 4 days of the demo runs, CMU "logged 12,500 autonomous miles" on the test course. Only 1% of the miles "were run in manual mode due to various system glitches", as for example an "over-cautious safety circuit."[522]

A comparative perspective can highlight some differences between the platoon and the free agent demonstration. While the platooning vehicles of PATH did not demonstrate obstacle detection, CMU integrated this function in their cars. In the PATH scenario, a human could not override the system while the vehicles moved in a platoon – its slow reflexes would be a source of danger. In the CMU demo, the system could be switched on and off at any moment, like a cruise control. While the PATH system aimed at high traffic density and needed a dedicated lane, the CMU system worked in mixed traffic.

Thorpe, Jochem and Pomerleau explain the utility of the different concepts by the different infrastructures around the research facilities: Most freeways around Berkeley in the San Francisco bay area have at least three lanes in each direction. Therefore, one lane could be dedicated for automated driving. In the less urban western end of Pennsylvania around Carnegie Mellon University, safety is a more important issue than congestion, claim the authors. An approach independent of infrastructure is more appropriate because most highways do not have three lanes in each direction and the highways are very long – infrastructure modifications would be very expensive. They conclude that each system "has its niche in which it may be more appropriate."[523] But from our perspective 25 years later we can resume that the Free Agent concept has gained much more acceptance.

At the end of this chapter we should also mention the third scenario that demonstrated the transition from steering by machine vision to steering by magnetic markers: The joint-venture of PATH and Honda had the aim to show the transition between operations in a high-density urban highway with magnetic markers in the road as lane reference and a lower-density rural environment where steering accuracy is not as critical, so the car steers by lane-marking detection, using stereo video cameras mounted inside the windshield. In both parts of the scenario, laser radar was used in addition to the stereo vision system to measure the distance to the car

---

520 Thorpe, Chuck, Jochem, Todd, and Pomerleau, Dean, Automated Highways and the Free Agent Demonstration, online in: https://www.ri.cmu.edu/pub_files/pub2/thorpe_charles_1997_1/thorpe_charles_1997_1.pdf, p. 6.

521 Ibid., p. 1.

522 Thorpe, Chuck. 1998. Mobile Robots and Smart Cars. In *Field and Service Robotics*, ed. Alexander Zelinsky, p. 4. Berlin, New York.

523 Thorpe, Chuck, Jochem, Todd, Pomerleau, Dean, Automated Highways and the Free Agent Demonstration, p. 7.

ahead and to detect obstacles. Two modified and one conventional Honda Accord – simulating a slow non-automated car – were used to show the transition between these two approaches (platooning, automated lane changes and obstacle detection and avoidance).

A fourth alternative guidance scenario of the Ohio State University under the direction of the electrical engineer Ümit Özgüner demonstrated radar-reflective tapes in the center of the lane, on the road surface, manufactured by 3M and tested with a different set of Honda Accords. The tape had an aluminum foil layer with slots in the middle that responded well to the 10 GHz automotive radar that detected the road and other vehicles on the side.[524] Together with a vision system, it was used for lateral control. Radar was also used for detecting vehicles at the side. A laser system was used for longitudinal (speed and braking) control.[525]

The other scenarios demonstrated infrastructure diagnostics by maintenance vehicles (Lockheed Martin), the use of driver assistance radars for heavy trucks (Eaton Vorad) and an evolutionary scenario for driver assistance (Toyota).

2.6.4 The Shift from Infrastructure to Vehicle-Based Solutions

It can be resumed that the approaches of PATH / University of California Berkeley and Carnegie Mellon University that were shown at Demo '97 represent the two main ideas to vehicle automation that had been developed throughout the twentieth century – the first based on support from the infrastructure (but with all control decision made in the vehicle), the second focused on solutions onboard the vehicle.[526]

A TRB-report highlights that at the beginning of the 1990s, the Department of Transportation pursued two "distinctly different research paths".[527] The NHTSA evaluated partially automated assistance systems (e.g. collision warning devices), following an *evolutionary* implementation of vehicle automation while the FHWA aimed at the demonstration of an automated highway system, following the idea of an *accelerated* research program that will lead to a prototype. The report emphasizes that both approaches coexisted until 1997, concluding that the program was not successful in reaching a consensus among stakeholders on what kind of highway technology should be introduced in the future.

From a historical perspective Demo 97 is on the one hand important because both approaches came together at the same event, showing "the potential of combining intelligent vehicles with intelligent highways" as a DOT-report highlights.[528] It

524 Ibid., p. 3.

525 NAHSC, Where the research meets the road, p. 10.

526 This view is also supported by Davies, p. 50.

527 Transport Research Board, 1998. *National Research Council, National Automated Highway System Research Program*, A review, Special Report 253, Washington, online in: https://onlinepubs.trb.org/onlinepubs/sr/sr253/sr25302.pdf, p. 18.

528 Auer, et al. p. 17.

can be seen as "an important ancestor of today's focus on automated and connected vehicles."

On the other hand, Demo '97 can be seen as a historical shifting point towards vehicle-based solutions. The ISTEA Initiative and the San Diego-Demo are significant from a historical perspective as they can be seen as the climax of the decades-old idea that the creation of an automated highway and an automated vehicle *prototype* might lead to an accelerated realization of a fully automated system. The TRB-report observes that this vision was very similar "to that of GM researchers in the 1950s and 1970s who sought to specify an optimal fully automated highway system for application before the turn of the century".[529] In this approach, the idea of a gradually emerging autonomous highway system through *incremental* advances in driver assistance systems was rejected.

After this successful demonstration of an AHS-system, the political priorities changed quickly. ISTEA expired on September 30, 1997 and was followed by the Transportation Equity Act for the twenty-first Century (TEA-21, signed into law on June, 9, 1998). The Department of Transportation cancelled the long-term National Automated Highway System Research Program (NAHSRP) on full automation in 1998 and abandoned the idea of an AHS-prototype. All funding was withdrawn from the NAHSC.[530] The idea of magnetic nails, detected by vehicle-based sensors, was substituted by an approach that used vehicle based sensors only.

After Demo '97, research efforts were recentered on near-term, vehicle-based safety technologies. Federal support moved to the Intelligent Vehicle Initiative (IVI), a private-public partnership that aimed at the more immediate goal to develop driving assistance systems. PATH also concentrated its activities more on "near-term, incremental research rather than in longer-term transformational research."[531] This was a historically important shift in concepts.

2.7 Conclusion

The presentation of the research efforts in the United States allowed to identify three main actors that were engaged in the field of automated and autonomous driving: In the 1950s, the automotive industry together with academic partners from electrical engineering started tests on vehicle automation, focusing on Automated Highway Systems. In the 1960s, academic research laboratories at Stanford and Carnegie Mellon University began the development of mobile robots, using concepts of

[529] Transportation Research Board 1998, p. 17.

[530] Thorpe, Jochem, Pomerleau, p. 8. Jochem comments that the decision of the federal government to stop funding was probably right because ten years later the automotive industry deployed some of these technologies and by 2015, Tesla "was deploying a system that was nearly equivalent to the best CMU systems for AHS on open highways." (E-mail of Todd Jochem to the author, May 10, 2020).

[531] Shladover 2007, p. 8.

artificial intelligence. With the purely military research programs of the following decades, a third actor entered the field.

An important finding of the chapter is that the two actors last mentioned were never clearly separated in the US – a big contrast to other regions. Especially in the robotics domain the relation between civil and military actors is very close. AI research is "at least in the United States, first and foremost a military affair", confirms Crevier.[532] From the 1960s on and in the following decades, all the research on mobile robots was funded by military agencies like DARPA. Without the creation of DARPA, triggered by Sputnik and the Space Race, the enormous financial resources for the basic research on technologies that play a crucial role for autonomous driving had not been made available. Without this military funding, these technologies would probably not exist or at least in another form.

At the beginning of the 1980s a shift from automobile-centered to military-centered technology development can be observed: While funding for Automated Highway Systems was completely stopped, important research projects on autonomous driving in the United States like the ALV and the NavLab were funded by military sources. The research on vision systems (ALV) received more than $10 million from the Strategic Computing Initiative.

Another observation specific to the US-chapter is that the research efforts of the automotive, robotics and military field began to overlap in the 1990s, at least at some occasions: As vehicles able to perform autonomous cross-country operations became a priority, civil and military domains collaborated on many occasions. For instance, the technology development of the NavLab vehicles contributed directly to military demands: CMU participated from 1990–1996 with NavLab2 in the DARPA-sponsored Demo II-program that focused on the military use case of Unmanned Ground Vehicles. Much of the perception technology used in this military research program had been developed by CMU. When CMU demonstrated at the San Diego-Demo 1997 autonomous driving with five vehicles, showing onboard decision making capabilities, military and civil technology development became undistinguishable on the level of the algorithms.

The second point to be highlighted here concerns the concepts of automated and autonomous driving. The United States were the first country where vehicle automation had been researched. A common element of all approaches was the aim to find an answer to the key question how a communication process between the road environment and the vehicle could be established. Two main answers were given, first mainly infrastructure-based second mainly vehicle-based. And sensors were always in the center of the task: The early automation concepts of RCA, inspired from railway concepts, depended on *sensors in the road*. Institutions like Ohio State University struggled with *missing in-vehicle sensors* and interferences of the guidance system with ferrous materials in the road. More robust vehicle based solutions became only possible with developments in the robotics field, putting the accent on *perception* and decision making capabilities. At the beginning of the 1960s, the idea

[532] Crevier, p. 313.

to use a camera for guidance purposes came up with the Stanford Cart. The step from remote control to computer vision controlling a moving machine was a big milestone. As it gave a machine the ability to see, it was a central step in the direction of machine vision. The efforts of the following decades – detect obstacles and avoid them – would not have been possible without vision concepts.

In a long-term perspective, the decades-long research on mobile robots were much more influential than the early guidewire-concepts from the electronics and automotive industry. We have seen that the early robots like Shakey and the Stanford Cart were not able to perceive, understand and move at the same time. While standing still, they first had to focus on perception, calculating and planning. Then they moved a certain distance and had to stop again. This changed in the 1980s when the digital processing power of small onboard computer systems made perception and movement in real time possible. But it was only at the end of the 1990s that the decades-old idea of an infrastructure-based automated highway had been abandoned.

Bibliography

Adams, James L. 1961. *Remote Control with Long Transmission Delays*, PhD in Mechanical Engineering. Stanford University.

Adam, John A. 1986, January. Aerospace and Military. *IEEE Spectrum*.

Albus, James S., and Alexander M. Meystel. 2001. *Engineering of Mind, An Introduction to the Science of Intelligent Systems*. New York.

Albus, James, and Anthony Barbera. 2006. 4D/RCS Reference Model Architecture for Unmanned Ground Vehicles. In *Intelligent Vehicle Systems: 4 D/RCS Approach*, ed. Raj Madhavan, Elena R. Messina, and James Albus, 1–26. New York.

Auer, Ashley, Shelley Feese, and Stephen Lockwood. 2016. History of Intelligent Transportation Systems, Performed by U.S. Department of Transportation, Intelligent Transportation Systems Joint Program Office, Report No. FHWA-JPO-16-329, Produced by Booz Allen Hamilton, May 2016.

Author Unknown. 1958. Highway of the Future. *Electronic Age (RCA-Journal)* 17 (1): 12–14.

———. 1970. Bendix Works on Automatic Vehicle Project. *Ann Arbor News*, June 19, 1970, online in: https://aadl.org/aa_news_19700619-bendix_works_on_automatic_vehicle_project. Accessed in June, 13, 2019.

———. 1972. Firm Devises Driverless Vehicle Controls. *Ann Arbor News*, June 30, 1972, online in: https://aadl.org/aa_news_1972060-firm_devises_driverless_vehicle_controls. Accessed on June 18, 2019.

———. 1991. Take Me, I'm Yours, The Autobiography of SAIL, 1991, online in: http://infolab.stanford.edu/pub/voy/museum/pictures/AIlab/SailFarewell.html. Accessed on April 10, 2019.

———. 2007. More Attacks, Mounting Casualties. *The Washington Post*, September 30, 2007, online in: https://www.washingtonpost.com/wp-dyn/content/graphic/2007/09/28/GR2007092802161.html. Accessed July 7, 2020.

———. 2009. The Inside Story of the SWORDS Armed Robot "Pullout" in Iraq. *Popular Mechanics*, October 1, 2009. https://www.popularmechanics.com/technology/gadgets/a2804/4258963/.

Bel Geddes, Norman. 1940. *Magic Motorways*. New York.

Bender, James G., Robert E. Fenton, and Karl W. Olson. 1971. An Experimental Study of Vehicle Automatic Longitudinal Control. *IEEE Transactions on Vehicular Technology* VT-20 (4): 114–123.

Bender, James G., and Robert E. Fenton. 1969. A Study of Automatic Car Following. *IEEE Transactions on Vehicular Technology* VT-18 (3): 134–140.

Bertozzi, Massimo, Alberto Broggi, Massimo Cellario, Alessandra Fascioli, Paolo Lombardi, and Marco Porta. 2002. Artificial Vision in Road Vehicles. *Proceedings of the IEEE* 90 (7): 1258–1269.

Bertozzi, Massimo, Alberto Broggi, and Alessandra Fascioli. 2000. Vision-Based Intelligent Vehicles: State of the Art and Perspectives. *Robotics and Autonomous Systems* 32 (1): 1–16.

Bishop, Richard, Kevin Dopart, and Steven Shladover. 2020. Commemorating the 20th Anniversary of the National Automated Highway Systems Consortium Demo '97 in San Diego, Powerpoint-Presentation, online in: https://path.berkeley.edu/sites/default/files/demo97foravs17v6.pdf. Accessed March 30, 2020.

Braisted, Paul. 1963. *Study of a Predictor for Remote Control Systems Operating with Signal Transmission Delays, PhD in Mechanical Engineering*. Stanford University.

Brooks, Rodney. 1986. A Robust Layered Control System for a Mobile Robot. *IEEE Journal of Robotics and Automation* 2 (1): 14–23.

California PATH, Platoon Demonstration, Demo '97-Flyer. http://www.path.berkeley.edu/sites/default/files/my_folder_76/NAHSC_Brochure- Demo%26Groups-971.pdf, pp. 14–17. Accessed June 17, 2018.

Carafano, James Jay, and Andrew Gudgel. 2008. The Pentagon's Robots: Arming the Future. In *Competitive Technologies for National Security*, ed. James Jay Carafano, Andrew Gudgel, and Alane Kochems, 27–32. Washington, DC: The Heritage Foundation. online in: https://www.heritage.org/node/15537/print-display. Accessed on June, 24, 2020.

Cassel, David. 2017. Remembering Shakey, the First Intelligent Robot. *The New Stack*, March 5, 2017, online in: https://thenewstack.io/remembering-shakey-first-intelligent-robot/.

Champeny-Bares, Lee, Syd Coppersmith, and Kevin Dowling. 1991. *The Terregator Mobile robot*. Pittsburgh: Field Robotics Center, The Robotics Institute, Carnegie Mellon University.

Cheon, Sanghyun. 2003. *An Overview of Automated Highway Systems (AHS) and the Social and Institutional Challenges They Face, UC*. Berkeley: University of California Transportation Center. online in: https://escholarship.org/uc/item/8j86h0cj, p. 23. Accessed on March, 28, 2020.

Chun, Wendell H., and Todd M. Jochem. 1994. Unmanned Ground Vehicle Demo II: Demonstration A. *Unmanned Systems* 12 (1): 14–20.

Congressional Research Service, Defense Primer, U.S. Policy on Lethal Autonomous Weapons Systems, December 19, 2019.

Crevier, Daniel. 1993. *The Tumultuous History of the Search for Artificial Intelligence*. New York.

Darrach, Brad. 1970. Meet Shaky, the First Electronic Person. *LIFE Magazine*, November 20, 1970, pp. 58A–58D, 61, 63–64, 65–66, 68.

Davies, Alex. 2021. *Driven, The Race to Create the Autonomous Car*, New York, London, Toronto, Sydney, New Delhi.

Defense Advanced Research Projects Agency, Next Generation Explosive Ordnance Disposal Remote Controlled Vehicle (NGEODRCV), Common Systems Architecture, Draft version 5, Arlington 2001.

Department of Defense, Unmanned Ground Vehicle Master Plan, July 1992. https://apps.dtic.mil/dtic/tr/fulltext/u2/a277757.pdf. Accessed June 25, 2020.

Department of Defense, Office of the Secretary of Defense. 2000. *Joint robotics program master plan*. Washington, DC.

Department of Defense, Directive, N° 3000.09, Autonomy in Weapon Systems, November 21, 2012, p. 2, online in: https://www.esd.whs.mil/Portals/54/Documents/DD/issuances/dodd/300009p.pdf. Accessed March 3, 2020.

De Vries, Marten Johannes, Pushkin Kachroo, and Kaan Ozbay, eds. 1998. Intelligent transportation systems. *Proceedings of the Society of Photo-Optical Instrumentation Engineers (SPIE)*, vol. 3207., 15–17 October 1997. Pittsburgh.

Dowling, Kevin, Robert Guzikowski, Jim Ladd, Henning Pangels, Sanjiv Singh, and William Whittaker. 1990. Navlab. An Autonomous Navigation Testbed. In *Vision and Navigation. The Carnegie Mellon Navlab*, ed. Charles E. Thorpe, 273–275. Boston/Dordrecht/London.

Dudek, Gregory, and Michael Jenkin. 2010. *Computational Principles of Mobile Robotics*. Cambridge, New York, Melbourne, Madrid: Cambridge University Press (first edition: 2000).

Earnest, Lester. Stanford Cart, 2018. https://web.stanford.edu/~learnest/les/cart.html. Accessed on February 3, 2019.

Engineering and Technology History Wiki, Milestones: SHAKEY: The World's First Mobile Intelligent Robot, 1972, last modified on May 24, 2017. http://ethw.org/Milestones:SHAKEY:_The_World's_First_Mobile_Intelligent_Robot,_1972. Accessed on December 5, 2019.

Fallon, Michael. 2019. *Self-Driving Cars, the New Way Forward*. Minneapolis.

Farivar, Cyrus. 2016. Dallas deployment of robot bomb to kill suspect is "without precedent". Ars technica, July 8, 2016, online in: https://arstechnica.com/tech-policy/2016/07/is-it-ok-to-send-a-police-robot-to-deliver-a-bomb-to-kill-an-active-shooter/. Accessed June 24, 2020.

Fenton, Robert E., Robert L. Cosgriff, Karl W. Olson, and Lyle M. Blackwell. 1968. One Approach to Highway Automation. *Proceedings of the IEEE* 56 (4): 556–566.

Fenton, Robert E., and Karl W. Olson. 1969. The Electronic Highway. *IEEE Spectrum*, pp. 60–66.

Fenton, Robert E. 1970. Automatic Vehicle Guidance and Control – A State of the Art Survey. *IEEE Transactions on Vehicular Technology* VT 19 (1): 153–161.

Fenton, Robert E., and Robert J. Mayhan. 1991. Automated Highway Studies at the Ohio State University – An Overview. *IEEE Transactions on Vehicular Technology* 40 (1): 100–113.

Fenton, Robert E. 1994. IVHS/AHS: Driving into the Future. *IEEE Control Systems Magazine* 14 (6): 13–20.

Flory, Leslie E. 1961. Electronic Control of Highway Vehicles. *Journal of the IEEE* 7 (77): 271–273.

Flory, Leslie E., et al. 1962. Electronic techniques in a system of highway vehicle control. *RCA Review* 23: 293–310.

Future of Life Institute. Autonomous Weapons: An Open Letter from AI & Robotics Researchers, online in: https://futureoflife.org/open-letter-autonomous-weapons/?cn-reloaded=1. Accessed March 5, 2020.

Gage, Douglas W. 1995. UGV History 101: A Brief History of Unmanned Ground Vehicle (UGV) Development Efforts, *Unmanned Systems, Special Issue on Unmanned Ground Vehicles* 13 (3): 2–7.

Gardels, K. 1960. Automatic car controls for electronic highways, General Motors Research Laboratories, *Rep. GMR-276*. Warren, MI.

General Motors. 1953. An Automatically Guided Automobile Cruised Along a One-Mile Check Road at General Motors Technical Center today.... Press release, USA.

Gershgorn, Dave. 2016. Police Used Bomb Disposal Robot to Kill a Dallas Shooting Suspect. *Popular Science*, July 8, 2016, online in: https://www.popsci.com/police-used-bomb-disposal-robot-to-kill-dallas-shooting-suspect/. Accessed June 24, 2020.

Hartford, Bill. 1967, April. You 'Fly' this Car with a Stick. *Popular Mechanics*, pp. 87–89 and p. 207.

Hartwich, Erwin, and Walter Zimdahl. 2012. Regelung von Straßenfahrzeugen: Ein Zeitzeugenbericht aus dem Institut von Prof. W. Oppelt, TH Darmstadt. *at-automatisierungstechnik* 60 (6): 338–343.

Hebert, Martial, Charles E. Thorpe, and Anthony Stentz. 1997. Introduction. In *Intelligent Unmanned Ground Vehicles, Autonomous Navigation Research at Carnegie Mellon*, ed. Martial H. Hebert, Charles E. Thorpe, and Anthony Stentz, 1–15. New York.

Ingraham, Joseph C. 1960. Electronic Roads Called Practical; New System of Guiding Cars Safely on Highways Is Shown at Princeton Fruit of 7 Years' Study R.C.A. and G.M. Jointly Conducted It – Full Use Seen 15 Years Away. *The New York Times*, June 6, 1960.

Ioannou, Petros A., ed. 1997. *Automated Highway Systems*. New York, London.

Jacobsen, Annie. 2015. *The Pentagon's Brain, An Uncensored History of DARPA, America's Top Secret Military Research Agency*. New York.

Jochem, Todd, and Dean Pomerleau. No Hands Across America General Information. *No Hands Across America Journal*. http://www.cs.cmu.edu/~tjochem/nhaa/Journal.html. Accessed on August 3, 2018.

Julia, Luc. 2019. *L'intelligence artificielle n'existe pas*. Paris.

Kanaris, Alexander, Petros Ioannou, and Fu-Sheng Ho. 1997. Spacing and Capacity Evaluations for Different AHS Concepts. In *Automated Highway Systems*, ed. Petros A. Ioannou, 125–167. New York, London.

Kay, Jennifer S. 1997, January. *STRIPE: Remote Driving Using Limited Image Data*, PhD thesis. Carnegie Mellon University, Computer Science Department.

Kilbon, Kenyon. 1960, Autumn. Tomorrow's Thruway is Here Today! *Electronic Age*, pp. 26–29.

Krishnan, Armin. 2009. *Killer Robots. Legality and Ethicality of Autonomous Weapons*. Surrey/Burlington.

Kröger, Fabian. 2016. Automated Driving in its Social, Historical and Cultural Contexts. In *Autonomous Driving, Technical, Legal and Social Aspects*, ed. M. Maurer, J.C. Gerdes, B. Lenz, and H. Winner, 41–68. Heidelberg.

Kuan, D., and U.K. Sharma. 1987. Model Based Geometric Reasoning for Autonomous Road Following. *Proceedings of the IEEE Int. Conf on Robotics and Automation*, Raleigh, NC, USA, March 31 – April 3, 1987, pp. 416–423.

Kuan, D., and G. Phipps. 1987. A-Shuan Hsueeh, Autonomous Land Vehicle Road Following. *Proceedings of the First Int. Conf. on Computer Vision*, London England, June 1987, pp. 557–566.

Kurman, Melba. 2017. Electronic Highway of the Future. Mechanix Illustrated. March 10, 2017, online in: https://mechanixillustrated.technicacuriosa.com/2017/03/10/electronic-highway-of-the-future/. Accessed May 25, 2021.

Lenzen, Manuela. 2018. *Künstliche Intelligenz, Was sie kann & was uns erwartet*. München.

Madhavan, R., T. Hong, and E. Messina. 2006. Temporal Registration of Sensed Range Images for Autonomous Navigation. In *Intelligent Vehicle Systems: 4 D/RCS Approach*, ed. Raj Madhavan, Elena R. Messina, and James Albus, 173–198. New York.

Main, Douglas. 2013. Robots to the Rescue. *Popular Science*, April 8, 2013, online in: https://www.popsci.com/trp-sponsored-article-slideshow/. Accessed July 7, 2020.

Mann, Martin. 1958. The Car That Drives Itself. *Popular Science*, No. 172, May 1958, pp. 75–79 and 226–227.

Matthies, Larry H. 1998. Robotic Perception for Autonomous Navigation of Mars Rover. *1998 Symposium on Frontiers in Engineering*, September 1998, pp. 1–6.

McCarthy, John, Marvin Minsky, Nathan Rochester, Claude Shannon. 1955. A Proposal for the Dartmouth Summer Research Project on Artificial Intelligence.

McCarthy, John. 2019. Computer Controlled Cars, (according to the author, the paper was written around 1968 and revised in the 1970s), online in: http://jmc.stanford.edu/commentary/progress/cars.pdf. Accessed October 1, 2019.

McKay, Shawn, Matthew E. Boyer, Nahom M. Beyene, Michael Lerario, Matthew W. Lewis, Karlyn D. Stanley, Randall Steeb, Bradley Wilson, and Katheryn Giglio. 2020. *Automating Army Convoys, Technical and Tactical Risks and Opportunities*. Santa Monica: RAND Corporation. in: https://www.rand.org/content/dam/rand/pubs/research_reports/RR2400/RR2406/RAND_RR2406.pdf. Accessed on June 24, 2020.

McMillan, Stuart I., and Jason G. McPhee. 2014. *An Analysis of the First Fifteen Years of the Department of Defense Framework for Unmanned Ground Systems*, Thesis, Naval Postgraduate School, Monterey, California.

Merrin, William. 2019. *Digital War, A Critical Introduction*. Oxon/New York.
Meystel, Alex. 1991. *Autonomous Mobile Robots, Vehicles with Cognitive Control*. Singapore.
Moravec, Hans. 1980, March. *Obstacle Avoidance and Navigation in the Real World by a Seeing Robot Rover*, Ph.D. thesis, Computer Science Department, Stanford University.
Moravec, Hans. 1982. The CMU rover. *AAAI Proceedings*, pp. 377–380.
———. 1988. *MIND Children, The Future of Robot and Human Intelligence*. Cambridge, London.
Moravec, Hans P. 1999. *Robot: Mere Machine to Transcendent Mind*. Oxford.
Morris, Jefferson. 2006. iRobot's Dyer Sees Missions Multiplying for PackBot. *Aerospace Daily & Defense Report* 217 (36).
Moye, William T. 1997, May. *The Genealogy of ARL*.
Murphy, Robin R. 2000. *Introduction to AI Robotics*. Cambridge, London: The MIT Press.
Mysliwetz, Birger. 1990. *Parallelrechner-basierte Bildfolgen-Interpretation zur autonomen Fahrzeugsteuerung*, Dissertation, Universität der Bundeswehr München, Fakultät für Luft- und Raumfahrttechnik. Institut für Systemdynamik und Flugmechanik.
NAHSC. 2018a. What is An Automated Highway System?, Demo '97-Flyer, in: http://www.path.berkeley.edu/sites/default/files/my_folder_76/NAHSC_Brochure-Demo%26Groups-971.pdf. Accessed June 17, 2018.
———. 2018b. Where the research meets the road, Demo '97-Flyer, in: http://www.path.berkeley.edu/sites/default/files/my_folder_76/NAHSC_Brochure-Demo%26Groups-971.pdf. Accessed June 17, 2018.
National Research Council. 2002. *Technology Development for Army Unmanned Ground Vehicles*. Washington DC.
Nilsson, Nils J. 2010. *The Quest for Artificial Intelligence, a History of Ideas and Achievements*. New York: Cambridge University Press.
Nocks, Lisa. 2007. *The Robot: The Life Story of a Technology*. Westport.
Nora, Simon, and Alain Minc. 1978. *L'Informatisation de la Société: rapport à M. le Président de la République*. Paris: La Documentation Française.
Nowacki, G. 2012. Development and Standardization of Intelligent Transport Systems. *International Journal on Marine Navigation and Safety of Sea Transportation* 6 (3): 403–411.
Özgüner, Ümit, Tankut Acarman, and Keith Alan Redmill. 2011. *Autonomous Ground Vehicles*. Boston, London.
Podnar, Gregg W. 1985a, February. The NEPTUNE Mobile Robot. *Annual Report, Technical Report CMU-RI-TR-86-04*, Robotics Institute, Carnegie Mellon University, pp. 123–125.
———. 1985b, February. The URANUS Mobile Robot. *Annual Report, Technical Report CMU-RI-TR-86-04*. Robotics Institute, Carnegie Mellon University, pp. 127–129.
Pomerleau, Dean A. 1993. *Neural Network Perception for Mobile Robot Guidance*. New York.
Quigg, Doc. 1960, June 7. Reporter Rides Driverless Car. *The Press-Courier*.
Rajamani, Rajesh, Han-Shue Tan, Boon Kait Law, and Wei-Bin Zhang. 2000. Demonstration of Integrated Longitudinal and Lateral Control for the Operation of Automated Vehicles in Platoons. *IEEE Transactions on Control Systems Technology* 8 (4): 695–708.
Rillings, Jim. 1997, February. *The National Automated Highway System: Cooperative Program*. National Automated Highway System Consortium.
Roland, Alex, and Philip Shiman. 2002. *Strategic Computing, DARPA and the Quest for Machine Intelligence*. Cambridge, London.
Schefter, Jim. 1985, October. Look, Ma! No Driver. *Popular Science*, pp. 94–95, 114–116.
Schmidt, Rodney. 1971. *A Study of the Real-Time Control of a Computer-Driven Vehicle*. PhD in Electrical Engineering. Stanford University.
Shladover, Steven E. 1992. The California PATH Program of IVHS Research and Its Approach to Vehicle-Highway Automation. *Proceedings of the Intelligent Vehicles '92 Symposium*, Detroit, pp. 347–352.
———. 1997. Reasons for Operating AHS Vehicles in Platoons. In *Automated Highway Systems*, ed. Petros A. Ioannou, 11–27. New York, London.

———. 2007. PATH at 20 – History and Major Milestones. *IEEE Transactions on Intelligent Transportation Systems* 8 (4): 584–592.

———. 2008. AHS Research at the California PATH Program and Future AHS Research Needs. *Proceedings of the 2008 IEEE International Conference on Vehicular Electronics and Safety*, Columbus, September 22–24, 2008, pp. 4–5.

Shladover, Steven E., Jane Lappin, Robert P. Denaro, and Bryant Walker Smith. 2018. Introduction: The Transportation Research Board's 2013 Workshop on Road Vehicle Automation. In *Road Vehicle Automation*, ed. Gereon Meyer and Sven Beiker, 1–12. Cham, Heidelberg, New York, Dordrecht, London.

Shoemaker, Charles. 2006. Development of Semi-Autonomous Robotic Ground Vehicles, DoD's Ground Robotics Research Programs: Demo I through Demo III. In *Intelligent Vehicle Systems: 4 D/RCS Approach*, ed. Raj Madhavan, Elena R. Messina, and James Albus, 283–315. New York.

Spofford, John R., Raymond D. Rimey, and Steve H. Munkeby. 1997. Overview of the UGV / Demo II program, online in: https://pdfs.semanticscholar.org/72fa/74930bb02e33e2c7d887f561dd78d4bf4271.pdf. Accessed on April 8, 2020.

Stone, Wesley L. 2005. The History of Robotics. In *Robotics and Automation Handbook*, ed. Thomas R. Kurfess, 1–12. Boca Raton, London, New York, Washington, DC.

Strachan, Jamie Robert. 2009. *NOVA: Nottingham Offroad Vehicle Architecture*, PhD thesis. University of Nottingham, in: http://eprints.nottingham.ac.uk/10842/1/Microsoft_Word_-_Complete_Thesis_-_issue_M.pdf.

Temple, David W. 2015. *Motorama, GM's Legendary Show and Concept Cars*. North Branch.

The Congress of the United States, Congressional Budget Office, High-Tech Highways: Intelligent Transportation Systems and Policy, October 1995, online in: https://www.cbo.gov/sites/default/files/104th-congress-1995-1996/reports/High-techHighways.pdf. Accessed March, 23, 2020.

Thielman, Sam. 2016. Use of Police Robot to Kill Dallas Shooting Suspect Believed to be First in US History. *The Guardian*, July 8, 2016, online in: https://www.theguardian.com/technology/2016/jul/08/police-bomb-robot-explosive-killed-suspect-dallas. Accessed June 24, 2020.

Thorpe, Charles E. 1990. Chapter 1: Introduction. In *Vision and Navigation: The Carnegie Mellon Navlab,* ed. Charles E. Thorpe, 1–8, Boston.

Thorpe, Charles, Martial H. Hebert, Takeo Kanade, and Steven A. Shafer. 1988. Vision and Navigation for the Carnegie-Mellon Navlab. *IEEE Transactions on Pattern Analysis and Machine Intelligence* 10 (3): 362–373.

Thorpe, Charles, Martial Hebert, Takeo Kanade, and Steven Shafer. 1991a. Toward Autonomous Driving: The CMU Navlab, Part I, Perception. *IEEE Expert* 6 (4): 31–42.

———. 1991b. Toward Autonomous Driving: The CMU Navlab, Part II, System and Architecture. *IEEE Expert* 6 (4): 44–52.

Thorpe, Chuck, Todd Jochem, and Dean Pomerleau. 2020. Automated Highways and the Free Agent Demonstration, online in: https://www.ri.cmu.edu/pub_files/pub2/thorpe_charles_1997_1/thorpe_charles_1997_1.pdf. Accessed September 11, 2020.

———. 1998. The 1997 Automated Highway Free Agent Demonstration. *Proceedings of Conference on Intelligent Transportation Systems 1997*, IEEE, S. 496–501.

Thorpe, Chuck. 2018. An oral history conducted in 2010 by Peter Asaro and Selma Šabanović, Indiana University, Bloomington Indiana, for Indiana University and the IEEE, in: http://ethw.org/Oral-History:Chuck_Thorpe#Interview. Accessed Mai 7, 2018.

Thrun, Sebastian, et al. 2006. Stanley: The Robot that Won the DARPA Grand Challenge. *Journal of Field Robotics* 23 (9): 661–692. online in: https://onlinelibrary.wiley.com/doi/epdf/10.1002/rob.20147. Accessed Mai 7, 2018), also In The 2005 DARPA Grand Challenge, the Great Robot Race ed. Buehler, Martin, Iagnemma, Karl, Singh, Sanjiv, Berlin, Heidelberg, New York 2007, pp. 1–44.

Torchinsky, Jason. 2019. *Robot, Take the Wheel: The Road to Autonomous Cars and the Lost Art of Driving*. New York.

Transport Research Board, National Research Council, National Automated Highway System Research Program, a review, Special Report 253, Washington, DC, 1998.

Turk, Mathew A., David G. Morgenthaler, Keith D. Gremban, and Martin Marra. 1988. VITS–A Vision System for Autonomous Land Vehicle Navigation. *IEEE Transactions on Pattern Analysis and Machine Intelligence* 10 (2): 342–361.

US Army Material Command, The U.S. Army's Report on International Cooperative Programs, Urbana-Champaign 1998.

US Department of Transportation, Request for Applications to Establish a National Automated Highway System Consortium, Contract number DTFH61-94-X-0000, December 15, The National Intelligent Transportation Systems Program: Where We've Been & Where We're Going, Report No. FHWA-JPO-97-0027, March 1997.

Velez-Green, Alexander. 2015. The Foreign Policy Essay: The South Korean Sentry—A "Killer Robot" to Prevent War, March 1, 2015, in: https://www.lawfareblog.com/foreign-policy-essay-south-korean-sentry—-killer-robot-prevent-war. Accessed March 3, 2020.

Weber, Marc. 2014. Where To? A History of Autonomous Vehicles, May 8, 2014, in: http://www.computerhistory.org/atchm/where-to-a-history-of-autonomous-vehicles/.

Weinberger, Sharon. 2018. *The Imagineers of War: The Untold History of DARPA, the Pentagon Agency That Changed the World*. New York.

Wetmore, Jameson M. 2003. Driving the Dream, The History and Motivations Behind 60 Years of Automated Highway Systems in America. *Automotive History Review*, Summer 2003, pp. 4–19.

———. 2020. Reflecting on the Dream of Automated Vehicles, Visions of Hands Free Driving Over the Past 80 Years. *Technikgeschichte* 87 (1): 69–94.

Wettergreen, David, Chuck Thorpe, and William Whittaker. 1993. Exploring Mount Erebus by Walking Robot. *International Conference of Intelligent Autonomous Systems*, pp. 72–81.

Zworykin, V.K., and L.E. Flory. 1958. Electronic Control of Motor Vehicles on the Highway. *Highway Research Board Proceedings* 37: 436–451.

Zworykin, Vladimir K. 1959. The Human Aspect of Engineering Progress. The Fourth Clerk Maxwell Memorial Lecture, July 1, 1959, Cavendish Laboratory, Cambridge. *Journal Brit. I. R. E.*, September 1959, pp. 529–544.

Chapter 3
History of the Research on Vehicle Automation in Japan

The following part of the book describes the historical context, the aims and technical concepts of the research on automated and autonomous driving in Japan. The research literature regularly mentions the Japanese research efforts, but only in more general summaries.[1] A more detailed presentation of the different stakeholders and their achievements in a chronological timeline is currently missing.

The different chapters follow a periodization that had been proposed by Sadayuki Tsugawa, one of the pioneers of machine vision in Japan. He classified the history of automated driving systems in four periods.[2] The first period (1950s and 1960s) featured systems based on cooperation between road infrastructure (inductive cables) and a system onboard the vehicle. The second period started with the 1970s and followed the goal to realize autonomous systems based on machine vision that function without any technology outside the vehicle. The third period begins in the mid-1980s when research focused on large-scale demonstrations of Intelligent Transport Systems (ITS). The fourth period consists of field operation tests which started in the year 2000. In a recent publication, a fifth period was added that is beyond the scope of this book. It started in 2010, when IT companies like Google and many automotive companies began to experiment with autonomous vehicles on public roadways.[3]

The presentation of the technical aspects of this chapter is mainly based on scientific articles of the laboratories that participated in the research. These primary sources have been completed in communication with Dr. Sadayuki Tsugawa, who

[1] Ioannou, Petros A., ed. 1997. *Automated Highway Systems*. New York, London, p. 7 and Özgüner, Ümit, Tankut Acarman, and Keith Alan Redmill. 2011. *Autonomous Ground Vehicles*, p. 4. Boston, London.

[2] Tsugawa, Sadayuki. 2011. Automated Driving Systems: Common grounds of Automobiles and Robots. *International Journal of Humanoid Robotics* 8 (1): 2.

[3] 保坂明夫，青木啓二，津川定之，自動運転(第2版):システム構成と要素技術，森北出版 (Translation of the book title in English: Hosaka, Akio, Aoki, Keiji, Tsugawa, Sadayuki, 2019. *Automated Driving, Systems and Technologies*, 2nd ed. Morikita Publishing, Tokyo).

F. Kröger, *From Automated to Autonomous Driving*, Archimedes 70,
https://doi.org/10.1007/978-3-031-49881-7_3

worked since 1973 in the Mechanical Engineering Laboratory. I thank him for his important insights and the time he spent giving me this advice.

3.1 Historical Context

As introduction to this chapter, the historical and economic contexts of the research on autonomous driving in Japan will be briefly considered, drawing on the work of scholars of different disciplines: historians, anthropologists, political scientists and economists. We decided to limit our presentation on two main factors[4]: The first paragraph details the Japanese industrial policy after World War II that was an important factor for funding research on driverless cars. The next paragraph explains how the economic boom, mass motorization and the increasing number of traffic accidents during the period of rapid growth motivated the quest for automated driving.

3.1.1 The Role of the Ministry of International Trade and Industry (MITI)

Most of the research on automated driving in Japan had been funded by the *Government Mechanical Laboratory* (later Mechanical Engineering Laboratory and currently National Institute of Industrial Science and Technology) that was part of the *Ministry of International Trade and Industry* (MITI). MITI was a powerful institution that had been created in 1949 to govern the interior and exterior commercial relations. During 40 years, it was the main actor of industrial policy, a term that signifies in the Japanese context the "protection of domestic industries, development of strategic industries, and adjustment of the economic structure in response to or in anticipation of internal and external changes".[5] MITI placed investments, controlled import and export, gave priority to special economic sectors and governed the economy together with private stakeholders. One of its main missions was the protection of the interior market from foreign competition.

After the end of World War II, the United States imposed on Japan a strong cooperation in the technology sector, and the obligation to import fundamental

[4] Demographic and geographic factors influence also the attractivity of autonomous driving for a country: First, Japan is an increasingly ageing society. Second, Japan is a mountainous country with little flatland. The population density and the car traffic are very high in urban areas (Tsugawa, Sadayuki. 1992. Concept of Super Smart Vehicle Systems and Their Relation to Advanced Vehicle Control Systems. *Transportation Research Record* 1358: p. 42).

[5] Chalmers, Johnson, 1982. *MITI and the Japanese Miracle: The Growth of Industrial Policy, 1925–1975*, Stanford 1982, p. 26.

knowledge from the United States.[6] The political scientist Chalmers Johnson states in his important study about MITI, that these technology transfers "were crucial to Japanese economic growth".[7] The importation of technology – "one of the central components of postwar Japanese industrial policy" – had been regulated by MITI. The ministry decided if foreign technology should be imported, whether a joint venture should be agreed upon and if patent rights should be bought.

According to the political scientist Guy Faure, the selection of technological projects by MITI during the 1960s and 1970s had been led by the goal of catching up with the United States.[8] Other scholars agree that until the 1970s, fundamental research had been done in the United States and the industrial development had been made in Japan.[9]

A study from the US-Japan Committee states on the one hand, that none of the most significant "hard technologies" had been developed in Japan during this time. The most important Japanese innovations were made in the field of management and system techniques.[10] On the other hand, the authors emphasize the creativity of the Japanese industry in the "application and modification of basic technologies imported from abroad".

First, they cite the example of Sony who had the idea to use the transistor, developed in 1948 by the US Bell Laboratories, to produce small radios. In 1955, Sony launched the first transistor radio.[11] The historian Tessa Morris-Suzuki emphasizes that Sony was able to conquer the transistor radio market only because it had the research capacity to adopt the early transistor technology to the "high frequencies necessary to broadcast the human voice".[12]

Second, the first microprocessor had been developed by Intel in 1969 – but only after a request from Busicom, a Japanese calculator company. Third, Japanese companies were also the first that replaced mechanical components in cameras with electronic ones. Like other scholars, the philosopher Alain-Marc Rieu summarizes

---

[6] Sabouret, Jean-François. 2007. Introduction, Japon: Le grand secret. In *L'Empire de l'intelligence: Politiques scientifiques et technologiques du Japon depuis 1945*, ed. Jean-François Sabouret, 9–21. Paris.

[7] Chalmers 1982, p. 16.

[8] Faure, Guy. 2007. Le METI et la science industrielle au Japon. In Sabouret, online in: http://books.openedition.org/editionscnrs/4945 (accessed on January 26, 2018).

[9] Rieu, Alain-Marc 2007, Le Japon comme "société de connaissance": quelles leçons pour la France? Politiques de recherche, innovation et réformes institutionnelles depuis la crise de l'énergie des années 1970. In Sabouret, pp. 185–216.

[10] Committee on Japan (ed.), 1997. Maximizing U.S. Interests in Science and Technology Relations with Japan, Committee on Japan Framework Statement and Report of the Competitiveness Task Force, National Academy Press, Washington, D.C., online in: https://www.nap.edu/read/5850/chapter/1#iii (accessed January 20, 2018).

[11] Morris-Suzuki, Tessa. 1994. *The Technological Transformation of Japan: From the Seventeenth to the Twentieth-first Century*, p. 171. Cambridge, New York, Melbourne.

[12] Ibid., p. 172.

that Japan was less dedicated to produce new knowledge; it was strong in transforming common knowledge into technology, into practical applications.[13]

This configuration began to change in the decade between the late 1960s and the late 1970s. In 1966, the Japanese government decided to finance expensive, long-term fundamental research projects that could produce an important economic impact in the future.[14] The main sectors of these investments had been chosen by MITI according to their significance for the Japanese society. The research had been planned by the Agency of Industrial Science and Technology (AIST), and the technology was developed by national research laboratories, including MEL. The economists Yuko Harama and Laurent Genod emphasize that the main goal of this policy was to "reduce the existing technological gap between the United States and Japan".[15]

In the 1970s, Japan became part of the "third wave" described by the futurist Alvin Toffler.[16] The postindustrial society became a society of information, based on new technologies of information and communication. According to Morris-Suzuki,[17] a lot of MITI reports from the late 1970s and early 1980s emphasized the importance of increasing Japan's capability to develop original technology. Concepts as *gijutsu rikkoku* (technology nation building or techno-nationalism) and the "advanced information society" were introduced. The main objective was to become less dependent on foreign technology. Since 1973, the economic growth stabilized around 4% per year and Japan now focused on its own development of technologies, not only based on imported technology. Science and technology politics became more independent from industrial politics.[18]

Most scholars agree that Japan reached the goal to catch up with the United States in the 1980s.

In 1980, the Japanese Government affirmed to pursue the objective of a "Nation Based on Science and Technology" as fundamental policy goal, aiming at a strong tripartite cooperation between Industry, University and State (national research laboratories). The budget for research and development was even higher than those of the United States, Great Britain and France.

But the economist Yveline Lecler highlights that Japan catched up only quantitatively, exploiting existing technologies imported from foreign countries. The qualitative development of own new technologies remained a challenge.[19]

Lecler shows also that most of the R&D efforts were funded by industry, not by the state. Until today, the public research sector does not play a very important role

---

[13] Rieu 2007.

[14] Harayama, Yuko, Genod, 2007. Laurent, La politique scientifique et technologique au Japon: une perspective historique. In Sabouret, pp. 23–49, online in: http://books.openedition.org/editionscnrs/4942 (Translation from French to English: F.K.) and Faure 2007.

[15] Harayama, Genod, 2007 (Translation French to English F.K.) and Faure 2007.

[16] Toffler, Alvin, *The Third Wave*, New York 1980.

[17] Morris-Suzuki 1994, pp. 209–219.

[18] Lecler, Yveline, 1991. La dynamique du progrès scientifique et technique au Japon. In *Gestion 2000, Management & Prospective*, No. 4, August–September, p. 155.

[19] Ibid., p. 157.

in the Japanese R&D activities. In 2010 82% of research was funded by private business.[20] As the industry is focused on fast commercialization, they do not finance much of fundamental research. This dominance of the industry could explain the weak efforts in basic research.[21]

3.1.2 The Boom Izanagi, Mass Motorization and Traffic Accidents in Japan

The era of mass motorization started in Japan in the 1960s, when cars became more affordable for the Japanese middle class. The period of economic growth (1965–1970) has been called *Boom Izanagi*, named after Izanagi, creator of the Japanese Islands and father of the goddess in the Iwato myth.[22] This economic boom had taken place after the Tokyo Olympic Games in 1964. Between 1963 and 1973, the average growth rates of the gross national product (GNP) per worker exceeded 8,7% each year.[23] In 1968, the economy of Japan became the second biggest of the world.

During this time, "three privileged items of mass consumption" became "signifiers of middle-class status" emphasizes Lukàcs.[24] In the late 1950s, refrigerators, washing machines and black-and-white television became the most important items of consumption. At the end of the 1960s, these objects of desire evolved into car, cooler (air conditioner) and color TV (3C). They were called the three sacred treasures (jap. (新・)三種の神器, *(Shin) Sanshu no Jingi*), "a parody of the three sacred treasures of the Imperial Household: the mirror, the sword, and the curved jewel."

During this period, car ownership grew rapidly: In 1966, 8 million vehicles – including passenger cars, trucks, buses and motorcycles – circulated on Japanese streets.[25] With about 16 million vehicles in 1970, this number doubled in only four years. It reached nearly 28 million vehicles in 1975.[26]

---

[20] Ijichi, Tomohiro. 2012. Innovation policies and reforms in Japan. In *The Dynamics of Regional Innovation, Policy Challenges in Europe and Japan, Series on Innovation and Knowledge Management*, ed. Yveline Lecler, Tetsuo Yoshimoto, Takahiro Fujimoto, vol. 10, 44. Singapore.

[21] Lecler 1991, p. 158.

[22] Saito, Mitsuo. 2000. *The Japanese Economy*. Singapore, New Jersey, London, Hong Kong, p. 223.

[23] Okimoto, Daniel I., and Thomas P. Rohlen (ed.). 1988. *Inside the Japanese System: Readings on Contemporary Society and Political Economy,* p. 144. Stanford.

[24] Lukács, Gabriella. 2010. *Scripted Affects, Branded Selves: Television, Subjectivity, and Capitalism in 1990s Japan*, p. 214. Durham and London: Duke University Press.

[25] Already in 1966, the number of road fatalities – 13.904 persons – were important.

[26] The numbers of vehicles: 8,123,096 (1966), 16,528,521 (1970), 27,870,475 (1975). Today (2017) there are 81,260,206 vehicles on the road in Japan (see: Automobile Inspection and Registration Information Administration, 自動車保有台数の推移 (軽自動車を含む), (in eng.: Changes in the number of car owners [including mini vehicles]), online in: http://www.airia.or.jp/publish/statistics/ub83el00000000wo-att/hoyuudaisuusuiihyou.pdf [accessed November 9, 2017].

The infrastructure also evolved in these years. In 1963 – one year before the Tokyo Olympic Games – the first national expressway in Japan was constructed between Nagoya and Kobe (Meishin Expressway). The second highway (Tomei Expressway) connected Tokyo to Nagoya from 1969 on. Today, Japan has only approximately 10.000 kilometers of expressways.

The boom of car ownership led to an increasing number of traffic accidents. The rapid increase of traffic fatalities started already at the end of the 1950s and reached its peak in 1970, with 16.765 deaths.[27] Taro Sekine from the *Department of Mechanical Engineering* at the Nihon University explains this development by insufficient roads and safety technologies and a gap between the skills of the drivers and "the rapid increase in private four-wheeled automobile ownership."[28] The anthropologist Joshua Hotaka Roth emphasizes that the road infrastructure in Japan is largely a shared space between cars, pedestrians and bicycles. For that reason, "more than 50 per cent of traffic deaths in Japan involve pedestrians and bicyclists."[29]

At the end of the 1960s, traffic safety became an important issue for state and government policies. Like in most western countries, legislation changed. In April 1969, the integration of safety belts on the driver's seat became mandatory for all new Japanese vehicles.[30] In 1970 the *Traffic Safety Policies Basic Act* was enacted. It promoted the separation of pedestrians and vehicles and acknowledged the importance of new car safety technologies. Traffic accidents also became a much-discussed topic in the Japanese media discourse. Finally, the annual road traffic fatalities began to fall in 1971 and reached 8.466 in 1979. According to Roth, this decline of the number of fatalities can be explained by improvements in safety technologies, the increased response of emergency services and a decline of the number of young drivers.[31] The new drivers became more acculturated to the reality of shared roads.[32]

[27] Statistics Bureau, Ministry of Internal Affairs and Communications, Chapter 29, Disasters and Accidents, Road Traffic Accidents 1924–2005, online in: http://www.stat.go.jp/data/chouki/zuhyou/29-11.xls (accessed November 9, 2017).

Note: At this point in history, only fatalities that occurred within 24 h after the accident became part of the statistic – today the limit is 30 days.

[28] Sekine, Taro, 2015. The image of automobile safety. In *International Association of Traffic and Safety Sciences*, Interdisciplinary wisdom of IATSS, online in: http://www.iatss.or.jp/common/pdf/en/publication/commemorative-publication/iatss40_theory_06.pdf, p. 62 (accessed November 9, 2017).

[29] Roth, Joshua Hotaka. 2017. The shared road: Cars, pedestrians and bicyclists in Japan. In *Cars, Automobility and Development in Asia: Wheels of Change*, ed. Arve Hansen, and Kenneth Bo Nielsen, 134. London/New York: Routledge.

[30] Sekine 2015, p. 62.

[31] Naikakufu, Heisei 27 Kotsu anzen hakusho [eng.: White paper on traffic safety], 2015, in: www8.cao.go.jp/koutu/taisaku/h27kou_haku/pdf/gaiyo/1-1-1-1.pdf (accessed February 2, 2018), p. 4, cit. in: Roth 2017, p. 135.

[32] Roth 2017, p. 135.

This historical background is important because increased highway capacity and improved safety are often cited as motivation for the research on automated driving in Japan.

3.2 First Period: Automatically Operated Vehicles Using Guidance Cables (1961–1968)

The first experiments on automated driving in Japan started at the beginning of the 1960s. They had been much influenced by the US-American research on the guide wire principle, the state-of-the-art reference of automatic steering technologies during this period (see Sect. 2.1).

3.2.1 The Test-Track in Higashi-Murayama

Early research on automated driving in Japan had been conducted by the Government Mechanical Laboratory, established in 1937. It was part of the Ministry of International Trade and Industry (MITI).

The main research facility was based in Igusa, Suginami-ku, a district located in the western part of Tokyo Metropolis. Another branch of the laboratory had been based in the city of Higashi-Murayama, also in the western part of Tokyo. This city was an ideal location for experiments on automated driving: In 1952, the first automobile test track in Japan had opened in Higashi-Murayama, financed by Nissan, Toyota and Isuzu.[33] The construction of the test track had already started in 1940 but was interrupted by World War II. The oval course was 2 kilometers long and the interior part offered rough roadways for test purposes (Fig. 3.1).

At this site, the Government Mechanical Laboratory tested imported cars and disassembled them to learn more about car technology.[34] This has been called *reverse engineering*, a method already practiced after the Meiji restoration 1868, when Japan opened itself to the world and started to import foreign technology. The term means an appropriation of technologies by disassembling an existing product that is available on the market, studying its conception and reassembling it with some small modifications explains Lecler.[35] The common steps of development, conception and fabrication are inverted. As reported by Lecler the accumulation of these small modifications is one of the key factors of the Japanese innovation system.

---

[33] Tombo, 日本で最初のテストコース 旧東村山自動車試験道路, July 2, 2009. http://blog.livedoor.jp/thetombo2/archives/51493535.html (accessed December 3, 2017).

[34] E-mail of Sadayuki Tsugawa to the author, November 28, 2017.

[35] Lecler 1991, p. 150.

Fig. 3.1 Aerial photography of the Higashi-Murayama proving ground, Japans first automobile test track. (Credits: AIST. Image Source: http://livedoor.blogimg.jp/thetombo2/imgs/3/c/3c84c480.jpg)

Reverse engineering means "more than simple copying" but an "understanding of underlying scientific principles" clarifies the historian Tessa Morris-Suzuki in her book about the technological transformation of Japan.[36] Imported technology "was translated into industrial strength only because it was combined with domestic innovation".[37] She understands it as a "complement" and "not a substitute" for local research and development. The reshaping of technology after transnational transfers had also been called "hybridization".[38]

3.2.2 The First Guide-Wire Experiments of the Government Mechanical Laboratory (1961–1968)

In 1961, Eiichi Kikuchi, Makoto Kimura and Shuntetsu Matsumoto from the Government Mechanical Laboratory started a research project on "automatic automobile driving". They were advised by the control engineer Prof. Oshima from the

[36] Morris-Suzuki 1994, p. 115.

[37] Ibid., p. 169.

[38] In theoretical discussions about "the transfer, adaptation, or convergence of productive models" in the manufacturing industry it has been argued that a "direct transplantation or imitation" is "rarely feasible or even desirable. Systems cannot be transferred without being significantly reshaped." Some authors privilege a perspective that focusses on the "process of 'hybridization'", that means the interaction of different models during the process of innovation (see Boyer, Robert, Elsie Charron, Ulrich Jürgens, and Steven Tolliday. 1998. Introduction: Between Imitation and Innovation: The Transfer and Hybridization of Productive Models in the International Automobile Industry. In *Between Imitation and Innovation, The Transfer and Hybridization of Productive Models in the International Automobile Industry*, ed. Robert Boyer, Elsie Charron, Ulrich Jürgens, and Steven Tolliday, 1. New York).

University of Tokyo.[39] The test track in Higashi-Murayama was used for driving experiments. On the one hand, the project was motivated by the aim to "increase the highway capacity and improve the safety" in the future.[40] On the other hand, automatic driving was also seen as useful for testing the dynamic behavior of cars at high speed.

The main lines of the concept were adapted from the system developed by Vladimir Zworykin from the Radio Corporation of America (RCA) in the 1950s and presented in Sect. 2.1.1. A research report of Matsumoto from 1965 made this knowledge transfer explicit, referencing publications of R. L. Cosgriff (Ohio State University) and L. E. Flory (RCA).[41] Both engineers developed Zworykins concept further.

The Government Mechanical Laboratory developed three systems to control the speed, the steering and to avoid collisions. First, the accelerator and the brake mechanism of the car had been controlled by a system called "automatic speed control". It electronically compared the actual speed of the car with the manually set desired speed. The deviation signal then activated a servo valve that adjusted a hydraulic motor which controlled accelerator and brakes. The position of throttle and brakes were fed back to an electronic amplifier to adjust the speed of the vehicle. The electronic devices were analog ones, because at that time on-board digital computers were not available.

Second, the "automatic steering" control had been realized with a cabtyre cable that was buried under the road surface of the two-kilometer test loop in Higashi-Murayama.[42] Two experimental cars – a Toyota Crown, Model 1963 and a Nissan, model unknown[43] – had been equipped with a pair of on-board sensors (pick-up coils) mounted on the front bumper. These devices had to "detect the deviation of the car from the guidance cable" (Fig. 3.2). To improve stability, the yaw angle of the car in relation to the guidance cable had also been "picked up by two coils mounted on the rear bumper". The electrical deviation signal of the four devices then had been "introduced to the servovalve whose output drives the hydraulic actuator connected to the tie rod of the steering mechanism", details the research paper.

Automatic steering had been tested on a circular course up to a speed of 37 km/h and then on a test loop that permitted automatic steering of the car up to a speed of 50 km/h.[44] The authors emphasize that the test loop in Higashi-Murayama had a radius of only 22,5 meter, limiting the potential speed to 60 km/h. Some years later,

---

[39] Oshima, Y., E. Kikuchi, M. Kimura, and S. Matsumoto. 1965. Control System for Automatic Automobile Driving. In *Proceedings of the International Federation of Automatic Control (IFAC) Tokyo Symposium on Systems Engineering for Control System Design*, ed. Kankuro Kaneshige and Keisuke Izawa, pp. 347–357.

[40] Oshima et al. 1965, p. 347.

[41] Ibid., p. 357.

[42] Ibid., p. 351.

[43] The Nissan vehicle followed the Toyota vehicle in the experiments for rear end collision avoidance.

[44] Ibid., pp. 353f., 356.

Fig. 3.2 The research vehicle – a Toyota Crown from 1963 – is following an inductive cable in the test loop in Higashi-Murayama. (Credits & Image Source: Tsugawa 2011, p. 3). Below, the on-board control unit is shown. (Credits & Image Source: Sadayuki Tsugawa)

the control speed was increased on a loop for high speed testing: On March 25, 1967, the test vehicle drove automated with a speed of 100 km/h[45] on a test track of the Japan Automobile Research Institute (JARI) in Yatabe, Tsukuba-gun, (currently a part of Tsukuba-shi), Ibaraki.

Third, the engineers developed also an "automatic collision prevention" system that permitted to stop behind a leading car in proper distance. It had been based on

[45] Matsumoto, Shuntetsu. 1971, February. Highway Communication System Using Guidance Cable. *IEEE Transactions on Vehicular Technology* VT-20 (1): 12, also confirmed in: Tsugawa, Sadayuki. 2011. Automated Driving Systems: Common grounds of Automobiles and Robots. *International Journal of Humanoid Robotics* 8 (1): 3.

the cooperation between vehicles and the road infrastructure.[46] 400 meter of the test track in Higashi-Murayama had been equipped with magnetic fields at intervals of 6 meters.[47] These inductance loops in the road were able to detect the presence of a car. To prevent collisions, the receiver coils of the research vehicles picked up the signal of the induction fields. The brakes were actuated if the manually set speed was larger than the speed of the car ahead. In this case, the car switched to a safety speed. "If the leading car decelerates or stops, the following car with such a device can automatically decelerate or stop at the proper distance from the leading car."[48]

It must be emphasized that both, the *lateral* (steering) and the *longitudinal* control (acceleration/deceleration) of the Toyota Crown were automated.[49] Otherwise the experiments with two vehicles on rear end collision avoidance could not have been realized.

But the infrastructure-based research vehicle had two disadvantages: First, it "could neither detect nor avoid obstacles on its path".[50] Only if the obstacle was another car, detected by the magnetic field, it was recognized. Second, the system depended on the installation of cables under the road, a current feeding into the guide wire and maintenance of this infrastructure.

But a more efficient approach was appearing at the horizon: In a comment on these experiments, Y. T. Lie from the MIT in Cambridge mentioned already in 1965 the idea of machine vision. He emphasized the need to have a "reliable equivalent" to the capability of the human eye and asked: "What can we do to replace the pattern recognition ability of our eyes"?[51] The experiments of Robert Fenton at Ohio State university struggled also with this problem of a missing sensor (see Sect. 2.1.2).

According to a chronology of the Ministry of Land, Infrastructure, Transport and Tourism the guidance cable experiments of the Government Mechanical Laboratory were conducted until 1967, the year of the 100 km/h driving test.[52] But another paper reports that the experiments continued for at least another year: In May 1968, Matsumoto started a new experiment with two parallel guidance cables, laid with 6 m distance along both shoulders of the test course of the Government Mechanical Laboratory in Higashi-Murayama.[53] Fed with two alternating currents, the cables

---

[46] Tsugawa, Sadayuki. 2008. *A History of Automated Highway Systems in Japan and Future Issues, Powerpoint-Presentation, Institute of Electrical and Electronics Engineers, International Conference on Vehicular Electronics and Safety*, Ohio, USA, September 22, 2008, p. 6.

[47] Oshima et al. 1965, pp. 348, 355.

[48] Ibid., p. 349.

[49] This was confirmed by Tsugawa (E-mail of Sadayuki Tsugawa to the author, April 14, 2018).

[50] Tsugawa, Sadayuki, Teruo Yatabe, Takeshi Hirose, and Shuntetsu Matsumoto. 1979. An automobile with artificial intelligence, *Proceedings of the 6th International Joint Conference on Artificial Intelligence*, vol. 2, p. 893. San Francisco.

[51] Oshima et al. 1965, p. 357.

[52] Ministry of Land, Infrastructure, Transport and Tourism, Road Bureau, ITS Homepage, Archive, ITS Handbook Japan 2008, AHS (Advanced Cruise-Assist Highway System), online in: http://www.mlit.go.jp/road/ITS/1998HBook/chapter3/3-3e.html (accessed December 12, 2017).

[53] Matsumoto 1971, p. 12–16.

Fig. 3.3 The Nissan Laurel 1800 C 30 (1968) research vehicle, used during the experiments with two guidance cables in Higashi-Murayama. (Credits & Image Source: Matsumoto 1971, p. 15)

created a magnetic field that was detected by the two coils of the research vehicle (Fig. 3.3), a Nissan Laurel 1800 C 30 (1968–1972). One year before, Robert Fenton from Ohio State University in the United States also made an experiment with two wires (see Sect. 2.1.2). These parallel works show that both countries were working at the same engineering problems during this period.

While the single-wire system was restricted by the distance between the two detecting coils of the car, the two-wire system had the advantage to extend the operating range of the system. The vehicle could be operated over the whole distance between the two cables. The twin-lead-system also had the potential to be used as communication system between road authority and automated vehicles.[54] A loudspeaker in the car could even receive messages. On November 14, 1968, "this system was successful in a public experiment", reports Matsumoto, without giving more details.[55]

3.3 MITI's Guideway-Based Automated System CVS (1970–1978)

Higashi-Murayama was not only the location of automated driving tests for automobiles, but also for track-based vehicles on rubber tires. Beside the guidance-cable tests and the Intelligent Vehicle that will be discussed in the next chapter, an

[54] Matsumoto 1971, p. 12.

[55] Ibid., p. 16.

automated Personal Rapid Transit system – the Computer Controlled Vehicle System (CVS) – was tested since 1970.

It consisted of small four-person, driverless, lightweight, rubber-tired, air conditioned and radio equipped electric vehicles running on special guideways. A consortium of eight Japanese industrial players[56] developed the project under the general direction of MITI and the Japan Society for the Promotion of Machine Industry. MITI funded the construction of test tracks in Higashi-Murayama with 10 million dollars. The CVS was motivated by the necessity to find alternatives to the automotive traffic system with its well-known problems – "traffic congestion, accidents, air-pollution, noise, vibration".[57]

The idea was to create a transport system that combines the advantages of public transport and individual automobile transport: The CVS was conceived like an "automatically controlled taxi". In contrast to transport by bus and train, the user is not obliged to share his cabin with many other people and to change vehicles. He buys a ticket that specifies his destination, inserts it in a reader, pushes a button and the cabin will take him directly to his destination. "The selection of the route and the operation of the vehicle will be handled automatically by the computer system."[58] Instead of fixed stations, flexible stops were planned at a minimum distance of 100 meters. Therefore, "better door-to-door service can be provided than by automobile".[59] With intervals of one second between each cabin, the CVS was also conceived to have more lane capacity than automobile traffic.[60] As freight traffic depended in Japan almost exclusively on road transportation, it was also an aim of CVS to develop an alternative. Dual-mode vehicles were also in development, able to be driven outside the concrete paved guideway network for rubber-tired vehicles.

Several projects of this type were developed since the 1950s in the US and since the 1960s in Europe.[61] It may be argued that these systems are on the margin of this

---

[56] Toyo Kogyo (currently Mazda), Mitsubishi Jukogyo (Mitsubishi Heavy Industries), Tokyo Shibaura Denki (Toshiba), Hitachi Seisakusho (Hitachi), Fujitisu, Sumitomo Denki Kogyo (Sumitomo Electric Industries), Nippon Denki (NEC) and Shin-Nihon Seitetsu (currently Nippon Steel Corporation).

[57] Ishii, Takemochi, Masakazu Iguchi, and Masaki Koshi. 1976. CVS: Computer-Controlled Vehicle System (trans: Yamashita, Akira). In *Personal Rapid Transit III*, ed. D.A. Gary, M.J. Gary, A.L. Kornhauser, and W.L. Garrard. Minneapolis: University of Minnesota, online in: http://staff.washington.edu/jbs/itrans/cvs%20prt%20paper%20with%20illustrations.pdf (accessed July 7, 2019), p. 79.

[58] Ibid.

[59] Ibid., p. 81.

[60] For safety reasons, the first tests were only conducted with headways of 5 s and a speed of 20–30 km/h. The four seats faced backwards and standing was not permitted (Schneider, Jerry, Description of the Japanese Computer-controlled Vehicle System (CVS) [derived largely from a report entitled Automated Guideway Transit: An Assessment of PRT and Other New Systems, U.S. Congress, Office of Technology Assessment, June 1975, November 11, 2014, online in: http://staff.washington.edu/jbs/itrans/cvs1.htm, accessed July 7, 2019]).

[61] Anderson, Edward, 1996. *Some Lessons from the History of Personal Rapid Transit (PRT)*, August 4, Conference Paper, online in: http://staff.washington.edu/jbs/itrans/history.htm (accessed on July 8, 2019).

Fig. 3.4 The test track of the Computer-Controlled Vehicle System in Higashi-Murayama. (Credits & Image Source: Ishii et al. 1976, p. 78)

book because they are based on special guideways. But it seems to be appropriate to mention them. They have in common with automated vehicles the absence of a human driver. All driving functions were planned to be operated by a computer system. Instead of following a guide-wire, lateral control was realized by a guideway. As the individualized cabins of a Personal Rapid Transit System imitate the private space of an automobile, they form a hybrid between train and car. The CVS shows that Japan addressed the limitations of an automobile centered traffic system with different concepts of future transport systems and focused not only on automated automobiles.

Planned since 1968, CVS started as a demonstration for a traffic game at World Expo '70 in Osaka, expanding 1971 into a large project with a test track on the automotive test facility of MITI in Higashi-Murayama (Fig. 3.4). A 200-meter track was opened in September 1972 and first CVS-test cars were run from October to December 1972. The full test course of 4.8 kilometers was completed in 1973 and the computer control of several vehicles, automated loading and unloading of freight vehicles, changing into different rail-lanes and automated overtaking at high speeds was tested.[62] Two kilometers of the guideway, operated by a Hitachi-Computer system (HIDIC-350), were designed for a maximum speed of 80 km/h, another one, operated by a Toshiba-system (TOSBAC-40), allowed speeds below 40 km/h. A Fujitsu-computer system (FACOM-230-359) functioned as a supervisory computer for the whole system.

[62] Schneider 2014.

A system with 60 vehicles was tested from February to October 1974 – the test track became "one of the largest urban transport system research facilities of the world".[63] Nevertheless, an installation in a real environment was finally not realized due to a lack of funding. In 1976, a second phase of testing with 60 cargo vehicles under complete computer control started. "Reliable and consistent one second headways were attempted" during this period.[64] Mechanisms for collision avoidance, obstacle detection and avoidance and an emergency braking system were also tested. But the project ended in 1978 when the Ministry of Land, Infrastructure and Transport refused to license it, arguing with safety concerns. The PRT-expert Edward Anderson mentions also that the "size, cost and visual impact" of the 3-meters wide guideway, problems with wet-and-icy weather traction, the rough ride and difficulties with adequate capacity in stations were other reasons to discontinue the project.[65]

The Mechanical Engineering Laboratory was not involved in the CVS-project. But beside the vision guided vehicle that will be presented in the next chapter, it conducted with MITI a second project during the 1970s.

Following a research proposal of Toyota in 1973, the Comprehensive Automobile Traffic Control System (CACS) was based on digital communication between infrastructure and vehicle. Its main function was a dynamic route guidance system that had the aim to increase "the efficiency of automobile traffic" and to decrease accidents and congestion.[66] A computer in a central control room collected traffic data that was transmitted to individual cars equipped with a receiver. The driver could decide to follow these instructions to avoid congested roads. A first test was conducted in 1975, "followed by pilot trials in Tokyo's Meguro-ku and Shibuya-ku from October 1977".[67] The system helped to manage traffic problems – but the calculating capacity of the early computers was not high enough to handle all data. For that reason, the project did not surpass the research status.[68]

---

[63] Ishii et al. 1976.

[64] Schneider 2014.

[65] Anderson 1996.

[66] Tsugawa 1992, p. 42.

[67] Toyota, A 75-Year History through Text, Chapter 2: Satisfying Public Needs, Section 2: Response to Increase in Traffic Accidents, Item 5: Research and Development of a Comprehensive Automobile Traffic Control System, in: http://www.toyota-global.com/company/history_of_toyota/75years/text/entering_the_automotive_business/chapter2/section2/item5.html (accessed January 3, 2018).

[68] Lanson, Florian, 2010. Les Systèmes de Transport Intelligents au Japon, Rapport d'Étude, Ambassade de France au Japon, Service pour la Science et la Technologie. p. 5f.

The system evolved in 1984 into RACS (Ministry of Construction) and in 1987 into AMTICS (National Police Agency). The two programs contributed to development of the basic functions of today navigation systems. In 1991 both systems had been combined in the Vehicle Information and Communication System (VICS) from the Ministry of Post and Telecommunication (Wootton, J.R., and A. García-Ortiz. 1995. Intelligent Transportation Systems. *Mathematical and Computer Modelling* 22 (4–7): 266).

CACS is not part of the history of autonomous driving but the history of navigation systems. Nevertheless, it was decided to mention it here: According to Tsugawa, it was the "first dynamic route guidance system in the world"[69] and one of the first Intelligent Transport Systems (ITS).[70] It may show that Japan "started working on ITS much earlier than many other countries around the world."[71] But two other systems were also developed in the 1970s, the Electronic Route Guidance System (ERGS) in the United States and the Autofahrer Leit- und Informationssystem in Germany.

3.4 Second Period: Intelligent Vehicles with Machine Vision (1974–1984)[72]

In 1970, the Government Mechanical Laboratory changed its name to Mechanical Engineering Laboratory (MEL). The MEL was the first laboratory worldwide that dealt with road vehicles guided by machine vision. At the end of the 1970s, it developed the first vision-based full-size automobile with Artificial intelligence.[73] This vehicle had the advantage to function without modification of the road infrastructure; however, it needed guiderails at the side of the road. It was also independent from off-board computing: Before, all the research on autonomous driving worldwide had been based on a separation of vehicle and computer units in the laboratory. The shift to vision based approaches introduced the second period of research on autonomous driving for road vehicles.

[69] Tsugawa, Sadayuki. 2009. A Survey on Effects of ITS–related Systems and Technologies on Global Warming Prevention. In *Proceedings of the 12th IFAC Symposium on Transportation Systems,* September 2–4, p. 337. Redondo Beach, CA, USA.

[70] Tsugawa, Sadayuki, Aoki, Masayoshi, Hosaka, Akio, Seki, Kaoru, A Survey on present IVHS activities in Japan. In *1996 IFAC, 13th Triennial World Congress*, p. 7844.

[71] Ministry of Land, Infrastructure, Transport and Tourism, Road Bureau, ITS Homepage, Comprehensive Plan for ITS in Japan, Chapter 1 Significance of Promoting ITS, 2008, in: http://www.mlit.go.jp/road/ITS/5Ministries/chap1.html (accessed December 12, 2017).

[72] This period is cited in: Ministry of Land, Infrastructure, Transport and Tourism, Road Bureau, ITS Homepage, Archive, ITS Handbook Japan 2008, AHS (Advanced Cruise-Assist Highway System), online in: http://www.mlit.go.jp/road/ITS/1998HBook/chapter3/3-3e.html (accessed December 12, 2017).

[73] Tsugawa et al. 1979, p. 893–895.

3.4.1 The First Phase Intelligent Vehicle: Visual Obstacle Detection (1974–1978)

In the first years of the 1970s, the Mechanical Engineering Laboratory decided to create an "Intelligent Vehicle"[74] that reacts quickly, keeps on the left lane and maintains a safe locomotion in road environments.[75] Shuntetsu Matsumoto, who led the early guidance-cable experiments during the 1960s, directed also the new research team until 1981. Matsumoto majored communication engineering and worked for a meteorological observatory as a radar engineer before he joined the laboratory. Drawing on this background, he had the idea to use a TV camera for vehicle guidance.[76]

During the years 1974/1975, MEL acquired a Toyota Crown S 60 / S 70 (model 1971/1972) passenger car that had to serve as research vehicle. According to Tsugawa, it was very difficult to integrate onboard control units and measurement equipment on a passenger car with its limited interior space. But at that time, the Crown was the most luxurious car in Japan – it had been chosen by Matsumoto because it was ideal to "show feasibility of automated driving to the society".[77] So the car was not only a scientific tool but also had a public function – its aim was to persuade the public about automated driving.

As the MEL had a research agreement with the Denso Corporation – a global automotive components manufacturer, owned to 25% by Toyota -, they asked Denso to customize the car as experimental vehicle. Its steering and speed mechanism should be controlled with electric (analog) signals. Denso equipped the car with speed sensors and an angle sensor for the steering wheel. To turn the steering wheel, it had also been equipped with a hydraulic actuator. An electric motor for the throttle valves and vacuum servo brakes were also added.

In the years before 1975, Matsumoto had already worked on an algorithm for the detection of obstacles with stereo vision. His colleague Taketoshi Nozaki conducted first basic tests of the algorithm with the Crown and published the results 1976 in a journal article in Japanese language.[78] As reported by him, the MEL-team used two vertically aligned cameras (Fig. 3.5) and installed equipment for pattern recognition

---

[74] The term "intelligence-car" was first mentioned by Taketoshi Nozaki, a staff member of Shuntetsu Matsumoto, but it was Matsumoto who coined the name "intelligent car" in 1971 (Nozaki, Taketoshi. 1976. Characteristics of an Automatic Automobile Steering System Using the Information of a Course Deviation at Some Distance Ahead of the Automobile (article in Japanese). *Transactions of the Society of Instrument and Control Engineers* 12 (6): 687–693, online in: https://www.jstage.jst.go.jp/article/sicetr1965/12/6/12_6_687/_pdf/-char/ja [accessed on June 27, 2019]).

[75] Tsugawa et al. 1979, p. 894.

[76] E-mail of Sadayuki Tsugawa to the author, April 14, 2018.

[77] Ibid.

[78] Nozaki 1976.

Fig. 3.5 This photo shows the first set of experimental cameras, installed on the hood of a Toyota Crown in the years 1974/1975. (Image Source: 武敏, 野崎, 自動車の自動運転のための通路パターン認識 [Nozaki, Taketoshi, 2016. Passage Pattern Recognition for Automated Vehicle Driving], Working Paper of MEL)

and recording in the back of the car. The system was tested in a road environment with rubber cones that had to be identified by the obstacle detection system. The objective of the work was lateral control of the vehicle with a deviation from the cones detected by the TV cameras at a single look-ahead distance. The control algorithm was based on an analog proportional–integral–derivative (PID) controller. But according to Tsugawa, Nozaki stopped the tests around 1975 and did not join the team that developed the first Intelligent Vehicle from 1978 on.[79]

In the summer of 1978 MEL continued the preliminary works of Matsumoto and Nozaki with the Toyota Crown in Higashi-Murayama and later at a new test track in Tsukuba that had been completed around 1983.

To realize a visual guidance system that is capable to detect obstacles in a two-dimensional field of view, the engineers replaced the cameras that Nozaki had used with another type of modified TV cameras (Ikegami Tsushinki, Model: CTC2500A) on the right-hand side of the front. While ordinary TV cameras use interlace scanning, the modified cameras used progressive scanning and vertical scanning, both of which were essential to the algorithm of obstacle detection based on parallax.[80] These special cameras were not commercially available.

[79] E-mail of Sadayuki Tsugawa to the author, April 14, 2018.

[80] Technical specifications of the camera: Tube: 2/3 inch Vidicon (at that time CCD cameras were not available), Resolution: 550 lines, Aspect ratio: 2:1, Scanning frequency: horizontally 30 Hz and vertically 15.75 Hz, Focus length of lens: 12.5 mm, Viewing angle: horizontally 40 deg. and vertically 32 deg. (E-mail of Sadayuki Tsugawa to the author, April 14, 2018).

Fig. 3.6 The research vehicle, a Toyota Crown S 60 / S 70 (first phase IV) at the test facility in Higashi-Murayama (1978). (Credits & Image Source: Sadayuki Tsugawa)

While the first cameras were both on the hood, the engineers now removed the front bumper of the Toyota Crown to fix the lower camera (Fig. 3.6). The ideal location of cameras may be the center of a vehicle front, but that is difficult to realize due to the hood structure. Thus, taking account of the Japanese "keeping left" rule, the cameras were put at the right front.

In 1979, the MEL-team presented the detection system for lateral guard rails in a conference article.[81] An algorithm detected differences in light intensity at the crossing point of the optical axes of both cameras. A "road pattern recognition unit" transformed the analog video signals into two digital pulse trains that indicated changes in the brightness and darkness of the two images. "One train is delayed for a fixed time corresponding to a distance of a range where an obstacle is to be searched. The delay compensated the shift between the images."[82] In comparing the pulse sequences of the two trains, elevated areas of objects – traffic cones and guard rails – could be detected within a field of view of 5–20 meters and a viewing angle of 40 degree in front of the vehicle. To realize a two-dimensional field of view, 16 processing units with different delays and parallel processing were employed. This data was then processed in a microcomputer system, the "problem solving unit" (Fig. 3.7).[83]

---

[81] Tsugawa et al. 1979.

[82] Ibid., p. 893.

[83] The microcomputer consisted of a micro CPU, RAMs (8 KB), ROMs (3 KB), I/O Interfacing LSIs and D/A and A/D converters.

Fig. 3.7 On-board computer equipment of the first phase Intelligent Vehicle (1978). (Credits & Image Source: Sadayuki Tsugawa)

According to Tsugawa, the MEL-team used an 8-bit microprocessor (Motorola M6800) to find a steering angle with the obstacle data of the pattern recognition unit. The outputs of the microprocessor were converted to an analog format to become input for the control unit of the vehicle. Since the two-dimensional road scene ahead of the vehicle was available in real-time, a table-look-up algorithm was developed taking account of the low computing capability of the 8-bit microprocessor. The algorithm classified five behaviors of the vehicle – the car is following the road, the car approaches obstacles, the car is turning right or left, the car is heading to a dead end. Taking into account a time span of 100 msec at most to get control, the car did run at a maximal speed of 36 km/h.[84] The exact distance that had been traversed at this speed is not mentioned in the paper.

It can be resumed that MEL created the world's first visually guided road vehicle that could record and process (on-board) pictures of obstacles via two cameras: It was not linked to a laboratory outside the vehicle; all computer equipment was in the back of the car. It was also the first system that functioned without electronic ground equipment.

Whether the system can be called *autonomous* is a question for discussion: The research paper claimed that "the vehicle is completely autonomous and can be driven without a human driver."[85] In an exchange with the author of this book Tsugawa explained that the vehicle had a lateral and longitudinal control system – it was able to steer, accelerate and brake autonomously.[86] According to Tsugawas

[84] Tsugawa et al. 1979, p. 894f.

[85] Tsugawa et al. 1979, p. 893.

[86] E-mail of Sadayuki Tsugawa to the author, April 14, 2018.

definition of autonomous driving that had been presented in the introduction, the vehicle can be called laterally autonomous as it functioned independently from guidance cables etc.

But it should be mentioned that the vehicle remained restricted to a certain type of environment: According to Tsugawa, the main innovation was the development of a stereo vision system that "detected guard rails, and the vehicle was steered to drive along the guard rails."[87] It also had no function of lane marking detection on the ground as white lines have no height.[88] This shows that autonomous driving depended in this case on special prerequisites – if the guardrail would suddenly end, how would the vehicle react? From a contemporary perspective it could be questioned if a vehicle can be called autonomous if it is limited to such a narrow scenario.

3.4.2 The Second Phase Intelligent Vehicle: Autonomous Navigation (1980–1984)

At the end of the 1970s, the Japanese government moved laboratories under many ministries including MITI from the Tokyo area to Tsukuba to create a science city,[89] an idea that dated back to 1962.[90] Therefore the test track facility in Higashi-Murayama[91] had been relocated to Tsukuba in 1979. One year later, the Mechanical Engineering Laboratory moved also to Tsukuba.[92] "Based on the experiments with the previous system",[93] MEL continued research with the Toyota Crown in Tsukuba during four years (1980–1984).

Due to difficult economic conditions after the oil crisis the construction of a new 3,2 kilometer-test track in Tsukuba – now without an inductive cable – was delayed and only completed in 1983.[94] For that reason, the second phase IV was tested in the laboratory yard (Fig. 3.8). The private roadways of the laboratory were good and big enough for the experiments.

---

[87] Tsugawa 2011, p. 4.

[88] Tsugawa et al. 1979, p. 893. Note: In many sources, road marking detection is erroneously attributed to the IV (e.g. Torchinsky, p. 65, Davies, p. 19).

[89] Bloom, Justin L., and Shinsuke Asano. 1981. Tsukuba Science City: Japan Tries Planned Innovation. *Science, New Series* 212 (4500): 1239–1247.

[90] Morris-Suzuki 1994, p. 180.

[91] Today, the former test site is a park, the Higashi-Murayama Central Park (see http://wikimapia.org/#lang=fr&lat=35.747731&lon=139.459730&z=16&m=b&show=/object/history/show/?object_type=&id=48095&lng=2&rev=2&rev2=3&search=Higashimurayama)

[92] E-mail of Sadayuki Tsugawa to the author, June 30, 2017.

[93] Tsugawa, Sadayuki, Takeshi Hirose, and Teruo Yatabe. 1984. An Intelligent Vehicle with Obstacle Detection and Navigation Functions. *Industrial Electronics Conference (IECON)* 1: 303–308.

[94] E-mail of Sadayuki Tsugawa to the author, April 14, 2018.

Fig. 3.8 The second phase intelligent vehicle Toyota Crown (model 1971/1972) at a roadway in the laboratory yard, Namiki, Tsukuba-shi, 1984. (Credits & Image Source: Tsugawa 2008, p. 7)

With the second phase Intelligent Vehicle, the MEL-team improved the obstacle detection system further that had been developed with the first phase Intelligent Vehicle. According to a research report from 1984, the two cameras now allowed to differentiate obstacles and figures on the ground: If there is no time interval between the signals of both cameras, the object is an obstacle with vertical extension. If the time interval is different, it is a figure on the ground, that means with a non-vertical surface.[95]

During an interview with the author of this book Sadayuki Tsugawa clarified that this capability to differentiate obstacles from figures should not be misinterpreted as full lane marking detection function. In 1992, Tsugawa published images of a lane marking detection system in an article, citing a publication of the MEL-team member Takeshi Hirose in Japanese.[96] This may have given the impression that the second phase IV successfully realized such a system. But according to Tsugawa, this was not the case. At the beginning of the 1980s, Hirose indeed worked on a lane-marking detection system, but he "stopped the research on white line detection before completion" and his work "did not contribute to the research of the second phase IV".[97]

[95] Tsugawa et al. 1984, p. 304.

[96] Hirose, Takeshi. 1989. Position Prediction of Automobiles – White Line Detection (in Japanese). *Proceedings of the 28th SICE Annual Conference* 1: 23–24 (cit. in: Tsugawa, Sadayuki, 1992. Concept of Super Smart Vehicle Systems and Their Relation to Advanced Vehicle Control Systems. *Transportation Research Record*, Issue 1358, p. 44).

[97] E-mail from Sadayuki Tsugawa to the author, April 14, 2018.

Fig. 3.9 On-board computers of the second phase Intelligent Vehicle. (Credits & Image Source: Sadayuki Tsugawa)

A difference to the experiments described in the paper from 1979 is that the engineers now used a faster processing unit with a digital "hard-wired logic instead of a microprocessor system" to directly process the video signals in real time.[98] While microprocessors are programmable, electronic components with a predefined set of functions are following a "hard-wired" logic that serves a single purpose only. These circuits are customized to a specific task only. Therefore, they are more efficient but less flexible.

The main aim of the second phase intelligent vehicle was the development of an on-board navigation system that could acquire route information. While most of the navigation systems that had been developed in the past depended on equipment on the ground, the team developed a system not needing ground equipment. The location of the vehicle was calculated with the measured velocities of both rear wheels. The difference of the velocities of both wheels indicated a change in the direction of the vehicle, called yaw rate.

The system controlled the steering and velocity of the vehicle along a predefined path, memorized in the minicomputer LSI 11-2 (16-bit processor) of the Digital Equipment Corporation (DEC). The LSI-11 was a microprocessor-version of the PDP-11, small enough to be used on-board an automobile. The microcomputer, the processing unit and the counting units had been installed in the back of the Toyota Crown (Fig. 3.9).

The steering angle was calculated by an arithmetic control algorithm, developed by Tsugawa. It contained a planned path, "represented by a sequence of points".[99]

[98] Tsugawa et al. 1984, p. 303.

[99] Ibid., p. 306.

The vehicle had to hit these target points one after another. "When an obstacle is detected, a temporary target point is installed beside the obstacle." In curves, the target points were set up densely, in straight parts of the route sparsely. With a velocity of about 10 km/h, the second phase IV was slower than the first phase IV because the mathematical calculation of location and steering angle took more time.[100]

The engineers concluded that the vehicle "was driven safely and autonomously from a starting point to a goal".[101] Nevertheless, they mention also "some defects": The obstacle detection system could be affected by "brightness and direction of light. An active system for obstacle detection using ultrasonic, laser or radar would be probably free from the defect."[102] Another problem was that the measurement of revolution counts of the rear wheels accumulated "errors caused by slip and skid." Therefore, the system was only useful at low velocity and on a smooth ground. The research on the second Intelligent Vehicle ended with the year 1984 when the MEL-team presented a more detailed report at the Industrial Electronics Conference (IECON).[103]

Studies on pattern recognition and image processing were becoming a very popular research topic at that time, but they required expensive computer systems. For that reason, Tsugawa stopped the machine vision study. After the end of the research on the Intelligent Vehicle, MEL started in 1984 a study on vehicle-to-vehicle-communication with the aim to control a group of autonomous vehicles.[104] As the budget was limited, two small automated vehicles used in factories were used as research objects. Infrared based vehicle-to-vehicle communication had been chosen to form a platoon of two vehicles: the second followed the trajectory of the first. According to Tsugawa, the DEMO 2000 demonstration with five passenger cars was based on this project (see Sect. 3.5.7).[105]

3.5 Third Period: AHS and ITS, Vehicle-Networks and Platooning (1985–2000)

As we have seen in the preceding chapters, during the thirty years from 1960 to 1990 most research efforts on automated driving in Japan were made by the Mechanical Engineering Laboratory of MITI. This domination can be explained by the fact that other laboratories could not afford a test vehicle and they had no test

[100] Ibid., p. 307 and e-mail from Sadayuki Tsugawa to the author, April 14, 2018.

[101] Ibid., p. 308.

[102] Ibid., p. 307.

[103] Tsugawa et al. 1984.

[104] Tsugawa 1992, p. 46.

[105] E-mail from Sadayuki Tsugawa to the author, April 14, 2018.

track at their disposition. According to Tsugawa, the test tracks in Higashi-Murayama and Tsukuba were a crucial advantage for MEL.[106]

The situation changed in the mid-1980s when the third period of research on autonomous driving began. The power of digital processors became more significant and the memory size augmented also. This made new research strategies possible, focused on large-scale demonstrations of what has been called after 1994 Intelligent Transport Systems (ITS). This concept is based on information transfer between "humans, roads and automobiles", according to a definition of the Japanese government corporation ITS Japan.[107] It was presented as a "fundamental solution" for "traffic accidents, congestion and environmental pollution." It must be emphasized that this definition includes the infrastructure (roads), an approach that was less acclaimed during the period of the first machine-vision systems since the end of the 1970s. But as we will see in this chapter, ITS can be divided into systems that are based on the infrastructure and systems that are based on vehicles only.

Beside MITI three other rival ministries and an agency entered the research on Intelligent Transportation Systems in Japan, now subjected to a strong competition: The most important were the Ministry of Construction (MOC), responsible for the public infrastructure – highways, bridges, roads – and the Ministry of Transport (MOT) charged with the regulation and certification of road vehicles. The Ministry of Post and Telecommunications (MPT), engaged in the development and regulation of wireless communications systems and the National Police Agency (NPA), responsible for traffic management in all cities also started to work in the field. These five public institutions had their own rivaling research branches, working on different projects independent from the other ministries. After 2001, MOC and MOT were integrated in the Ministry of Land, Infrastructure, Transport and Tourism (MLIT) but they continued to act independently as Roadway Bureau (MOC) and Automobile Bureau (MOT).[108]

The following chapter presents the activities of the three main ministries since the end of the 1980s: The Ministry of Construction developed the cooperation between vehicles and infrastructure in the frame of its Automated Highway System (AHS)-project. The Ministry of Transport worked with its Advanced Safety Vehicle (ASV) on vehicle based sensing and active safety assistance systems. MITI continued its investigations on machine vision and worked on vehicle-to-vehicle communication. Compared to the other ministries, it made "only modest investments" in

---

[106] Interview of the author with Sadayuki Tsugawa, Tokyo, May 21, 2018.

[107] ITS Japan, About ITS, in: http://www.its-jp.org/english/about_e/ (accessed January 30, 2018).

[108] In the United States, a similar structure can be identified: The National Highway Traffic Safety Administration (NHTSA) is in charge of vehicles (e.g. research on assistance systems) and the Federal Highway Administration (FHWA) is in charge of highways (e.g. research on highway signs). Both administrations are part of the Department of Transportation. They cooperated on big projects like the AHS Demo 97 (see Sect. 2.6.2).

Intelligent Transport Systems but conducted "long-term conceptual studies" on fully automated driving on automated highways.[109]

3.5.1 The Personal Vehicle System (1987–1991) of MITI

During the shift of most research projects to vehicle-networks, MITI and an auxiliary organization engaged in bicycle racing – the Mechanical Social Systems Foundation – started in 1987 a new experimental vehicle with machine vision called Personal Vehicle System (PVS).[110] The decision of MITI to start this project can be explained by progress in visual real-time road and lane recognition that had been reported by US-American (Carnegie Mellon) and European universities (UniBwM) on several conferences since 1985. It was also a reaction to the ALV-program of DARPA (see Sect. 2.4).

According to Tsugawa, the PVS was a successor of the Personal Rapid Transit (PRT) project that MITI developed in the 1970s under the name Computer Controlled Vehicle System (CVS) (see Sect. 3.3). While the CVS was a rail based system, the PVS had the aim to become a non-railed personal rapid transit system.

The technical realization came from Nissan and Fujitsu who developed a special computer for image processing. As high speed computers had important physical dimensions at the end of the 1980s, the system had to be integrated in a large Nissan Civilian bus (model W40 from 1984). The computer needed so much electric power that the team even had to install a power generator and an oil tank on the vehicle (Fig. 3.10).

The first difference to the Intelligent Vehicle presented in 1984 was that the number of cameras had been increased: The bus had been equipped with three cameras for lane detection and two stereo television cameras for obstacle detection. Ultrasonic ranging sensors were also added.[111] The second innovation was a laser-based obstacle detection system (LIDAR).

Like the second phase Intelligent Vehicle, the PVS used a dead-reckoning system based on a differential odometer to locate its own position.[112] According to Tsugawa, it allowed the bus to choose an optimal route from a starting point to a goal. The bus drove autonomously and could create a map and explore a road network. "In 1989,

[109] VanderWerf, Joel, Steven Shladover, and Mark A. Miller. 2004. *Conceptual Development and Performance Assessment for the Deployment Staging of Advanced Vehicle Control and Safety Systems*, California PATH Research Report UCB-ITS-PRR-2004-22, August 2004, online in: https://pdfs.semanticscholar.org/3476/e3f1195402060c9398e95674f7c2c6d0fa49.pdf (accessed on July 8, 2019), p. 19 (p. 60 in pdf).

[110] Taniguchi, M., et al. 1991. The Development of Autonomously Controlled Vehicle, PVS. In *Proceedings of the 1991 Vehicle Navigation and Information Systems Conference*, pp. 1137–1141 (cit. in: Tsugawa 1992, p. 45).

[111] Tsugawa 2008, p.13.

[112] Tsugawa 1992, p. 45.

Fig. 3.10 The 1987 PVS (Personal Vehicle System) Nissan Civilian bus on a test track in Hakone near Mount Fuji. (Credits & Image Source: Akio Hosaka)

the PVS ran at 30 km/h along straight lines and 10 km/h along curved lines on a special proving ground and at 60 km/h along a straight line" on the test track in Tsukuba.[113] Its vision system detected painted markings at the sides of the lanes and steered around a parked vehicle in its lane. That means it avoided a collision by lane changing, not by braking.[114] In 1990 and 1991, the three cameras for lane detection had been replaced by a single camera on a turntable in the interior.

3.5.2 MITIs Super Smart Vehicle System – A Vehicle Information Network (1990–1992)

Influenced by European projects like PROMETHEUS, the Association of Electronic Technology for Automobile Traffic and Driving (JSK) – today the Japan Automobile Research Institute (JARI) – started in 1990 a MITI-supported study on an information system for drivers, called *Super Smart Vehicle System (SSVS)*. It was independent from the infrastructure and used only minimal in-vehicle sensing.[115] The research project had the aim to create vehicle-oriented intelligent transport systems including

[113] Tsugawa 1992, p. 46. Note: I replaced km/hr. (in the original sentence) by km/h.

[114] Shladover, Steven E., and Advanced Vehicle Control Systems. 1990. *SAE Technical Paper* 901129, p. 108.

[115] VanderWerf et al. 2004, p. 20 (p. 61 in pdf).

a network of vehicles that communicate with each other and the road to increase safety and efficiency. It covered not only passenger cars but also cargo and public transportation.

According to Tsugawa, JSK had already started to study vehicle-to-vehicle communication in 1981.[116] He emphasizes that "Japan was at the leading edge of the inter-vehicle communications research", beginning before Europe and the US with this kind of experiments.[117] The SSVS was based on experiences with the CACS mentioned in the last chapter, but also the Intelligent Vehicle and the PVS. It resulted in cooperative driving systems with inter-vehicle communications by infrared (1997) and by DSRC (2000).

3.5.3 Japan's First AHS-Demonstration (1995) of the Ministry of Construction

The Japanese Ministry of Construction (MOC) is "one of the most powerful and wealthiest ministries in Japan" and is considered by some authors as an "engine of economic growth".[118] As most of the infrastructures – bridges, tunnels, highways – were already built, it searched in the 1990s for new fields of investment. As the ministry had strong relationships with the electronics industry and had a large budget, it focused on investments in the electronic infrastructure, not the vehicles. Equipping roadways with ITS-electronics became an important part of this new mission.

MOC had already started in 1989 a R&D program on an Automated Highway System as reported by the Road Bureau of the Ministry of Land, Infrastructure, Transport and Tourism.[119] The research focused on information technologies and assistance systems for vehicle control but had also the goal to realize fully automated driving.

In 1995, MOC initiated an Automated Highway System (AHS)-program. The fierce international competition in the field of AHS-development accelerated these ambitions. It should be kept in mind that only some months before, in December 1994, the US-American Federal Highway Administration (FHWA) funded the National Automated Highway System Consortium (NAHSC), responding to the demand of the ISTEA-legislation. As we have learned in Sect. 2.6 its mission was to develop a federal AHS program and the prototype of an automated test track until 1997. After a platooning experiment in 1992, PATH performed the world's first

[116] Tsugawa 1992, p. 46f.

[117] Tsugawa, Sadayuki. 2005. Issues and recent trends in vehicle safety communication systems. *IATSS Research* 29 (1), p. 9.

[118] VanderWerf et al. 2004, p. 17 (p. 58 in pdf).

[119] Ministry of Land, Infrastructure, Transport and Tourism, Road Bureau, ITS Homepage, Archive, AHS (Advanced Cruise-Assist Highway System), online in: http://www.mlit.go.jp/road/ITS/1998HBook/chapter3/3-3e.html (accessed December 12, 2017).

public platooning demo with four cars in 1994 in San Diego. This certainly influenced the Japanese research policy to accelerate its efforts.

The qualitative and quantitative dimension of the 1995 demo in Japan was impressive: According to Tsugawa, it was only "in the spring of 1995" that MOC "had asked automotive companies like Toyota, Nissan, Honda, Mitsubishi and Mazda to make automated vehicles in half a year".[120] The results were presented with 19 vehicles during the first Japanese AHS-demonstration in November 1995 at a 3-kilometer test course of the Public Works Research Institute (PWRI), the research arm of MOC in Tsukuba.[121] The demonstration with many visitors was held in the context of the ITS World Congress that took place in Yokohama the same year. It had the aim to show "vehicle control systems for longitudinal and lateral behavior."[122]

At the end of 1996, MOC changed the definition of the acronym AHS to "Advanced Cruise-Assist Highway System", emphasizing information, warning and control assistance systems rather than autonomous driving.[123] The ministry now focused on "infrastructure-based sensing and intelligence for collision warning and avoidance, with communication from infrastructure to vehicles"[124] – systems that were later demonstrated at Demo 2000 in Tsukuba. "This change in focus was coupled with the formation of a public-private partnership between the MOC and the private industry", institutionalized 1996 in the Advanced Cruise-Assist Highway Systems Association (AHSRA). A similar development can be observed some years later in the United States: In 1994, the public-private partnership National Automated Highways System Consortium (NAHSC) was created but after the San Diego Demo 1997 the US-Department of Transportation started also to move away from Automated Highway Systems (see Sect. 2.6).

3.5.4 *The Second AHS-Demo with Magnetic Markers and LCX-Cables (1996)*

Guided by the Ministry of Construction, a second AHS-test run took place in September 1996 on a new segment of the Joshinetsu Expressway before public use (from Komoro Junction to Tohbu-Yunomaru Junction).[125] Its objective was to dem-

[120] E-mail from Sadayuki Tsugawa to the author, April 14, 2018.

[121] The PWRI has several international agreements with the FHWA in the US and the Highways Agency in UK, exchanging regularly staff members (VanderWerf et al. 2004, p. 18 (p. 59 in pdf).

[122] Ministry of Land, Infrastructure, Transport and Tourism, Road Bureau, ITS Homepage, Archive, AHS (Advanced Cruise-Assist Highway System), online in: http://www.mlit.go.jp/road/ITS/1998HBook/chapter3/3-3e.html (accessed December 12, 2017).

[123] VanderWerf et al. 2004, p. 17 (p. 58 in pdf).

[124] VanderWerf et al. 2004, p. 20 (p. 61 in pdf).

[125] Ministry of Land, Infrastructure, Transport and Tourism, Road Bureau, ITS Homepage, Archive, ITS Handbook Japan 2008, AHS (Advanced Cruise-Assist Highway System), online in: http://www.mlit.go.jp/road/ITS/1998HBook/chapter3/3-3e.html (accessed December 12, 2017).

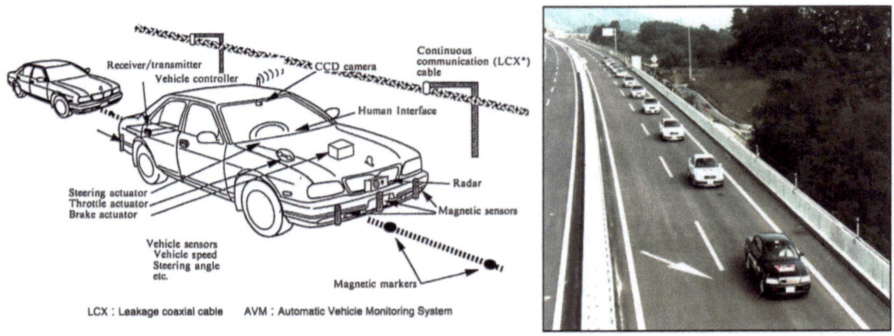

Fig. 3.11 The AHS-test run based on magnetic markers and LCX-cables on the side of the road, Joshinetsu Expressway, September 1996. (Credits & Image Source: Tsugawa 2008, p. 15)

onstrate the driving maneuvers of the 1995-demo (PWRI-testtrack) in a real-road environment.[126] Eleven cars – four Toyota Crown, three Nissan Cima, two Honda Accord and two Mitsubishi Diamante – followed each other autonomously on a 11-kilometer round-trip (Fig. 3.11). The event served also as test run for the demonstration in San Diego 1997. At this large-scale demonstration of autonomous driving systems in the US, "it drew attention as Japan's own concept on AHS development."[127]

The platoon concept was promoted with the promise to reduce energy costs – a group of vehicles can move with less aerodynamic drag per vehicle than a single autonomous vehicle, as Tsugawa explains.[128] It was also often mentioned that a platoon could reduce the occupied road space and this may avoid congestion – by the way a promise that is vulnerable to some contradictions, as Chuck Thorpe argues: If the concept is based on a merge lane next to the platoon lane, more road space will be consumed. Additional platoon-to-platoon lanes at interchanges with human-driven cars would even increase the infrastructure to 8 layers of roadways.[129]

---

[126] "Key Demo features included tests of vehicle control on roads involving lateral and longitudinal inclines, the relationship between in-vehicle devices and various external disturbances such as power lines and overpasses, adaptive cruise control, platoon formation driving, longitudinal control (speed and braking control), lane departure warning using magnetic lane markers embedded on the road, laser radar tracking control function for vehicle spacing in platoon operation, obstacle/roadway hazard detection using onboard laser radar, and incident detection and warning function using roadside sensors." VanderWerf et al., p. 27 (p. 68 in pdf).

[127] Ministry of Land, Infrastructure, Transport and Tourism, Road Bureau, ITS Homepage, Archive, ITS Handbook Japan 2008, AHS (Advanced Cruise-Assist Highway System), online in: http://www.mlit.go.jp/road/ITS/1998HBook/chapter3/3-3e.html (accessed December 12, 2017).

[128] Tsugawa, Sadayuki. 2013. An Overview on an Automated Truck Platoon within the Energy ITS Project. *7th IFAC Symposium on Advances in Automotive Control, The International Federation of Automatic Control (IFAC)*, September 4–7, 2013, p. 42. Tokyo, Japan.

[129] Mail of Chuck Thorpe to the author, March 2, 2020.

Platoon driving has been envisioned mostly for trucks because the reduction of energy costs, the reduction of the workload for the driver and the increased safety should have the most impact in truck traffic. Nevertheless trucks with big heavy Diesel engines and air brakes will respond slowly, "so in practice it is hard to accurately control the headway between trucks without lots of throttle and brake action", says Thorpe.[130] An improvement of fuel efficiency is therefore not proven in every case. But experiments in the US and Japan with a platoon of heavy trucks have shown fuel efficiency as the control system avoided the micro accelerations mentioned by Thorpe.[131]

The control system combined infrastructure based equipment and on-board systems: The *lateral* control of the vehicles had been realized by a series of magnetic markers[132] on the road surface (Toyota, Nissan, Honda) and a lane marking detection system that used computer vision (Mitsubishi).[133] The approach to use magnetic markers and the concept of coordinated platoons may be seen as an adoption from prior works of the California PATH project[134] of the UC Berkeley (USA) (see Sect. 2.6.1 in this book).

The *longitudinal* control was realized with on-board LIDAR that measured the gaps between the vehicles. Vehicle speed had been controlled by leaky coaxial cables (LCX) that were originally developed for train communication systems in a tunnel for Tokaido and Sanyo Shinkansen, the bullet train: The outer conductors of LCX-cables have regularly positioned slots that can emit and receive radio waves, they function like antennas. With this technology, speed commands were sent from the roadway to the vehicles and the vehicles did send back a signal to the roadway about their location. Cables have the advantage of offering a more stable communication than radio communication systems.[135]

From a historical point of view, this concept is significant as it reinvented the guidance-cable approach that had been mostly abandoned – with the exception of Automated Guided Vehicles in factories – in the United States and Europe: While during the 1950s and 1960s cables were mostly used for lateral guidance, the Ministry of Construction used the cables now for longitudinal control. Installed along the roadway and not buried into the ground, the system also presented a solution to reduce the high costs of the early guide-wire concept.

---

[130] Ibid.

[131] Tsugawa, Sadayuki, Sabina Jeschke, and Steven E. Shladover. 2016. A Review of Truck Platooning Projects For Energy Savings. *IEEE Transactions on Intelligent Vehicles* 1 (1): 68–77.

[132] Tsugawa 2011, p. 5f.

[133] Tsugawa 2008, p. 14.

Car models added according to Tsugawa (E-mail from Sadayuki Tsugawa to the author, April 14, 2018).

[134] VanderWerf et al. 2004, p. 17 (p. 58 in pdf).

[135] Chatani, Yoshiyuki. 2008. Antennas for the Bullet Train. In *Mobile Antenna Systems Handbook*, ed. Kyohei Fujimoto, 3rd ed., 417. Boston, London.

3.5.5 *The Advanced Safety Vehicle of the Ministry of Transport (1991-Today)*

The Ministry of Transport (MOT) is a "much more inward-looking" ministry than the MOC and participates only rarely in international projects. It launched in June 1991 the *Advanced Safety Vehicle* (ASV) project, a public-private partnership with 24 major private companies from the automobile industry.

The ASV-project focused on assistant systems for automobiles – danger warning systems, vehicle recognition and rear-end collision prevention systems.[136] Its main goal was active safety and the reduction of damages after accidents. All technology development was funded by the private companies themselves, not by the ministry and each company promoted its own ASV-prototype. A research report from PATH emphasizes that all ASV-systems were "based on autonomous vehicle sensing technologies, with no cooperation with the infrastructure or other vehicles." The report claims that these technologies were "highly sophisticated and innovative, comparable to or ahead of the leading technologies in Europe and North America."[137] The ASV-concept influenced the US-American research efforts, but slowly: Only six years later, in 1997, a similar project was launched by the US-Department of Transportation: The Intelligent Vehicle Initiative (IVI).

The first phase (1991–1995) of the ASV-project had the aim to study the technical possibilities of active safety systems for passenger vehicles, define goals and verify the efforts of accident reduction.[138] It "focused on concept design, research on component and system technologies, development of a prototype vehicle and demonstration of this vehicle".[139]

The second phase of the ASV-project (1996–2000) concentrated on research and development for market introduction; as well as the definition of design principles and guidelines. The main development work focused on "sensing for collision warning and avoidance" for autonomous vehicles.[140] In addition to the passenger vehicles from the first phase, trucks, buses and motorcycles now became part of the research, involving all major Japanese automotive manufacturers (Daihatsu, Hino,

[136] Ministry of Land, Infrastructure, Transport and Tourism, Road Bureau, ITS Homepage, Archive, AHS (Advanced Cruise-Assist Highway System), online in: http://www.mlit.go.jp/road/ITS/1998HBook/chapter3/3-3e.html (accessed December 12, 2017).

For a full list of all systems see Tsugawa, Sadayuki, Aoki, Masayoshi, Hosaka, Akio, Seki, Kaoru, A Survey of present IVHS activities in Japan. In *1996 IFAC, 13th Triennial World Congress*, p. 7845f. San Francisco, USA.

[137] VanderWerf et al. 2004, p. 18 (p. 59 in pdf).

[138] Ministry of Land, Infrastructure, Transport and Tourism, Study Group for the Promotion of ASV, Seeking Even Greater Traffic Accident Reductions through Vehicle Advancements, October 2017, in: http://www.mlit.go.jp/jidosha/anzen/01asv/resourse/data/asv6pamphlet-e.pdf (accessed July 13, 2017).

[139] VanderWerf et al. 2004, p. 19 (p. 60 in pdf).

[140] Ibid.

Fig. 3.12 JSK-demo with four cars – one Toyota Soara, two Honda Legend, one Mazda Eunos 800 – communicating with infrared signals, 1997. (Credits & Image Source: Tsugawa 2005, p. 10)

Honda, Isuzu, Mazda, Mitsubishi, Nissan, Subaru, Suzuki, Toyota). The ASV-project continued with a third (2001–2005), fourth (2006–2010), fifth (2011–2015) and a sixth phase (2016–2020) that is focused on autonomous driving.

3.5.6 Cooperative Driving Demos (1997 and 2000)

The trend to vehicle-based automation concepts was obvious: In 1997, the Association of Electronic Technology for Automobile Traffic and Driving (JSK) tested a cooperative driving system with four cars that used infrared signals to communicate, aiming at driver assistance and smooth traffic flow. The photo (Fig. 3.12) shows how the car in the third position from the top semi-automatically merges into the gap between the second and the fourth vehicle. The distance between the vehicles was measured by "triangulation between a pair of infrared markers on the roof of the vehicles".[141] One of them functioned as a transmitter.

After the 1997-demo JSK started new inter-vehicle communications research with dedicated short range communication (DSRC). From November 21–27, 2000 MITI and JSK held a "Cooperative driving demo" with five automated vehicles on the MEL-test track in Tsukuba.[142] The vehicles drove in a platoon (Fig. 3.13) with speeds between 40 and 60 km/h and with gaps in a range of 20–34 meters. They performed splitting and merging a platoon and lane changing.[143]

[141] Tsugawa 2005, p. 10.

[142] Kato, S., et al. 2002. Vehicle Control Algorithms for Cooperative Driving with Automated Vehicles and Inter-vehicle Communications. *IEEE Transactions on Intelligent Transportation Systems* 3 (3): 155–161.

[143] Ohmine, Hiroyuki. 2008. Antennas for ITS. In *Mobile Antenna Systems Handbook*, ed. Kyohei Fujimoto, 429. Boston, London: Artech House.

Fig. 3.13 The five vehicles – four Mazda Eunos 800 and one Toyota Celsior (Lexus LS 400) – showing the merging of two platoons (3 vehicles and 2 vehicles) during the "Cooperative driving Demo" of MITI in the year 2000 in Tsukuba. (Credits & Image Source: Sadayuki Tsugawa)

In contrast to the platoon that had been tested in 1996, this system was independent from the road infrastructure (cables and magnetic markers) and was based on inter-vehicle communication only.[144] The cars were linked with 5.8 GHz dedicated short-range communication (DSRC) systems that allowed the exchange of data with high-speed. The DSRC-system had been developed to allow drivers to pass and pay at an electronic toll collection (ETC) gate without stopping the car.[145] The vehicles also shared their positions via highly accurate D-GPS and were equipped with a LIDAR-system for obstacle detection.

3.5.7 The Demo 2000 Smart Cruise 21 Test Run

Only one day later, two other test runs started: From November 28 to December 1, 2000, Japan's Ministry of Transport and the Ministry of Construction jointly conducted a test run called "Smart Cruise 21 – Demo 2000" together with private

[144] Özgüner, Ümit, Tankut Acarman, and Keith Alan Redmill. 2011. *Autonomous Ground Vehicles*, p. 6. Boston/London.

[145] Ibid.

industry partners on the PWRI and JARI test tracks in Tsukuba.[146] The demonstrations showed infrastructure-based devices (Ministry of Construction) and vehicle based sensing devices (Ministry of Transport):

The Ministry of Construction (MOC), the major organizer behind this event, showed infrastructure-based sensing and communication devices ("SmartWay") on the PWRI test course.[147] US-American colleagues from the PATH-program in California participated with three scenarios (automatic steering control, driver lane guidance, and lane departure warning).[148]

Based on the development of the Advanced Safety Vehicle, the Ministry of Transport showed vehicle-based sensing at the JARI test track. According to a PATH-report, "this part of the demonstration was almost invisible to international visitors, but it actually contained more advanced technologies and more impressive capabilities than the demos at PWRI (attended by international visitors)."[149]

According to the report, the split into three demonstrations can be explained by the "deep-seated rivalries" among the different ministries. The report identified this concurrence as the "primary weakness" of the Japanese innovation system.[150]

3.5.8 *The Intelligent Multimodal Transit System (IMTS) of Toyota (1990–2008)*

In Japan, most of the research on automated driving had been organized by state institutions. The automotive industry participated often in the research programs, but always under the umbrella of programs directed by the different government ministries.

The Japanese automotive industry also worked independently on automated driving systems but most of these projects remained confidential. On very few occasions, scientific papers were published. Tsugawa also estimates that the automotive industry imitated experiments of the MEL, never admitting that it had been influenced by the public research laboratories.[151]

---

[146] Ministry of Land, Infrastructure, Transport and Tourism, Road Bureau, ITS Homepage, ITS Handbook Japan 2000, Section 4 ITS deployment progress in Japan, Toward practical use of the Smart Cruise System. http://www.mlit.go.jp/road/ITS/2000HBook/chapter4/4-19e.html (accessed December 12, 2017).

[147] VanderWerf et al. 2004, p. 17 (p. 58 in pdf).

[148] VanderWerf et al. 2004, p. 26 (p. 68 in pdf).

[149] VanderWerf et al. 2004, p. 18 (p. 59 in pdf). According to this PATH-report, the ASV-demo included forward collision warning and avoidance, adaptive cruise control with brake control, side collision-, lane departure-, roadway condition- and drowsy driver warning systems, rear vehicle proximity monitoring and head-up displays (VanderWerf et al. 2004, p. 28 (p. 69 in pdf).

[150] VanderWerf et al. 2004, p. 19 (p. 60 in pdf).

[151] Interview of the author with Sadayuki Tsugawa, Tokyo, May 21, 2018.

Fig. 3.14 The Intelligent Multimodal Transit System (IMTS) of Toyota. (Credits & Image Source: Tsugawa 2008, p. 18)

During the 1990s, research on autonomous driving by private car companies in Japan became more visible.[152] Around 1991, Toyota used the test track of the Higashi-Fuji Technical Center – opened in 1966 – on the slopes of Mt. Fuji (Susono, Shizuoka Prefecture) for automated driving tests.[153] Toyota also developed a dual mode bus system with manual and automated driving mode (Fig. 3.14). The Intelligent Multimodal Transit System (IMTS) assured the lateral control of the vehicles with magnetic markers and the longitudinal control with millimeter wave radar and Inter-Vehicular Communication (IVC). Experiments on a test track with a platoon of three buses were performed with 30 km/h and a gap of 5 meters between the vehicles. Later, the system had been used for a bus system on a theme park on Awaji Island in 2001 and during the world fair 2005.[154] It proved to be a real public service as it carried ten million passengers around the fair grounds during the 6 months period of the fair in Nagoya.

[152] Bommel, Nathan. 2015. Toyota testing self-driving cars 'since the 1990s'. *USA Today*, October 6, 2015, in: https://eu.usatoday.com/story/money/cars/2015/10/06/toyota-driverless-cars-automated-highway-driving/73448028/ (accessed July 13, 2019).

[153] Mail of Chuck Thorpe to the author, March 2, 2020.

[154] Ibid., p. 17.

3.6 Conclusion

Japan was the first non-western country that tested automated driving systems for automobiles. As in western societies, the problem of growing traffic accidents at the end of the 1960s became a major driving force behind the idea to develop automated cars. Driven by the aim to catch up with the United States, the Japanese activities started since 1961 with reverse engineering, drawing on the guidance-cable concept from RCA. In 1967, a vehicle with automatic lateral and longitudinal control drove on a test track, reaching a speed of 100 km/h. At the end of the 1960s, the United States and Japan worked on the same engineering problems.

First concentrated on the adaptation of US-American concepts, Japan became an innovation leader in the 1970s. The transformation of the postindustrial society towards a society of information and the aim to become independent from technology imports accelerated the research efforts in the mobility sector. The Mechanical Engineering Laboratory was the first laboratory worldwide that used a camera for guidance purposes on a road vehicle. Working on an obstacle detection algorithm already before 1975, MEL presented the world's first visually guided autonomous vehicle for roads with guard rails in 1978. Independent from other ground equipment, it drove along these guard-rails and reached a speed of 36 km/h. Japan worked also on an automated Personal Rapid Transit-system and a traffic information system for automobile navigation purposes.

Already in 1981, Japan worked on vehicle-to-vehicle communication. In 1984, an autonomous research vehicle of MEL was able to distinguish between elevated obstacles and figures on the ground, reaching a speed of 10 km/h. Work on general-purpose machine vision started with the Personal Vehicle System in 1987, reaching a speed of 60 km/h in 1989. Until 1991, obstacle detection had been realized with more cameras, ultrasonic sensors and laser-based systems.

In the 1990s, Japan developed infrastructure-based and vehicle based approaches in parallel and was very advanced in the research on assistance systems: Since 1991, the Advanced Safety Vehicle-project of the Ministry of Transport researched vehicle-based active safety systems, earlier than the United States that started a similar initiative only in 1997. The work on vehicle networks was also especially strong in Japan during the 1990s.

From the mid-1990s on, Japan reached its goal to catch up with the United States and became part of an international run towards Intelligent Transportation Systems (ITS). In 1995, the Ministry of Construction started an Automated Highway System-project, and organized in the same year Japans first AHS-demo focusing on the cooperation between vehicles and infrastructure. It followed the approach of the Californian PATH-program that made the first platooning experiments with magnetic markers for lateral control in 1992. In 1996, Japan organized a platooning demo on an unopened highway, based on magnetic markers and LCX-cables.

Both approaches, infrastructure based and vehicle based sensing systems were demonstrated in the year 2000. The automotive industry started own programs on autonomous driving without the participation of the state.

Following a comparative perspective, two major non-technical cultural differences between autonomous driving in Japan on the one hand and the United States and Europe on the other hand should also be highlighted.

First, for historical reasons the research on autonomous driving in Japan had not been funded by military institutions – a sharp contrast to the United States. After the drop of two atomic bombs on the cities of Hiroshima and Nagasaki during World War II, atomic and aircraft research had been banned in Japan since the end of 1945.[155] Following the American occupation, Japan laid down in its constitution that it should not have an army and should not support military efforts.[156] Article 9 of the Constitution specifies that Japan "forever renounce war as a sovereign right of the nation and the threat or use of force as a means of settling international disputes."

The economist Yveline Lecler emphasizes the influence of this decision on the industrial structure and the development of science and technology. In most other countries, the military plays an important role in the development of technical innovations and the private industry often benefits from military funding. Japan followed another path – its definition of "national security" concentrated on the protection and development of civil key industries that may play a vital role in the future. These long-term visions, based on prospective analyses, were given an institutional frame with several ministries and agencies for science and technology, all under the cabinet of the Prime Minister.[157]

This institutional structure is also reflected in the position of individual researchers who "were definitively negative to receive money from the military" as Sadayuki Tsugawa emphasizes.[158] As many Japanese know about the disaster of the Second World War, researchers were "more conscious" about what they are doing in what context. For instance, Japanese Universities decided not to participate in the DARPA Challenges. Only recently, this situation changed: According to Tsugawa, the Japanese Ministry of Defense started to sponsor university research on innovative high technologies in 2015. While some universities decided not to apply for this kind of funding, some universities apply and receive money from the military. One reason is that the budget of the Ministry of Education and the number of excellent papers from Japan is annually decreasing.

Second, in contrast to the United States and Germany, none of the research vehicles that were used during forty years of research on automated driving became a part of Science and Technology-Museums. Instead of becoming museum artifacts, the research vehicles were scrapped: The two 1963 Toyota Crown that were used for tests with guidance cables got abandoned in 1970.[159] The modified Toyota Crown S 60 / S 70 – the first vision based automated road vehicle in the world – went to the scrapyard at the end of the 1990s. All these research vehicles did not start a second

[155] Morris-Suzuki 1994, p. 162.

[156] Lecler 1991, p. 151.

[157] Ibid., p. 152.

[158] Interview of the author with Sadayuki Tsugawa, Tokyo, May 21, 2018.

[159] E-mail from Sadayuki Tsugawa to the author, April 14, 2018.

career as an exhibition object like the Mercedes 500 SEL (VaMP) research vehicle from UniBwM in Germany, now exposed in the traffic exhibition of the Deutsches Museum in Munich or the Stanford Cart, exhibited at Googles Computer History Museum in California. Today, the most important part of the Miraikan (National Museum of Emerging Science and Technology) in Tokyo, opened in 2001, is dedicated to the great number of Japanese innovations in the robotics field. A live performance of the humanoid robot ASIMO from Honda, playing soccer with a human, is the main attraction of the museum.

Bibliography

Anderson, Edward. 1996. Some Lessons from the History of Personal Rapid Transit (PRT), August 4, 1996. Conference Paper, online in: http://staff.washington.edu/jbs/itrans/history.htm. Accessed July 8, 2019.

Bloom, Justin L., and Shinsuke Asano. 1981. Tsukuba Science City: Japan Tries Planned Innovation. *Science*, New Series 212 (4500): 1239–1247.

Bommel, Nathan. 2015. Toyota testing self-driving cars 'since the 1990s'. *USA Today*, October 6, 2015. https://eu.usatoday.com/story/money/cars/2015/10/06/toyota-driverless-cars-automated-highway-driving/73448028/. Accessed July 13, 2019.

Boyer, Robert, Elsie Charron, Ulrich Jürgens, and Steven Tolliday. 1998. Introduction: Between Imitation and Innovation: The Transfer and Hybridization of Productive Models in the International Automobile Industry. In *Between Imitation and Innovation, The Transfer and Hybridization of Productive Models in the International Automobile Industry*, ed. Robert Boyer, Elsie Charron, Ulrich Jürgens, and Steven Tolliday, 1–20. New York.

Chalmers, Johnson. 1982. *MITI and the Japanese Miracle: The Growth of Industrial Policy, 1925–1975*. Stanford.

Chatani, Yoshiyuki. 2008. Antennas for the Bullet Train. In *Mobile Antenna Systems Handbook*, ed. Kyohei Fujimoto, 3rd ed., 417–425. Boston, London.

Committee on Japan (ed.). 1997. *Maximizing U.S. Interests in science and technology relations with Japan, Committee on Japan framework statement and report of the competitiveness task force*. Washington, D.C., online in: https://www.nap.edu/read/5850/chapter/1#iii. Accessed January 20, 2018.

Faure, Guy. 2007. Le METI et la science industrielle au Japon. In *L'Empire de l'intelligence: Politiques scientifiques et technologiques du Japon depuis 1945*, ed. Jean-François Sabouret. Paris, online in: http://books.openedition.org/editionscnrs/4945. Accessed on January 26, 2018.

Harayama, Yuko, and Laurent Genod. 2007. La politique scientifique et technologique au Japon: une perspective historique. In *L'Empire de l'intelligence: Politiques scientifiques et technologiques du Japon depuis 1945*, ed. Jean-François Sabouret, 23–49. Paris, online in: http://books.openedition.org/editionscnrs/4942. Accessed January 20, 2018.

Hirose, Takeshi. 1989. Position Prediction of Automobiles – White Line Detection (in Japanese). *Proceedings of the 28th SICE Annual Conference* 1: 23–24.

保坂明夫，青木啓二，津川定之，自動運転(第2版)：システム構成と要素技術，森北出版 (Translation of the book title in English: Hosaka, Akio, Keiji Aoki, Sadayuki Tsugawa. 2019. *Automated Driving, Systems and Technologies*, 2nd ed. Tokyo: Morikita Publishing).

Ijichi, Tomohiro. 2012. Innovation Policies and Reforms in Japan. In *The Dynamics of Regional Innovation, Policy Challenges in Europe and Japan, Series on Innovation and Knowledge Management*, ed. Yveline Lecler, Tetsuo Yoshimoto, and Takahiro Fujimoto, vol. 10, 41–71. Singapore.

Ioannou, Petros A., ed. 1997. *Automated Highway Systems*. New York, London.

Ishii, Takemochi, Masakazu Iguchi, and Masaki Koshi. 1976. CVS: Computer-Controlled Vehicle System, translated by Akira Yamashita. In *Personal Rapid Transit III*, ed. D.A. Gary, M.J. Gary, A.L. Kornhauser, and W.L. Garrard, 77–83. Minneapolis: University of Minnesota, online in: http://staff.washington.edu/jbs/itrans/cvs%20prt%20paper%20with%20illustrations.pdf. Accessed July 7, 2019.

Kato, S., et al. 2002. Vehicle Control Algorithms for Cooperative Driving with Automated Vehicles and Inter-vehicle Communications. *IEEE Transactions on Intelligent Transportation Systems* 3 (3): 155–161.

Lanson, Florian. 2010. Les Systèmes de Transport Intelligents au Japon, Rapport d'Étude, Ambassade de France au Japon, Service pour la Science et la Technologie.

Lecler, Yveline. 1991. La dynamique du progrès scientifique et technique au Japon. In *Gestion 2000, Management & Prospective*, No. 4, August–September 1991, pp. 149–178.

Lukács, Gabriella. 2010. *Scripted Affects, Branded Selves: Television, Subjectivity, and Capitalism in 1990s*. Japan, Durham and London.

Matsumoto, Shuntetsu. 1971. Highway Communication System Using Guidance Cable. *IEEE Transactions on Vehicular Technology* VT-20 (1): 12–16.

Ministry of Land, Infrastructure, Transport and Tourism, Road Bureau, ITS Homepage, ITS Handbook Japan 2000, Section 4 ITS deployment progress in Japan, Toward practical use of the Smart Cruise System, 2000. http://www.mlit.go.jp/road/ITS/2000HBook/chapter4/4-19e.html. Accessed December 12, 2017.

Ministry of Land, Infrastructure, Transport and Tourism, Road Bureau, ITS Homepage, Comprehensive Plan for ITS in Japan, Chapter 1 Significance of Promoting ITS, 2008. http://www.mlit.go.jp/road/ITS/5Ministries/chap1.html. Accessed December 12, 2017.

Ministry of Land, Infrastructure, Transport and Tourism, Road Bureau, ITS Homepage, Archive, ITS Handbook Japan 2008, AHS (Advanced Cruise-Assist Highway System), 2008. http://www.mlit.go.jp/road/ITS/1998HBook/chapter3/3-3e.html. Accessed December 12, 2017.

Ministry of Land, Infrastructure, Transport and Tourism, Study Group for the Promotion of ASV, Seeking Even Greater Traffic Accident Reductions through Vehicle Advancements, October 2017. http://www.mlit.go.jp/jidosha/anzen/01asv/resourse/data/asv6pamphlet-e.pdf. Accessed July 13, 2017.

Morris-Suzuki, Tessa. 1994. *The Technological Transformation of Japan: From the Seventeenth to the Twentieth-First Century*. Cambridge/New York/Melbourne: Cambridge University Press.

Nozaki, Taketoshi. 1976. Characteristics of an Automatic Automobile Steering System Using the Information of a Course Deviation at Some Distance Ahead of the Automobile (article in Japanese). *Transactions of the Society of Instrument and Control Engineers* 12 (6): 687–693, online in: https://www.jstage.jst.go.jp/article/sicetr1965/12/6/12_6_687/_pdf/-char/ja. Accessed on June 27, 2019.

Ohmine, Hiroyuki. 2008. Antennas for ITS. In *Mobile Antenna Systems Handbook*, ed. Kyohei Fujimoto, 427–457. Boston, London: Artech House.

Okimoto, Daniel I., and Thomas P. Rohlen, eds. 1988. *Inside the Japanese System: Readings on Contemporary Society and Political Economy*. Stanford.

Oshima, Y., E. Kikuchi, M. Kimura, and S. Matsumoto. 1965. Control System for Automatic Automobile Driving. In *Proceedings of the International Federation of Automatic Control (IFAC), Tokyo Symposium on Systems Engineering for Control System Design*, ed. Kankuro Kaneshige and Keisuke Izawa, 347–357.

Özgüner, Ümit, Tankut Acarman, and Keith Alan Redmill. 2011. *Autonomous Ground Vehicles*. Boston, London.

Rieu, Alain-Marc. 2007. Le Japon comme "société de connaissance": quelles leçons pour la France?, Politiques de recherche, innovation et réformes institutionnelles depuis la crise de l'énergie des années 1970. In *L'Empire de l'intelligence: Politiques scientifiques et technologiques du Japon depuis 1945*, ed. Jean-François Sabouret, 185–216. Paris.

Roth, Joshua Hotaka. 2017. The Shared Road: Cars, Pedestrians and Bicyclists in Japan. In *Cars, Automobility and Development in Asia: Wheels of Change*, ed. Arve Hansen and Kenneth Bo Nielsen, 134–151. London/New York.

Sabouret, Jean-François. 2007. Introduction, Japon: Le grand secret. In *L'Empire de l'intelligence: Politiques scientifiques et technologiques du Japon depuis 1945*, ed. Jean-François Sabouret, 9–21. Paris.

Saito, Mitsuo. 2000. *The Japanese Economy*. Singapore, New Jersey, London, Hong Kong.

Schneider, Jerry. 2014. Description of the Japanese Computer-Controlled Vehicle System (CVS) (derived largely from a report entitled Automated Guideway Transit: An Assessment of PRT and Other New Systems, U.S. Congress, Office of Technology Assessment, June 1975), November 11, 2014, online in: http://staff.washington.edu/jbs/itrans/cvs1.htm. Accessed July 7, 2019.

Sekine, Taro. 2015. The Image of Automobile Safety. In *International Association of Traffic and Safety Sciences, Interdisciplinary Wisdom of IATSS.*, 2015, online in: http://www.iatss.or.jp/common/pdf/en/publication/commemorative-publication/iatss40_theory_06.pdf, p. 62. Accessed November 9, 2017.

Shladover, Steven E., Advanced Vehicle Control Systems. 1990. *SAE Technical Paper* 901129: 103–112.

Statistics Bureau, Ministry of Internal Affairs and Communications, Chapter 29, Disasters and Accidents, Road Traffic Accidents 1924–2005, online in: http://www.stat.go.jp/data/chouki/zuhyou/29-11.xls. Accessed November 9, 2017.

Taniguchi, M., et al. 1991. The Development of Autonomously Controlled Vehicle, PVS. *Proceedings of the 1991 Vehicle Navigation and Information Systems Conference*, pp. 1137–1141.

Toffler, Alvin. 1980. *The Third Wave*. New York.

Tombo, 日本で最初のテストコース 旧東村山自動車試験道路, July 2, 2009, in: http://blog.livedoor.jp/thetombo2/archives/51493535.html. Accessed December 3, 2017.

Tsugawa, Sadayuki, Teruo Yatabe, Takeshi Hirose, Shuntetsu Matsumoto. 1979. An Automobile with Artificial Intelligence. *Proceedings of the 6th International Joint Conference on Artificial Intelligence*, Vol. 2, San Francisco, pp. 893–895.

Tsugawa, Sadayuki, Takeshi Hirose, and Teruo Yatabe. 1984. An Intelligent Vehicle with Obstacle Detection and Navigation Functions. *Industrial Electronics Conference (IECON)* 1: 303–308.

Tsugawa, Sadayuki. 1992. Concept of Super Smart Vehicle Systems and Their Relation to Advanced Vehicle Control Systems. *Transportation Research Record* 1358: 42–49.

———. 2005. Issues and recent trends in vehicle safety communication systems. *IATSS Research* 29 (1): 7–15.

———. 2008. A History of Automated Highway Systems in Japan and Future Issues, Institute of Electrical and Electronics Engineers. *International Conference on Vehicular Electronics and Safety*, Ohio, USA, September 22, 2008, p. 2–3.

———. 2009. A Survey on Effects of ITS–related Systems and Technologies on Global Warming Prevention. In *Proceedings of the 12th IFAC Symposium on Transportation Systems*, Redondo Beach, USA, September 2–4, 2009, pp. 334–341.

———. 2011. Automated Driving Systems: Common Grounds of Automobiles and Robots. *International Journal of Humanoid Robotics* 8 (1): 1–12.

———. 2013. An Overview on an Automated Truck Platoon within the Energy ITS Project. *7th IFAC Symposium on Advances in Automotive Control, The International Federation of Automatic Control (IFAC)*, September 4–7, 2013, Tokyo, Japan, pp. 41–46.

Tsugawa, Sadayuki, Sabina Jeschke, and Steven E. Shladover. 2016. A Review of Truck Platooning Projects Foe Energy Savings. *IEEE Transactions on Intelligent Vehicles* 1 (1): 68–77.

VanderWerf, Joel, Steven Shladover, and Mark A. Miller 2004. Conceptual Development and Performance Assessment for the Deployment Staging of Advanced Vehicle Control and Safety Systems, California PATH Research Report UCB-ITS-PRR-2004-22, August 2004, online in: https://pdfs.semanticscholar.org/3476/e3f1195402060c9398e95674f7c2c6d0fa49.pdf. Accessed on July 8, 2019.

Wootton, J.R., and A. García-Ortiz. 1995. Intelligent Transportation Systems. In *Mathematical and Computer Modelling* 22 (4–7): 259–268.

Chapter 4
History of the Research on Vehicle Automation in Europe

After the first two chapters about the research on automated and autonomous driving in the United States and Japan we will now present the development efforts in a third geographical space, Europe – with an emphasis on Germany and France.

It would go beyond the scope of our investigation to give a complete overview of all individual research projects in all European countries. Instead of quantity, we focused on quality – only projects that illustrate some general development trends are presented. Priority is also given to institutions that developed test vehicles and demonstrated their capabilities in international demo programs.[1]

The first chapter starts with the early experiments on automated driving in Europe: The guidance-cable research of the Road Research Laboratory in England (1960–1975), the semi-automated test-vehicle of Continental (1968) and the guide-wire directed trucks of MAN and Bosch (1971) in Germany are presented. In France, Peugeot used guidance cables only for test drives.

The second chapter presents the shift to vision systems in Europe at the end of the 1970s. Like in the US and Japan, machine vision became a more promising approach to vehicle automation than guidance-cables. At the end of the 1970s, the mobile robot HILARE was created in France. In Germany, Ernst Dieter Dickmanns started at UniBwM to work on dynamic machine vision with a simulation loop. At the beginning of the 1980s the automotive industry became interested in visual guidance experiments (Volkswagen Opto-Pilot).

The third chapter is dedicated to the concepts, research vehicles and milestones of UniBwM in Germany. Starting with the vision-system of the research vehicle VaMoRs and the 4-D approach (1985–1991) of Dickmanns, the first outdoor-runs of VaMoRs (Skidpan-demo at Daimler Stuttgart) in 1986 are presented, followed by a

[1] However, some events and projects had to be skipped due to a lack of resources or time, for example the Automated Vehicle Guidance Demo'98 in Rijnwoude, Netherlands and the research efforts of Hans-Hellmut Nagel at the *Fraunhofer Institut für Informations- und Datenverarbeitung* (IITB).

F. Kröger, *From Automated to Autonomous Driving*, Archimedes 70,
https://doi.org/10.1007/978-3-031-49881-7_4

section on the Autonomous Mobile Systems-project with Daimler (1987–1988). A main milestone of the 1980s was VaMoRs' fully autonomous high speed test on the *Autobahn* (1987).

The fourth chapter is concentrated on the PROMETHEUS-program of the European Union (1988–1994). The Knowhow-Transfer of UniBwM-technology onto the VITA-research van of Daimler is presented, followed by a portrait of the research vehicles VITA-II and VaMP with the second-generation of the UniBwM-vision system based on transputers. After a paragraph about the second generation vision-system of VaMoRs, the performances of VaMP and VITA-II at the final PROMETHEUS-demo in Paris (1994) are detailed, followed by an overview about the state of the art of other research projects in France, the UK and Germany during this time.

The fifth chapter is focused on the research efforts that were undertaken after PROMETHEUS (1995–2000). The French projects Praxitèle, Cycab and La Route Automatisée are presented. The next paragraphs will highlight two important demonstrations, VaMP's long-distance drive from Munich to Odense (1995) and the Mille Miglia in Automatico of the University of Parma in Italy (1998).

The sixth chapter will go back in the chronology and present the defense related activities in Europe. Starting with the French military project DARDS, the experimental robotics-program PRIMUS in Germany is presented together with the AutoNav-agreement (1997–2003) and the third-generation vision system (EMS) of UniBwM.

4.1 Early Guidance-Cable Experiments in Europe

As in Japan, early experiments on automated driving in Europe were influenced by the guidance-cable concept of RCA and GM in the United States.

4.1.1 Guidance Cable Research at the Road Research Laboratory in England (1960–1975)

In Europe, the earliest research on automatic vehicle guidance was effectuated in 1960 by Cyril George Giles, Head of Surface Characteristics Section of the Road Research Laboratory (RRL)[2] in Harmondsworth, Middlesex (England).[3] The Laboratory was founded in 1933 as a government agency under the Department of Scientific and Industrial Research (DSIR), charged with road materials and construction. After the Second World War, the RRL focused on road safety issues,

[2] In 1972, the Road Research Laboratory became the Transport and Road Research Laboratory (TRRL). In 1996 it was privatized and is now called Transport Research Laboratory (TRL).

[3] Giles, C.G., and J.A. Martin. 1961. *Cable Installation for Vehicle Guidance Investigations in the New Research Track at Crowthorne, Report RN/40 57/CGG*. Crowthorne: Road Research Lab (cit. in: Ioannou, p. 9).

supported by the Ministry of Transport, the National Physical Laboratory and the DSIR. Johnson highlights that the RRL was at this time the most important and best endowed road research facility in the world because it could run “full-scale, full-speed experiments.”[4]

To avoid motorway congestion, it also started to investigate traffic control by a central computer – it was estimated that such a system could increase the road capacity by 50% and prevent 40% of the accidents. First experiments were effectuated since 1960 at a new test track that opened in Crowthorne, Berkshire.[5] After successful tests, four miles of the M4 between Slough and Reading were equipped with two guidance cables beneath its surface. Experiments started before the road was opened to the public.[6]

The system automated *both, longitudinal and lateral control*: The cables transmitted electronic signals that were picked up by sensors which actuated steering and brake mechanisms of the vehicles.[7] These experiments were called „drive-by-wire”[8] and four test vehicles were used. As in the RCA-experiment in Nebraska 1957 (see Sect. 2.1.1) the windscreen of a manually driven vehicle – a Standard Vanguard estate – was obscured with a cardboard and the driver had to follow steering commands that were displayed on the dashboard. A system called RITA (Road Information Transmitted Aurally) informed the driver acoustically via radio about the road signs.

After the Vanguard, an automated Citroën DS 19 was used for experiments. Today, the vehicle is presented in the Science Museum’s Road Transport Collection at Wroughton as „the world’s first completely automatically-controlled road vehicle”.[9] Its hydraulic system was modified to actuate steering, brakes and throttle. Following magnetic cables under the road bed with the help of sensors on the front and the rear, the Citroën circulated at the test track with speeds up to 130 km/h (80 mph) on dry surfaces. It operated even on snow covered surfaces at speeds up to 100 km/h (60 mph), as a research paper from 1970 reports.[10] But the engineers had no success to make the vehicle change lanes.[11]

---

[4] Johnson, p. 40.

[5] Reuters, U.K.: New Road Research track opened at Crowthorne, October 25, 1960, in: https://reuters.screenocean.com/record/547949 (accessed June 25, 2019).

[6] Torchinsky, p. 60.

[7] Reynolds, John. 2001. Cruising into the Future. *Telegraph.*, May 26, 2001, online in: https://www.telegraph.co.uk/motoring/4750544/Cruising-into-the-future.html (accessed June 12, 2019).

[8] Today, the term drive-by-wire is not used to describe a car following a cable buried in the roadbed. Instead it means the substitution of mechanical links between components like the gas pedal and the motor by electronic wiring.

[9] Reynolds, John. 2001. Cruising into the Future. *Telegraph.*, May 26, 2001, online in: https://www.telegraph.co.uk/motoring/4750544/Cruising-into-the-future.html (accessed June 12, 2019).

[10] Cardew, K.H.F. 1970. The Automatic Steering of Vehicles: An Experimental System Fitted to a DS 19 Citroen Car, Driver Aids and Abilities Section, Road Research Laboratory, Crowthorne, Berkshire, RRL report LR 340 (1970), online in: https://trl.co.uk/sites/default/files/LR340.pdf (accessed June 22, 2019).

[11] Waugh, Rob. 2013. How the first “driverless car” was invented in Britain. *Yahoo News UK*, July 17, 2013, online in: https://uk.news.yahoo.com/how-the-first%2D%2Ddriverless-car%2D%2Dwas-invented-in-britain-in-1960-093127757.html (accessed July 26, 2019).

As the hydraulic system was too complex for common use, an Austin Mini and a Ford Cortina with automatic transmission were equipped for automated driving experiments in 1969. The Ford Cortina can be seen in a video from the year 1971, probably filmed during one of the Public Open Days of the Laboratory.[12] The film shows that the RRL experimented at this time with radar sensors for distance keeping – like MAN and Bosch in Germany (Sect. 4.1.3). Further scenes are documented in a BBC-film about the Road Research Laboratory.[13] The film also mentions an automatic toll collect system, functioning with a black box allowing to identify each vehicle that activates a wire in the pavement while entering the city center.

Influenced by the oil crisis and the economic recession of the mid-1970s, the UK-government considered the long-term research in the field of automatic guidance as "unnecessary and unaffordable". All funding was withdrawn.[14] The use of guidance cables got abandoned due to the high costs of road modifications.[15] A similar retreat can be observed some years later in the United States (see Sect. 2.1.2).

4.1.2 *Semi-Automated Test Drives of Continental and Siemens in Germany (1968)*

Proving ground experiments on automated driving in (Western-) Germany started at the end of the 1960s. Initiated by the rubber and truck industry, the technology was developed by major electronics companies and technical universities. Vehicle automation was foremost seen as useful for dull tasks like test and truck driving.

At the end of the 1960s, the German automotive manufacturing company Continental developed together with Siemens a semi-automated radio guided Mercedes 250 (Fig. 4.1) for tire tests that was used until 1974.[16] The technical universities of Darmstadt and Munich also participated in the experiments. According to the current state of research, this was the first automated road vehicle in Germany. Continental was interested in the automatization of test procedures because preprogrammed test runs would allow more precise measurements and would relieve test drivers. The test runs started on September 11, 1968 at the new 2.8-km long test facility Contidrom that was opened one year before near Jeversen in the woodland of the *Lüneburger Heide*.

---

[12] British Movietone, 2015. Early Driverless Car in 1971, July 2, online in: https://www.youtube.com/watch?v=5ocvNxjN3dc (accessed on June 16, 2019).

[13] DRIVERLESS CARS and the FUTURE OF MOTORING | BBC News | Retro Transport | BBC Archive (Video Title), BBC Broadcast of May 18, 1971, in: BBC Archive (Youtube-Account), Uploaded on May 5, 2022, online in: https://www.youtube.com/watch?v=Y2Lw3ZvrEdk (accessed May 5, 2022).

[14] Reynolds, John. 2001. Cruising into the Future. *Telegraph.*, May 26, 2001, online in: https://www.telegraph.co.uk/motoring/4750544/Cruising-into-the-future.html (accessed June 12, 2019).

[15] Torchinsky, p. 62.

[16] Continental AG, Controlled by Electronics: Continental Launched Its First Driverless Vehicle 50 Years Ago, Press Release September 11, 2018, online in: https://www.continental.com/en/press/press-releases/controlled-by-electronics-144932 (accessed September 25, 2020).

Fig. 4.1 The Mercedes 250 test vehicle of Continental driving under automated lateral control, 1968. (Credits & Image Source: Continental AG, Contidrom Electronic test, online in: https://www.continental.com/de/presse/pressemitteilungen/50-jahre-fahrerloses-fahren-144612 [accessed July 4, 2019])

The Mercedes was equipped with "electro-mechanical steering, an electro-mechanical throttle regulator and a radio system for reporting measurements".[17] The control electronics and an electro-pneumatic braking system were housed in the trunk.

As in the early US-American experiments, the magnetic field of a guide wire in the road was used to automate lateral control, but also to receive speed commands. Three coils at the front and the rear bumpers of the vehicle measured the intensity of the magnetic field.[18] The antennas on the left and the right side were linked to the steering system. When the car followed the guidance cable symmetrically, the voltage measured by both coils was identical. If the car left the guidance cable, the measured tensions started to differ and the electronics in the trunk of the car adjusted the steering mechanism. While the lateral control of the vehicle was automated, the longitudinal control – acceleration and braking – was remote controlled from a control room. The desired speed was determined by modulating the current in the guide-wire with an audio-frequency. The third antenna in the center received these commands for longitudinal control.

[17] Ibid.

[18] Continental AG, 50 Jahre fahrerloses Fahren: Pressematerial, online in: https://publicarea.admiralcloud.com/p/a49d3d8ba92f3cdbfa864f (accessed September 25, 2020a).

This system was developed by Siemens, the US-American company Westinghouse and researchers at the technical universities of Munich and Darmstadt.[19] In the early 1960s Westinghouse developed a driverless technology for the Skybus-Project in Pittsburgh, a driverless transit system with rubber-tires on a busway.

Continental board member Adolf D. Niemeyer qualified the vehicle as the "world's first teleguided vehicle for tire tests".[20] But it should be remembered that two years before, the US-American manufacturer Rubber Co. had already started tire tests with automated vehicles on a five-mile test-track near Laredo (Texas), United States.[21] The vehicles were equipped with a guidance boom, extended two feet to the right of the front bumper. The detector coil on the boom followed the path of a guidance cable.[22]

4.1.3 The Semi-Automated Truck of MAN and Bosch in Germany (1971)

In 1971 the German truck company MAN (Maschinenfabrik Augsburg Nürnberg) and the technology company Bosch started a collaboration in the field of automated driving.[23] The project was called *Automatische Fahrzeugführung* (AFF) – automated vehicle guidance. It was motivated by the idea to relieve truck drivers from the driving task, especially in enclosed areas like quarries. Two other use-cases were container loading in harbors and bus services on closed tracks. Automated trucks on the motorway remained a more long-term vision.[24]

A big MAN 16.320 F truck served as test vehicle. Both sides of the vehicle displayed in big letters the slogan *Vollautomatisch ohne Fahrer* – fully automated without driver (Fig. 4.2). According to the MAN-engineer Hans Hagen, the truck had been equipped with an automatic transmission, ABS, an electric motor for steering control, take-up reels and an electronic Diesel-injection pump that was controlled "via radio, cable or the accelerator pedal".[25]

The pavement of a closed MAN-test track in Munich-Allach was equipped with a guidance-cable, fed with an alternating current (AC) signal of 10 kHz, 15 kHz and 20 kHz. A control tower next to the test track transmitted the signal (1 W / km) to

[19] Continental AG, The "Ghost Car", online in: https://www.continental.com/en/company/history/the-ghost-car-145316 (accessed September 25, 2020b).

[20] Meyer, Wolfgang. 1968. Ein Auto ohne Fahrer rast mit 120 Sachen über die Beton-Piste. *Bild*, September 13, 1968.

[21] Dunne, Jim. 1966. This 140-m.p.h. Test Car is Driven by Remote Control. *Popular Science*, November, p. 105.

[22] Ibid.

[23] MAN Bus & Trucks, Controlled by an invisible hand, January 28, 2020, online in: https://www.mantruckandbus.com/en/moments/1971.html (accessed September 25, 2020).

[24] Hagen, Hans, Lastwagen ohne Fahrer, VDI-Nachrichten, April 19, 1974.

[25] Hagen 1974.

Fig. 4.2 Automated lateral guidance tested by MAN and Bosch (1971). (Credits: Credits & Image Source: MAN Bus & Trucks, Controlled by an invisible hand, January 28, 2020, online in: https://www.mantruckandbus.com/en/moments/1971.html [accessed September 25, 2020])

the cable.[26] It was captured by the take-up reels in front and back of the truck that were fixed with a distance of 300 mm to the ground. As in the experiments of Continental, the truck was started, monitored and stopped from a control tower (Fig. 4.3).

The first experiments of Bosch were effectuated in 1971. In summer 1973, MAN and Bosch started experiments with an active guidance system in the highway – meaning that the vehicle followed commands from the cable.[27] *Lateral* control (steering) was automated up to a speed of 50 km/h by following the guidance cable. As in the experiments of Continental, *longitudinal* control – braking and accelerating – was realized by teleguidance from a control tower. The on-board electronics received start-stop and speed commands via the cable. The semi-automated truck could also be teleguided like a toy car – a corporate film from MAN shows an employee standing nearby the truck using a steering device with a long cable to control the vehicle.

The experiments of MAN and Bosch went further than those of Continental and Siemens: The engineers also worked on distance warning devices. As reported by Hans Hagen, obstacles were detected with a radar eye on the left side of the front bumper. Press reports mentioned also infrared-devices – one of them can be seen in the MAN-film (Fig. 4.4).[28] With this device, the truck was able to respect a safety

[26] Ibid.

[27] A passive system would use existing lane markings, avoiding important infrastructure changes. But as important changes in vehicle technology would be necessary, this approach was not explored further at this time (Hagen 1974).

[28] Firgau, Walter. 1973. Ohne Fahrer auf richtigem Kurs. *Süddeutsche Zeitung*, September 6.

Fig. 4.3 The control tower next to the MAN-test track. (Credits: Credits & Image Source: MAN Bus & Trucks 2020)

Fig. 4.4 Infrared device at the front of the MAN truck. (Credits: Credits & Image Source: MAN Bus & Trucks 2020)

distance to other vehicles. It also stopped in front of obstacles – but steering maneuvers to avoid them had to be realized by a human driver. The MAN-film shows the truck braking after being overtaken by another car cutting into its lane.[29] But Hagen explains that the sensor was not able to distinguish properly between an obstacle on its own lane and an overtaking vehicle on the adjacent lane.[30] For that reason it was planned to use the system within a third highway lane for automated trucks only.

[29] Man Bus & Trucks 2020, min 1:51.

[30] Hagen 1974.

4.1.4 Guidance-Cables for Test Drives – Peugeot in France (1979)

In France, guidance cables were first used at the end of the 1960s at a test circuit of Peugeot near Besançon.[31] At the end of the 1970s the system was used to test new vehicles at the proving ground of Peugeot in Belchamp near Sochaux.

The first lanes of this test track had been created in 1959. A high speed test track was added in 1968. In addition to the 5.4 km long test circuit, Peugeot created in 1979 an automated track of 1.4 km with very bad surface characteristics, called *Piste Automatique Pavée* (PAP). Guided only by two cables in the road, prototypes circulated without a driver with a speed of 35 km/h for several days non-stop.[32]

Peugeot used the guidance cable-approach only to automate test vehicles, a different use case than automation of automobile road traffic. Like the experiments of Continental in Germany at the end of the 1960s, the guidance cable was only used to relieve human test drivers from their exhausting task. Like in other European countries, the automotive industry in France did not yet develop automated driving as a concept for the automobile future.

4.2 The Shift to Vision Concepts in Europe

During the 1970s the guidance cable approach lost not only in the United States but also in Japan and Europe its status as most promising concept for an automated traffic system of the future. It became a *niche technology* that was only seen as useful for automated test-drives of research vehicles in closed areas. An international shift from guide-wire concepts to visual guidance concepts can be observed at the end of the 1970s. Japan abandoned guide-wires already at the end of the 1960s and started then to experiment with visual guidance (Sect. 3.4.1). The United States abandoned active wire-following systems in 1977 (Sect. 2.2). England stopped funding research on guidance-cables in the mid-1970s. In France, first steps with vision approaches had been started in 1977 with the mobile robot HILARE in Toulouse. In Germany, Ernst Dieter Dickmanns (UniBwM), observing the development rate of microprocessors, selected in 1977 machine vision as long-term research goal. Experiments of the automotive industry began 1982 with the Opto-Pilot at Volkswagen.

[31] Hartwich, Erwin, and Walter Zimdahl. 2012. Regelung von Straßenfahrzeugen: Ein Zeitzeugenbericht aus dem Institut von Prof. W. Oppelt, TH Darmstadt. *at-automatisierungstechnik* 60 (6), p. 340.

[32] Barbaza, Pierre. 1984. Profession Pilote Essayeur. *Echappement* (Vol. number unknown), p. 90.

4.2.1 The Mobile Robot HILARE of LAAS Toulouse (1977)

A better possibility than the guidance cable to synchronize a vehicle with its environment arose in the robotics field. Influenced by the mobile robots Shakey and Stanford Cart in the United States, academic research institutes in Europe and especially in France joined the research on mobile robots in the late 1970s. As we have already mentioned, the main tasks of robots and autonomous vehicles are identical: They must first be able to observe and understand their environment and second they have to decide and realize their own reactions in this environment.[33] Therefore the development of robots is directly relevant for the topic of this book.

This section will present the mobile robot HILARE, one of the first robots in Europe with the mission to avoid obstacles by sensing the environment. It was developed by LAAS in Toulouse, one of the biggest laboratories of the *Centre National de la Recherche Scientifique* (CNRS), the main public research agency in France.[34] LAAS had been created in 1967 by Jean Lagasse[35] – one of the pioneers of automatization in France – and started its work at the climax of social unrest in May 1968. Until 1973, the acronym LAAS meant *Laboratoire d'automatique et de ses applications spatiales*. As its name indicated, the first steps of LAAS were linked to space exploration. Today, it is widely known for its contributions to the science of information and communication (robotics, AI, automatization, systems control).

Driven by a growing demand for manufacturing robots, the industrial sector became more interested in robotics research at the beginning of the 1970s, as Claude Laurgeau outlines.[36] During this time, the French automotive company Renault started to work on industrial robots. These first generation robots were conceived as simple mechanical tools with a leg and a hand, able to capture objects, but without sophisticated perception capabilities.[37] To reinforce its activities in the field, Renault searched for academic expertise.

---

[33] Rachid Alami from LAAS describes these tasks in a definition of Artificial Intelligence: "La capacité pour une machine, un robot, d'appréhender son environnement, de le comprendre, d'anticiper son action et de la réaliser dans des bonnes conditions, donc une capacité de raisonner sur l'environnement et sur soi-même pour réaliser une tache." (Rachid Alami, responsable Robotique et AI, in: LAAS, Video: Robotique et Intelligence Artificielle, in: https://www.laas.fr/vod/public/films_Georges/5-robotique_IA_720p.html [accessed November 19, 2020]).

[34] Chatila, Raja. 1996. Autonomous Mobile Robots in France: Some Projects and Scientific Approaches. In *Autonome Mobile Systeme 1996*, ed. Günther Schmidt and Franz Freyberger, 2–7. München.

[35] The LAAS was created by merging teams of the *École nationale supérieure d'électrotechnique, d'électronique, d'informatique, d'hydraulique et des télécommunications* (ENSEEIHT) and the department Génie électrique of the *Institut National des Sciences Appliquées* (INSA) in Toulouse (Martinez, Augustin, La centrale technologique microélectronique, in: Lettre du LAAS, N° 37, October 2008, p. 39, online in: https://www.laas.fr/LettreDuLaasStatic/content/download/793/4835/version/3/file/La-Lettre-n37-5.pdf

[36] Laurgeau, Claude. 2009. Le siècle de la voiture intelligente, p. 8. Paris.

[37] Giralt, Georges. 1990. Robots et intelligence des machines. *Les Cahiers du MURS* n° 22, 4ème trimestre, p. 28.

Responding to this industrial demand, a robotics research group was created at LAAS. As a tribute to the evolution of its activities, LAAS changed its name in 1973 to *Laboratoire d'automatique et d'analyse de systèmes.*[38] Coming back from a research stay at the university of Berkeley, the French roboticist George Giralt (1930–2013) created in 1974 the robotics and artificial intelligence group at LAAS, combining mechanical and AI-approaches.

At the end of 1977, Giralts team started to work on the mobile robot HILARE (*Heuristiques intégrés aux logiciels et automatismes dans un robot évolutif*).[39] This robot is of special interest for the purpose of our publication because it was conceived to understand path finding and obstacle detection processes, tasks that are also crucial in autonomous driving concepts.

The three wheeled vehicle was powered by 24 Volt batteries, had a weight of 400 kg and operated in the laboratory only.[40]

On the perception side, HILARE used a laser range finder, a TV-camera on a pan/tilt platform, 14 ultrasonic sensors to detect nearby objects and an odometer to measure the distances traversed. HILARE processed the information from these sensors on four on-board Intel 80286 processors working on locomotion and sonar range processing. But like Shakey and the Stanford Cart, the most important computing power was located off-board: The robot functioned only with a radio link to a local mini-computer (SEL 77/80) for coordination, control, trajectory planning and navigation. This computer was linked to another bigger computer outside the laboratory (IBM 30/33).[41]

The robot was conceived to find the shortest path beyond fixed, movable and mobile obstacles to reach a goal. Its sensors gave HILARE the capability to perceive and move between obstacles in a blocks world like Shakey, as Giralt precises.[42] During the exploration of its environment – it reached a speed of 1 m/s[43] – HILARE

[38] In 1994 the meaning of the acronym was changed to *Laboratoire d'analyse et d'architecture des systèmes* (CNRS, La science de la machine intelligente, Interview with Georges Giralt, interviewed by Patrick Danès, in: La lettre du LAAS, N° 37, October 2008, p. 58, online in: https://www.laas.fr/LettreDuLaasStatic/content/download/793/4835/version/3/file/La-Lettre-n37-5.pdf).

[39] Giralt, Georges, R.P. Sobek, and Raja Chatila. 1979. A multi-level planning and navigation system for a mobile robot; a first approach to HILARE. *Proceedings of 6th IJCAI*, Tokyo, Japan.

[40] http://homepages.laas.fr/matthieu/robots/hilare.shtml

[41] Giralt, Georges, Raja Chatila, and Marc Vaisset. 1990. An Integrated Navigation and Motion Control System for Autonomous Multisensory Mobile Robots, p. 420–443, here: p. 423, see also CNRS, HILARE et les autres, Aspects de la robotique au LAAS, film réalisé par Jacques Brissot, France 1981, (televised on August 10, 1983, Antenne 2), 00:25 min.

[42] CNRS, La science de la machine intelligente, Interview with Georges Giralt, interviewed by Patrick Danès, in: La lettre du LAAS, N° 37, October 2008, online in: https://www.laas.fr/LettreDuLaasStatic/content/download/793/4835/version/3/file/La-Lettre-n37-5.pdf, p. 44 (accessed October 26, 2020).

[43] Giralt, Georges, Raja Chatila, and Marc Vaisset. 1990. An Integrated Navigation and Motion Control System for Autonomous Multisensory Mobile Robots. In *Autonomous Robot Vehicles*, ed. Ingemar J. Cox and Gordon T. Wilfong, p. 421. New York, Berlin, Heidelberg, London, Paris Tokyo, Hong Kong.

memorized the position of the obstacles it had detected. But it was also able to make decisions when confronted with non-repetitive events.

The kinematics of the three-wheeled robot with two wheels mounted fix was limited – like an automobile, it was a *non-holonomic* vehicle: As direct lateral motions were not possible, the robot had to plan several steps to move laterally. Like humans who have to learn how to steer their car into a parking lot with a limited number of maneuvers only, HILARE had to optimize the number of its motions, explains Giralt.[44]

At the beginning of the 1980s, Giralt started and directed the program *Automatisation et robotique avancée* (ARA), that recentered the activities on manufacturing robotics. HILARE was followed in 1986 by Junior, used until 2001.[45] In 1989, the project *Véhicule autonome planétaire (VAP)* was developed together with the *Centre National d'Études Spatiales (CNES)* in Toulouse. It focused on autonomous systems, driving architectures, planning algorithms and perception systems.[46] The goal of this autonomous robot project was to substitute humans in dangerous environments (exploration of planets, polluted environments, warehouses etc.). With a focus on perception, reasoning and mobility, the project surpassed the immediate industrial interests in manipulation robots for factory assembling. HILARE 2, created in 1990, was equipped with a pan/tilt zoom color camera and two pan/tilt black and white cameras.[47] Hilare 2bis was built in 1992.[48]

Since 2006, the first HILARE is shown in the *Musée des Arts et Métiers (CNAM)* in Paris (Fig. 4.5). Cyrille Foasso of CNAM resumes that „the successive versions of this research project led to significant progress in modelling the environment and the autonomy, reactivity and programming of mobile robots."[49] He emphasizes that HILARE is today "regarded as the first French mobile robot capable of autonomously negotiating an unknown environment."

4.2.2 *Early Research on Machine Vision at University of Bundeswehr Munich (1977)*

In the same year that LAAS started to work on mobile robots, Ernst Dieter Dickmanns at UniBwM in Germany selected machine vision for vehicles of any kind as long-term research goal. In the following decades this university became one of the most successful research facilities exploring autonomous driving.

[44] Giralt, Georges. 1990. Robots et intelligence des machines. *Les Cahiers du MURS* n° 22, 4ème trimestre, p. 44.

[45] http://homepages.laas.fr/matthieu/robots/junior.shtml

[46] Giralt 1990, p. 33.

[47] http://homepages.laas.fr/matthieu/robots/h2.shtml

[48] http://homepages.laas.fr/matthieu/robots/h2bis.shtml

[49] Foasso, Cyrille, Hilare autonomous mobile robot, in: https://artsandculture.google.com/asset/hilare-autonomous-mobile-robot-laboratory-of-analysis-and-architecture-of-systems/UwFV2X6hPn0NjA

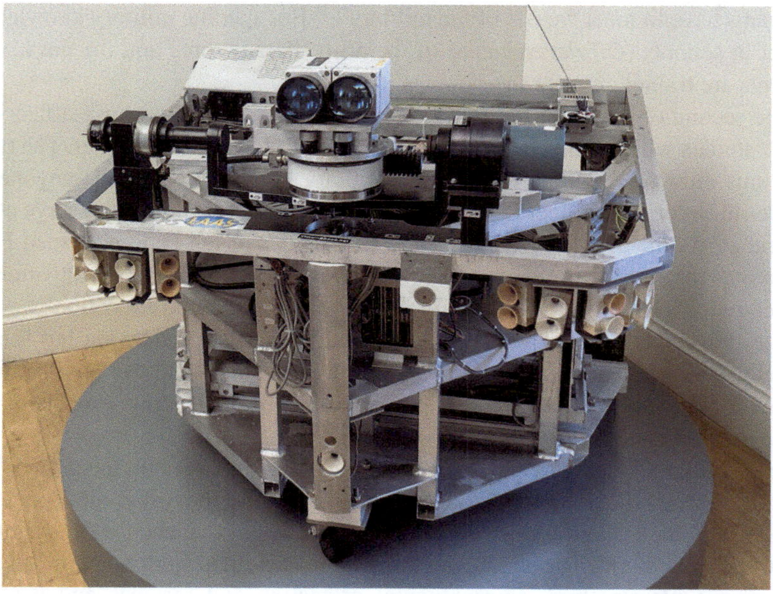

Fig. 4.5 HILARE – the mobile robot of LAAS – as it is today shown at the *Conservatoire des Arts et Métiers (CNAM)* in Paris. (Credits: F. Kröger, October 27, 2022)

For a better understanding of the institutional frame of these research efforts, we will start with a short introduction of the university. Located in Neubiberg in the south of Munich in Bavaria, UniBwM is one of two federal universities that are part of the *Bundeswehr*, a term that designates the armed forces of Germany and their civil administration. The second one is the Helmut Schmidt University / University of the German Federal Armed Forces in Hamburg.[50] The creation of the universities goes back to an initiative of Helmut Schmidt who was federal Minister of Defense (1969–1972) before he became Chancellor of Germany in 1974. After a decision of the Bundestag in 1972 both universities were founded in 1973. Until 1985 they were called *Hochschulen der Bundeswehr* (HsBw).

The universities were created with two objectives: First, a higher qualification for military officers was desired. Unlike military academies like Saint-Cyr in France, Sandhurst in Great-Britain or West Point in the US, the universities of the German Federal Armed Forces were therefore conceived as *civil* universities for *military* personnel. They offer courses with civilian contents, without direct relation to the military and corresponding to courses at regular German universities. In contrast to other schools like École Polytechnique in France, all professors of the Bundeswehr-universities are civilians. However, UniBwM confirms that the degree

[50] The university in Hamburg is called Helmut Schmidt University since 2003 (Uzulis, André. 2005. *Die Bundeswehr: Eine politische Geschichte von 1955 bis heute*, Hamburg, Berlin, Bonn).

programs have to meet "needs of the military".[51] But specific military knowledge is taught in schools (*Offiziersschulen*) of the army, the air force and the navy, where students are trained prior to their university studies.

The second objective of the Bundeswehr-universities is to propose officers academic studies with civilian degrees independent of their military career. The degree programs must be "suitable for the post-military life".[52] In Germany, anyone who decides to become an officer in the military service commits himself to serve at least 13 years.[53] As the degrees granted by the universities of the Federal Armed Forces are equivalent to those of German state universities,[54] officers can switch after their service in the military into a private job requiring an academic education.[55]

The universities are primarily concentrated on academic training but they are also engaged in research projects. It is important to understand the "civil-military" character of the universities because their name can be misleading, especially when it comes to research programs.

As already mentioned above, the Bundeswehr consists of both, the armed forces and the civil administration. The armed forces include the army, air force, navy, armed forces base (*Streitkräftebasis*), central medical service and the organizational area of cyber and information space. The civil areas include equipment, information technology, infrastructure, environmental protection and services and personnel (together they form the Bundeswehr administration), military pastoral care and the administration of justice. The strict organizational separation between the armed forces and the civil staff of the administration – including the armaments area – is prescribed in the constitution (*Grundgesetz*) of the Federal Republic of Germany. This organizational structure had been decided to avoid a concentration of power and is a lesson learned from the history of the German armed forces.

Reporting about the pioneering role of UniBwM in the development of machine vision and autonomous driving, a media report claimed: "Who invented the robot car? The Bundeswehr!"[56] On the one hand, this is correct because both universities were since their creation part of the armed forces base (*Streitkräftebasis*) that

[51] „Studiengänge müssen nicht nur einen Bedarf der militärischen Seite decken und gleichzeitig tauglich für die nach-militärische Lebensspanne sein." (Universität der Bundeswehr München, Präsidialabteilung, Struktur- und Entwicklungsplan der Universität der Bundeswehr München 2015–2020, München 2015, online in: https://www.unibw.de/universitaet/universitaet/leitung/struktur-und-entwicklungsplan-unibw-m-2015-2020.pdf, p. 26 [accessed May 7, 2020]).

[52] Ibid.

[53] Bundeswehr, Die Laufbahn der Offiziere (M/W/D), 2020. https://www.bundeswehr.de/de/ueber-die-bundeswehr/dienstgrade-laufbahnen-bundeswehr/laufbahn-offiziere (accessed May 8, 2020).

[54] Due to 'trimester'-terms (3 instead of 2 lecture periods per year) the regular study at UniBwM can be finished in 3½ years.

[55] See also Ellwein, Thomas, Achatz von Müller, and Harro Plander. 1974. *Hochschule der Bundeswehr zwischen Ausbildungs- und Hochschulreform*, p. 11–13. Opladen.

[56] Vieweg, Christof. 2015, July 31. Wer hat das Roboterauto erfunden ? Die Bundeswehr! In *DIE ZEIT*.

belongs to the military organizational area of the Bundeswehr.[57] They are funded by the Federal Ministry of Defense and they have no civil clause (*Zivilklausel*).[58] On the other hand, this headline could give the impression that autonomous driving came out of a military research project. That is wrong. As we will see in the next chapters, only very few research programs were funded by the Federal Ministry of Defense and the *Bundesamt für Wehrtechnik und Beschaffung* (BWB).[59] Most of the work on autonomous driving at UniBwM received funding from civilian sources – the Federal Ministry for Research and Technology (BMFT), the German Research Foundation (DFG), the European Union (EU) and the automotive industry.[60] Research on autonomous driving serves military interests but at least in Germany it was never a military research project only.

After this short introduction, we will present the first steps towards not only automated but autonomous driving in Germany that were undertaken at UniBwM under the direction of Dickmanns. As the concept of vehicle guidance with computer vision goes back to him – his lifetime project was to teach machines the sense of vision – a short biographical sketch will be given first, highlighting his professional background:

Dickmanns, born in 1936 in Niederkassel, grew up near Cologne during World War II. As a child, he witnessed the power of aircraft technology when a bomb exploded not far away from the bunker where he and his family took shelter. Witnessing this event was one of several factors that sparked his interest in airplane technology. As teenager he worked with horses on a farm right across the street of his home. When horses were exchanged against tractors he noted that the large

---

[57] Since April 2012 (Dresdner Erlass), the Bundeswehr universities are no longer part of the military area but belong to the civil administration (*Bundeswehrverwaltung*): They remain assigned (*zugeordnet*) to the Federal Ministry of Defense – the highest authority of all military and civil areas of the Bundeswehr – but they are now subordinated to its Human Resources department. That means they remain a department (*Dienststelle*) of Bundeswehr, but they are a "civilian service outside the armed forces" (Universität der Bundeswehr, Unterstellungsverhältnis, in: https://www.unibw.de/milbereich/allg-info/inhalt-allg-info/unterstellungsverhaeltnis). Strictly speaking it is therefore since 2012 imprecise to translate University of Bundeswehr as University of the Federal Armed Forces – as the university is no longer part of the armed forces.

[58] A civil clause is a voluntary commitment by scientific institutions to do research exclusively for civil purposes. The first civil clause came into force at the University of Bremen in 1986 and several German universities have introduced them today. Civil clauses exist only in a few other countries, especially in Japan.

[59] Today, BWB is called Bundesamt für Ausrüstung, Informationstechnik und Nutzung der Bundeswehr (BAAINBw).

[60] From 1982–1984, research on "models to improve the interpretation of dynamical scenes" in the field of visual autonomous road vehicle guidance was funded by the Program for Information Technology of the Federal Ministry of Research and Technology (BMFT). This was followed by two contracts from the German Science Foundation (DFG). The first contract (1984–1987), useful for satellite docking and road vehicle guidance, had the title "Perception and control of motion in a technical environment by processing of image sequences". The second contract (1987–1991) investigated the guidance of ground vehicles and autonomous visual landing approaches and was named "4-D-Scene recognition with integral spatiotemporal models".

increase in power also introduced a disadvantage with respect to human attention necessary. While horses learned to move along the edge of the road or along the boundary of vegetation on a field, the tractor had to be controlled all the time by a human. So the sense of vision for vehicles would be a desirable feature for the future.

After his school days he decided in 1956 to study Aero-Space Engineering at the Rheinisch-Westfälische Technische Hochschule (RWTH) in Aachen. After finishing his diploma thesis in 1961 he was hired at the Institute for Flight Mechanics of the Deutsche Versuchsanstalt für Luftfahrt (DVL) in Mülheim/Ruhr. A NASA student-fellowship allowed him in 1964/65 to study flight dynamics and control engineering at Princeton University, New Jersey. Back home, he wrote his dissertation on optimal trajectories of spaceships when they reenter the atmosphere of planets. After achieving his doctoral degree in 1969 from RWTH Aachen he became head of the flight performance group and soon vice director of the Institute for Systems Dynamic and Flight Mechanics. In 1971/72 he spent a one-year Postdoctoral Research Associateship with the NASA Marshal Space Flight Center in Huntsville, Alabama, investigating optimal re-entry trajectories for the planned Space Shuttle Orbiter. After return, his research centered on trajectories for the ascent and positioning of the first European communication satellite *Symphonie* and of spaceships reentering the atmosphere of the Earth.

In December 1974, after the launch of *Symphonie*, Dickmanns became acting director of the DFVLR-research center Oberpfaffenhofen. In this position, he came into contact with the analysis of satellite images at several institutions in the center. An important research field was the interpretation of high-resolution snapshots taken from airplanes and spacecrafts.

In 1975 he received a call from the new *Hochschule der Bundeswehr München* where he became Professor of Control engineering (*Steuer- und Regelungstechnik*), a position he held until 2001. Dickmanns founded the Institute for Systems Dynamic and Flight Mechanics (ISF) of the *Fakultät für Luft- und Raumfahrttechnik* (LRT), one of the largest aviation and aerospace faculties in Germany. From the beginnings, the institute worked, besides continuing trajectory optimization, on the goal of using technical image sequences for the guidance of vehicles. It aimed to provide the sense of vision for technical vehicles and machines.

In 1977, Dickmanns defined the new field of real-time computer vision *(Rechnersehen)* as long-term research goal.[61] In the long run, he planned to work on autonomous aircrafts and spaceships also; but as motion control on the ground is less complex than in aero-space applications, it was decided to focus on ground vehicles first, reports Dickmanns.[62] In 1978, "studies on road vehicle guidance by computer vision" were started.[63]

---

[61] Dickmanns, Ernst Dieter. 1980. Untersuchungen und mögliche Arbeitsschritte zum Thema "Künstliche Intelligenz: Rechnersehen und -steuerung dynamischer Systeme", p. 3.

[62] Dickmanns, Ernst Dieter. 1993. Interview, Der Lotse geht an Bord, p. 6. Hochschulkurier.

[63] Dickmanns, Ernst Dieter. 1987. PROMETHEUS, 11 170 Integrated Approaches, State of the art review, March 1987, p. 3.

At the end of the 1970s and the beginning of the 1980s, one of the main challenges in the field of computer and machine vision was real-time vision, meaning a system that enables a robot "to perceive changes in its environment while they are occurring" and soon enough to react properly to these changes.[64] Such a system should therefore respect a delay of not more than 100 milliseconds in reporting an event or providing data on a scene observed. But at this period real-time vision systems with digital microprocessors for on-board use in vehicles were not available. Most of the research groups worldwide analyzed single-frame images which took "from a few minutes to a few hours on a powerful computer", reports Graefe.[65] With the available computer technology it was impossible to read and evaluate full video images in real-time. It can be resumed that mobile robots at this time could only analyze the environment *or* move, but *not understand the environment and move through it at the same time* – as the US-chapter has shown. The tools were too weak and the task too hard.

Despite these difficulties, Dickmanns selected image sequence processing in real-time as long-term research goal. His vision was to merge the digital analysis of images with technologies in control engineering. His decision was inspired by the forecast of an ever increasing computer power described in 1964 by the American physicist Gordon Moore who observed that since the invention of integrated circuits in 1958 the number of transistors on an integrated circuit doubled every year.[66] In the years till the end of the 1970s, the prediction of Moore's law proved to be correct – the computing power of digital microprocessors increased by a factor of 10 every 4 to 5 years. Drawing on this development, Dickmanns estimated that until his retirement 25 years later, digital microprocessors would be one million times more powerful if they continue the path observed by Moore (see Fig. 4.6).

But instead of waiting until more powerful processors for real-time vision-systems would be available in the future, Dickmanns and his colleagues at UniBwM decided to simplify the task. They worked on methods that would reduce the complexity of image analysis, allowing at the same time to stick to the goal of real-time operation.[67]

To achieve his goal, Dickmanns started in 1978 to develop the *window-concept for real-time image sequence processing*.[68] It was based on two ideas: First, instead of working on the complete perception and analysis of the environment, camera images had to be analyzed only partially, focusing on the one hand on areas (rectangular "windows") with high contrast and on the other hand on areas in motion

[64] Graefe, Kuhnert, p. 13.

[65] Graefe Kuhnert, p. 13, see also Mysliwetz, p. 7.5.

[66] Moravec, Hans. 1988. *MIND Children, The Future of Robot and Human Intelligence*, p. 68. Cambridge, London.

[67] Dickmanns, Ernst Dieter. 1988. An Integrated Approach to Feature Based Dynamic Vision. *Proceedings of the Computer Society Conference on Computer Vision and Pattern Recognition*, p. 820. Ann Arbor.

[68] Dickmanns, Ernst Dieter, and Volker Graefe. 1988. Dynamic monocular machine vision. *Machine Vision and Applications* 1: 223–240.

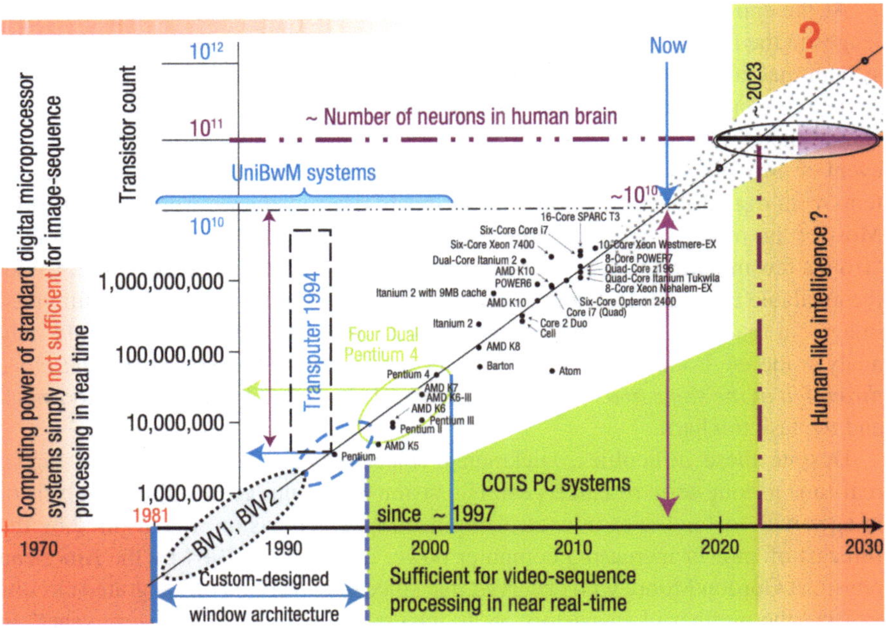

Fig. 4.6 Increase in number of transistors of digital microprocessors over time. (Credits: Ernst Dieter Dickmanns. Image Source: Dickmanns E.D. 2017: Developing the Sense of Vision for Autonomous Road Vehicles at UniBwM. IEEE-Computer, Dec. 2017. Vol. 50, no. 12. Special Issue: SELF-DRIVING CARS, pp. 24–31)

(Fig. 4.7). This reduced the amount of data. Second, with prediction-error-feedback methods from control engineering (Kalman-Filter) – that will be explained later in Sect. 4.3.1 – it was avoided to store pixels from previous images. This allowed UniBwM-researchers to assure that the image evaluation frequency would not drop below 10 Hz. That means an image should be analyzed in less than 100 milliseconds, allowing the development of the real-time systems mentioned above.

Based on this concept from Dickmanns, Volker Graefe from the *Institut für Meßtechnik* of UniBwM realized in the years 1979/1980 a new image sequence processor, called *Bildvorverarbeitungssystem* (BVV) or image preprocessing system. It used 8-bit microprocessors (Intel 8085) and was placed between a camera and a computer (PDP 11/60, VAX 750, Perkin Elmer 3220). As already mentioned, its task was to reduce the complexity of the data received from the camera by extracting only relevant information – as road objects – from the pixel data. Only inclined edge features with adjacent average grey values of the smoothly curved road, of lane markings and of moving objects like bicycles, pedestrians or other vehicles – representing less than 10% of the image area – were analyzed.[69] All other parts of the image, for instance the background, were skipped.

[69] Graefe, Volker, and Klaus-Dieter Kuhnert. 1992. Vision-Based Autonomous Road Vehicles. In *Vision-Based Vehicle Guidance*, ed. Ichiro Masaki, p. 16. New York.

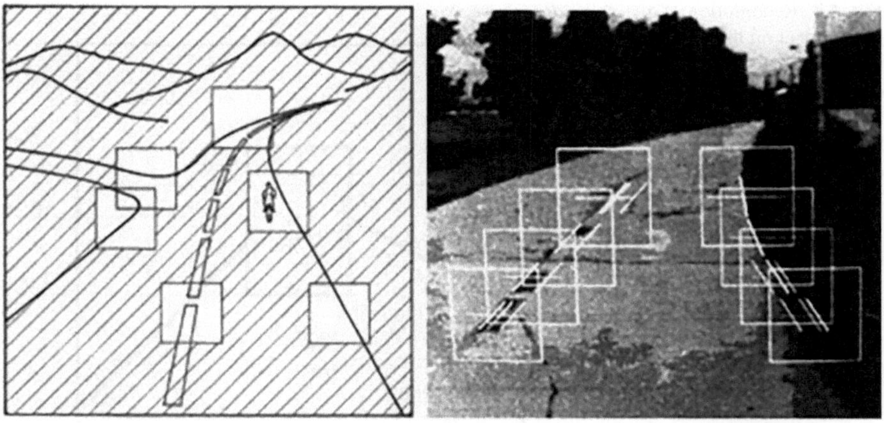

Fig. 4.7 The most important information of an image can be found in a few windows. (Credits & Image Sources: Graefe Kuhnert, p. 16, Dickmanns 2007, p. 225)

The next image (Fig. 4.8) shows in more detail how these edge features were used for object recognition on a road in front of the vehicle.

The figure shows a pair of features on horizontal and vertical search paths centering each other over time. They yield the hypothesis of an object if the average gray values on the inner sides (hashed) are within a small bandwidth of intensity. In order to determine the position of the object in the road environment, the inclined left (yBRl) and right (yBRr) boundaries of the road are determined by a pair of inclined edge features in the image line where the object touches the ground (zBOu). Assuming a flat surface and knowing the position and angles of the camera, the image line determines the distance to the object by inversion of perspective projection. Thus, edge feature extraction allows perceiving the essential elements of road scenes. More details on the evolution of this approach over two decades may be found in a book of Dickmanns.[70]

After establishing the window concept and the use of edge features in 1978, UniBwM started in 1979 laboratory experiments on the visual guidance of road vehicles with a 3-axis-motion-simulator (*Dreiachsen Bewegungs-Simulator DBS*). Developed in the field of aero-space engineering, these simulators – here a CARCO 452 R – (Fig. 4.9) are normally used to train human pilots on the ground how to fly airplanes using visual feedback.[71] Taking advantage of this technology available, the ISF used the simulator to design a simulation loop for road vehicle guidance by machine vision and for systems capable to fly aircrafts.[72] It was also called

[70] See Sect. 5.2 in: Dickmanns, E.D., 2007. *Dynamic Vision for Perception and Control of Motion*, Heidelberg.

[71] Carco Electronics in Menlo Park, California, was founded in 1961 to provide simulation facilities to the Aerospace and Defense industries. In 2005, it was acquired by ACUTRONIC.

[72] In France, a similar simulation facility was used by IRISA Rennes for indoor autonomous vehicle simulation Rives, P., and L. Marce. 1985, October. Use of Moving Vision Sensors to Control

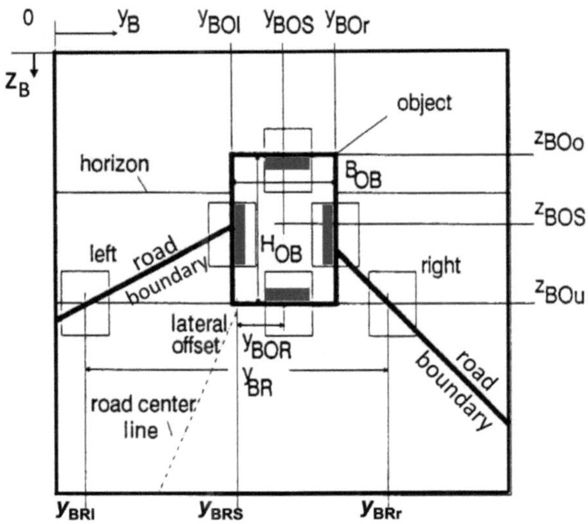

Fig. 4.8 Recognition of a large object on the road ahead by sets of special edge features with adjacent average gray values that help assigning features to objects. (Credits & Image Source: Dickmanns)

Hardware-in-the-loop (HIL)-simulator because it included real hardware – sensors, cameras and computers – for the simulated test drives.

This experimental dispositive used media techniques – camera, screen, projector – and reflections on the physiological function of the human visual apparatus for the automation of the driving task. Using a mechanical servo system of high dynamics, the machine simulated the motions of instruments and cameras in three rotational degrees of freedom (pitch, yaw, roll) and synchronized these movements with a computer-generated visual display of the outside world.

A block diagram (Fig. 4.10) shows the different elements of the simulator. The calculations for the production of perspective images were made by a special computer.[73] The result was then sent to a calligraphic display system[74] that produced a computer generated line image of the road environment outside. This synthetic graphical image was then displayed onto a cylindrical screen in front of the simulator where it was captured by two CCD-TV-cameras – wide angle and tele – in the center of the simulator.

Like a real car, the simulator adapted its movements to the changing image of the road that was given to him as visual input. Installed on a two-axis gaze platform in the center of the simulator, the cameras followed the features on the screen, realizing movements similar to the human eye. The images captured by the cameras were

Robots in an Unknown Universe. In *Proceedings of the 5th International Conference on Robot Vision and Sensory Controls*, 165–176. Amsterdam.

[73] PDP 11 / 60 (later VAX 11/750) from DEC. All information on the evolution of the computer equipment was taken from an image of the simulation loop – dated August 29, 1983 – that was added to the pdf-version of the following research paper (Dickmanns, Ernst Dieter, 1980. Untersuchungen und mögliche Arbeitsschritte zum Thema “Künstliche Intelligenz: Rechnersehen und -steuerung dynamischer Systeme”, p. 20).

[74] Picture System 2 from Evans & Sutherland.

Fig. 4.9 The CARCO 3-axis motion simulator with two cameras (center) and a video projector (above). (Credits & Image Source: Ernst Dieter Dickmanns)

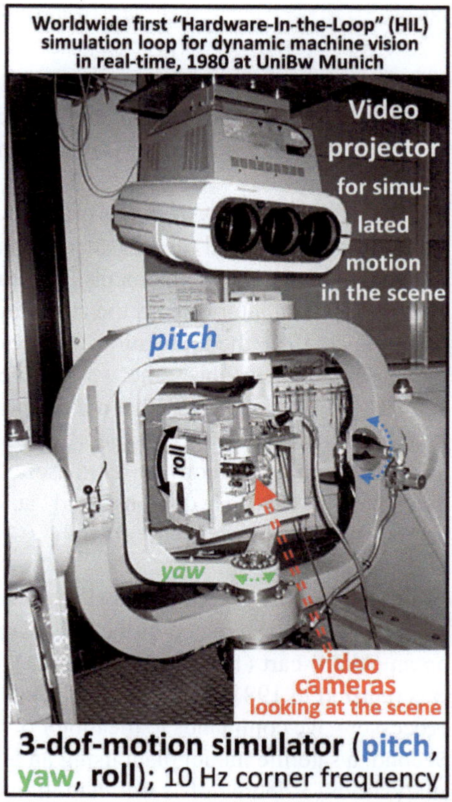

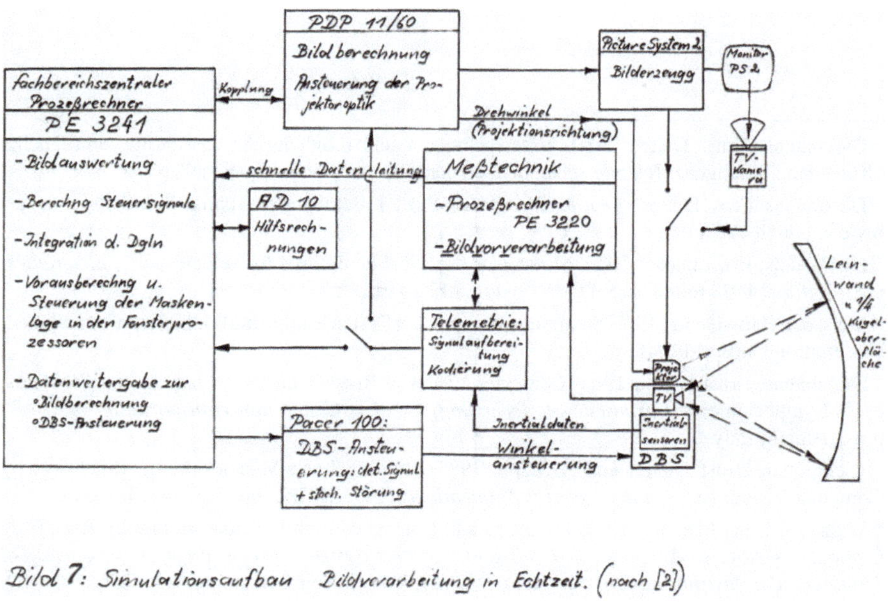

Fig. 4.10 Block diagram of the simulator. (Credits: Dickmanns. Image Source: Dickmanns, Ernst Dieter, Untersuchungen und mögliche Arbeitsschritte zum Thema "Künstliche Intelligenz: Rechnersehen und -steuerung dynamischer Systeme", 1980, p. 18)

then analyzed with the image preprocessing system BVV. Other computers interpreted the images and steered the simulator.[75] This information was sent to the graphics computer to project the current scene onto the screen and the loop was closed.

This experimental system was crucial for UniBwM because it allowed "to investigate the effects of real sensors under actual motion conditions in a laboratory environment", emphasizes Dickmanns.[76] As mechanical movements can disturb image sensors, the simulator offered the advantage that these disturbances could be simulated. Since 1981 the simulation loop was used to test the sensors and computer systems in "exactly the same way as they would be used later on in a real test vehicle".[77] For a laboratory that did not possess a test vehicle at this time, the simulator filled a crucial gap. Hans-Georg Meissner, the first PhD-student of the Institute, used the simulation loop for his dissertation about visual road vehicle guidance, published in 1982.[78] He demonstrated "automatic lane keeping and speed control on curved roads at speeds up to 60 km/h".[79] The results from the simulation loop were published in 1985.[80]

During this period, the first real-world research vehicles – of purely experimental nature – were also created at the ISF: First, an experimental rod-balancing system on an electro-cart (1-degree-of-freedom), controlled by real-time computer vision, was realized in 1981. The BVV analyzed the images of the camera and one special processor gave commands to the motor of the electro-cart, stabilizing the pendulum. Second, a satellite model plant using an air-cushion vehicle was used since 1983 to study a slow 3-degree-of-freedom motion process. Both experimental systems were based on the detection of edge and corner features in the images.[81]

---

[75] Dickmanns, Ernst Dieter. 1980. Untersuchungen und mögliche Arbeitsschritte zum Thema "Künstliche Intelligenz: Rechnersehen und -steuerung dynamischer Systeme", p. 19.

[76] Dickmanns, Ernst Dieter. 1987. PROMETHEUS, 11 170 Integrated Approaches, State of the art review, March 1987, p. 3.

[77] Dickmanns, Ernst Dieter, 2020. May a pair of 'Eyes' be optimal for vehicles too?, *Electronics*, 9(5), 759, pp. 1–28 (cited here: Draft-Version 6.12.2019, p. 12).

[78] Meissner, Hans-Georg. 1982. Steuerung dynamischer Systeme aufgrund bildhafter Informationen, Dissertation UniBwM/LRT.

[79] Dickmanns, Ernst Dieter. 1986. Computer Vision in Road Vehicles – Chances and Problems. ITCS-Symposion on "*Human Factors Technology for Next-Generation Transportation Vehicles*", p. 16. Amalfi, Italy.

[80] Dickmanns, Ernst Dieter, and A. Zapp. 1985. Guiding Land Vehicles Along Roadways by Computer Vision. *Proceedings Congrès Automatique 1985*, AFCET, pp. 233–244. Toulouse.

[81] Wünsche, Hans-Joachim. 1986. Detection and Control of Mobile Robot Motion by Real-Time Computer Vision. In *Advances in Intelligent Robotics Systems, Proceedings of the Society of Photo-Optical Instrumentation Engineers (SPIE)*, ed. Nelson Marquina, vol. 727, 100–109, Wünsche, Hans-Joachim. 1987. *Bewegungssteuerung durch Rechnersehen*, Dissertation. Universität der Bundeswehr München, Fakultät für Luft- und Raumfahrttechnik, Institut für Systemdynamik und Flugmechanik.

4.2.3 *Automated Lateral Guidance by Vision: VW Opto-Pilot (1982–1986)*

While UniBwM had already started to work on machine vision for vehicle guidance at the end of the 1970s, most of the automotive industry remained reluctant to the idea of a "seeing" automobile. Volkswagen was the first German automotive company that became interested in visual guidance systems. In 1982, a test project on "automatic steering with computer vision" called Opto-Pilot was started, led by Walter Zimdahl, head of the Future Research Department at Volkswagen.[82] In 1986 the results were presented in a conference paper.[83]

To understand the genealogy of this initiative, we have to go back to the 1960s: Zimdahl had already explored vehicle automation in his diploma thesis in 1963. The research topic was suggested to him by Winfried Oppelt from the Institute of Control Engineering at the Technical University in Darmstadt.[84] Beside mechanical and electrical steering systems, he also mentioned optical steering systems, based on colored lines on the ground detected by photo cells.[85] But in 1963, Zimdahl did not give any future to mechanical and optical steering systems.[86] Only electrical steering with guidance cables seemed to be useful for the future.

A short version of his diploma thesis, published in 1965, attracted the interest of Siemens.[87] The electronics company asked the Institute in Darmstadt to give advice with the creation of a guidance-cable project for Continental (see Sect. 4.1.2). On the occasion of a press-presentation Zimdahl together with his colleague Erwin Hartwich visited in 1968 the guidance-cable experiments at the Contidrom in Hanover. They documented their visit with a Super 8 camera, a film that Zimdahl made available to the author of this book.[88]

---

[82] Volkswagen Newsroom, 2018. Automatically bound for the future, The history of autonomous driving at Volkswagen, August 24. online in: https://www.volkswagen-newsroom.com/en/stories/automatically-bound-for-the-future-4118 (accessed March 17, 2020).

[83] Zimdahl, W., I. Rackow, and T. Wilm. 1986. OPTOPILOT – ein Forschungsansatz zur Spurerkennung und Spurführung bei Straßenfahrzeugen. *VDI Berichte Nr.* 162: 49–60

[84] Zimdahl, Walter. 1963. *Die Bewegung des Straßenfahrzeugs auf vorgeschriebener Bahn*, Diploma Thesis, 1963, Institut für Regelungstechnik, Technische Hochschule Darmstadt (supervised by Fritz Frank).

[85] Ibid., p. 75.

[86] "Abschließend ist zu sagen, daß sowohl die mechanischen als auch die optischen Systeme keine Aussicht auf Verwirklichung in der Straßenverkehrspraxis haben, da sie zu störanfällig gegen Verunreinigungen der Fahrbahn (Laub, Schnee und Eis) sind. Am aussichtsreichen, da relativ unempfindlich und störunanfällig, sind die elektrischen Systeme mit einem Leitkabel unter der Mitte der Fahrbahn." (Ibid., p. 77).

[87] Zimdahl, Walter. 1965. Führungsverhalten des vierrädrigen Straßenfahrzeugs bei Regelung des Kurses auf festgelegter Bahn. *Regelungstechnik, Zeitschrift für Steuern, Regeln und Automatisieren* 13 (5): 221–268, online in: https://doi.org/10.1524/auto.1965.13.112.221 (accessed October 1, 2020).

[88] The VHS-film of Zimdahl can be found on the Youtube-channel of the author: https://www.youtube.com/channel/UCMXD25D-_izEXLy21dUIU9Q/videos

In 1969 Zimdahl terminated his dissertation on the guidance cable approach. As in his diploma thesis, he concluded that mechanical and optical systems had no chance to be realized in the future as they could be disturbed by environmental conditions. Only the guidance cable seemed promising to him.[89]

More than one decade later, he changed his mind: Zimdahl remembers that he thought about visual guidance systems for the first time in 1982 when a small electronic camera from Panasonic was presented at the technology fair Hannover Messe.[90] The guidance-cable approach of General Motors[91] was now seen by him as "unsuitable" for private transport because of the expensive adaptation and maintenance of the road infrastructure.[92] Progress in microprocessor and sensor development encouraged Zimdahl to substitute the human eye with a camera, the human brain with microprocessors and the human hand with a steering motor, as he explains in a research report.[93] Two students, Ingo Rackow und Thomas Wilm from the Technical University in Braunschweig were then hired by Volkswagen to equip a vehicle with a camera, microprocessors and a steering servo.[94]

A miniature CCD camera with a wide angle was attached behind the windshield at the height of the rear view mirror, scanning a zone of 4–25 meters in front of the vehicle (Fig. 4.11).[95] Five lines of the TV-signal coming from the camera were digitally evaluated using standard 80186 microprocessors from Intel. The computer detected the "contrasts in brightness between the lane markings and the road surface" and converted them into lateral control signals.[96] An update about the course of the road was received every 80 ms.[97] According to Zimdahl, the very simple configuration of the system only allowed a detection of lanes until 25 m in front of the vehicle and only on roads with low curvatures.[98]

As electromechanical steering systems were not available at this time, an electric motor was installed parallel to the steering column and "linked directly to the steering using gear reduction".[99] Tests were effectuated with a VW Bus and a VW Golf at closed test tracks of the Volkswagen proving ground in Ehra-Lessien 25 km northwest of Wolfsburg that had opened in 1968.

---

[89] Zimdahl, Walter. 1969. *Regelungstechnische Probleme bei dem Leitkabelgeführten Kraftfahrzeug*, Dissertation. Fakultät für Elektrotechnik, Technische Hochschule Darmstadt.

[90] E-Mail from Walter Zimdahl to the author, September 29, 2020.

[91] Zimdahl cites papers of J. B. Bidwell, Head Engineering Mechanics Dept., General Motors, Warren, Michigan (Zimdahl, Rackow, Wilm, p. 60).

[92] Zimdahl, Rackow, Wilm, p. 51.

[93] Ibid.

[94] Hartwich, Erwin, and Walter Zimdahl. 2012. Regelung von Straßenfahrzeugen: Ein Zeitzeugenbericht aus dem Institut von Prof. W. Oppelt, TH Darmstadt. *at- automatisierungstechnik* 60 (6), p. 340.

[95] Zimdahl et al. 1986, p. 53.

[96] Volkswagen Newsroom 2018.

[97] Zimdahl et al. 1986, p. 54.

[98] Ibid., p. 60.

[99] Volkswagen Newsroom 2018.

Fig. 4.11 Zimdahl testing the automatic steering system in a Golf MK 2. (Credits: Volkswagen AG. Image Source: Volkswagen Newsroom, Automatically bound for the future, The history of autonomous driving at Volkswagen, August 24, 2018, online in: https://www.volkswagen-newsroom.com/en/stories/automatically-bound-for-the-future-4118 [accessed March 17, 2020])

The Opto-Pilot allowed semi-automated driving only: While a human driver controlled the longitudinal movements of the car, the camera system controlled the lateral movements. According to Zimdahl, the automatic lateral control had been tested for the first time in 1985 "at speeds up to 100 km/h.[100] In 1986, the system was presented at the VDI conference "Electronics in motor vehicles".[101] The research of UniBwM that was effectuated at the same time was unknown to him until the Prometheus-project.[102]

4.3 Concepts, Vehicles and Milestones of University of Bundeswehr Munich

After this presentation of the Opto-Pilot we will come back to the Institute for Systems Dynamic and Flight Mechanics (ISF) of University of Bundeswehr Munich (UniBwM) where most of the pioneering work in the field of machine vision for autonomous driving in Germany had been done.

With regard to the developments that had been presented in the US-chapter it should be mentioned that the research on "fully on-board autonomous vision systems for vehicle guidance with digital microprocessors" started in the 1980s "completely independent" in Germany and the US, as emphasized by Dickmanns.[103] Only in 1984 both communities on each side of the Atlantic learned to know of each

[100] Zimdahl et al. 1986, p. 59 and Hartwich and Zimdahl 2012, p. 340.

[101] Zimdahl et al. 1986, p. 49–60.

[102] Telephone interview with Zimdahl by the author, October 8, 2020.

[103] Dickmanns, Ernst Dieter. 2002. The Development of Machine Vision for Road Vehicles in the Last Decade. *Intelligent Vehicle Symposium* 1, p. 268.

other at a NATO Advanced Research Workshop on Vision and Image Understanding in Erice, Sicily.[104]

4.3.1 The 4-D Approach to Real-Time Vision

As we have seen in the US-chapter, the field of image sequence interpretation was dominated since its start in the 1960s by concepts of Artificial Intelligence and Computer Science. This scientific field had the aim to retrieve the three-dimensional structure of the world from images.[105] Since the early 1970s it was given the name Computer Vision.[106] In contrast to the activities in this domain, Dickmanns coined the term *dynamic machine vision* for his approach. The next paragraph will highlight the differences between both concepts.

The AI/CS-approach – sketched in the upper part of Fig. 4.12 – focused in a first step on locations of features in a sequence of single images. As in each single image the range – direct spatial depth – for each pixel-point is lost, several images of a sequence had to be compared for recovering depth, applying inverse perspective projection. By doing this, spatial objects and their motion with time in the real world were recognized. But the reconstruction of complex spatial scenes from individually analyzed images – relying on a complete capture of the environment – required high storage capacities and extended the time needed for analysis.

When Dickmanns began to explore the field of machine vision with control engineering methods in 1977, he set himself apart from these approaches in computer science and artificial intelligence.

He recognized that the objects of the outside world do not only exist in the three spatial dimensions – length, width and height – that compose the edges of a 3D geometric shape. The outside world is four-dimensional – it includes not only space but also time. Dickmanns concluded that the internal computer model should also be four-dimensional. This is the basic idea of the 4-D approach for image sequence analysis, developed and tested during 10 years at UniBwM since 1983.[107] The

[104] Dickmanns, Ernst Dieter, 2015. Contributions to Visual Autonomous Driving, Part I, online in: http://dyna-vision.de/main/Text_PDF/2015%20ContribPart%20I%20to%20Visual%20Auton.%20Driving%20EDD.pdf (accessed October 6, 2020).

[105] Szeliski, Richard. 2010, Computer Vision: Algorithms and Applications, September 3, draft, online in: https://www.cs.ccu.edu.tw/~damon/tmp/SzeliskiBook_20100903_draft.pdf, p. 11 (accessed February 2, 2021).

[106] Rosenfeld 1998, Szeliski, p. 11.

[107] Dickmanns, Ernst Dieter. 1988. An Integrated Approach to Feature Based Dynamic Vision. *Proceedings of the Computer Society Conference on Computer Vision and Pattern Recognition*, p. 824. Ann Arbor.

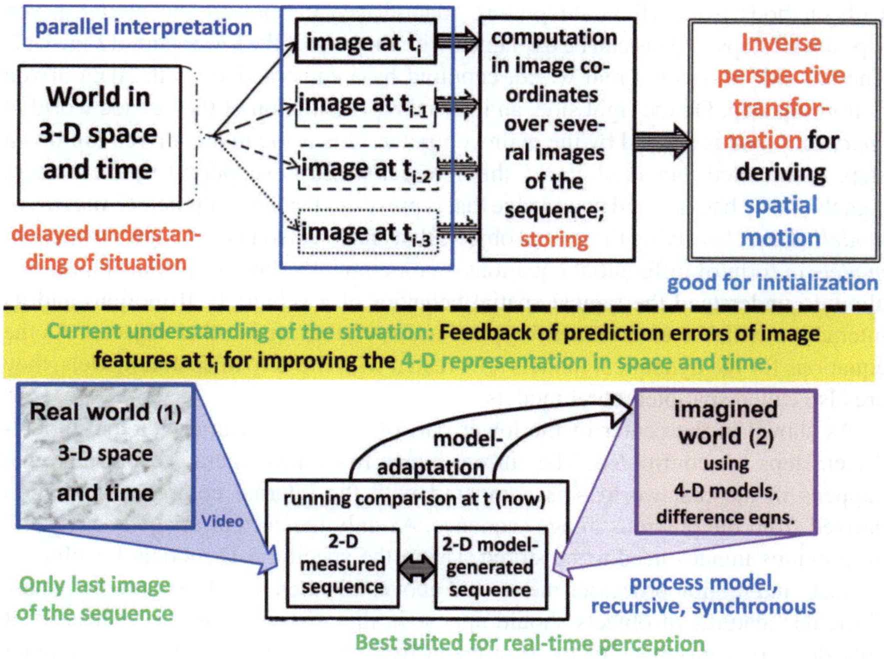

Fig. 4.12 Two different ways of image sequence interpretation: On the top the CS/AI-approach for 3-D scene understanding, on the bottom the UniBwM-approach for 4-D scene understanding. (Credits & Image Source: Dickmanns)

method had been dubbed 4-D approach[108] in contrast to the 2.5-D approach of David Marr.[109]

Instead of inferring from the *two* dimensions of the image plane to three-dimensional objects in 3D *space* – as done in the AI/CS approach –, Dickmanns proposed to take three dimensions (2D of the image plane plus time) simultaneously into account to identify moving 3D-objects in *space and time*. "Motion processes of objects are modeled instead of objects" only, as it is the case in the computational vision approach.[110]

To integrate time in the visual perception process, Dickmanns decided to combine "expectation-based methods from systems dynamics and control engineering

[108] Dickmanns, Ernst Dieter. 1988. 4-D-Dynamic Scene Analysis with Integral Spatio-Temporal Models. In *Proceedings of the Fourth International Symposium on Robotics Research, Santa Cruz 1987*, ed. Robert C. Bolles and Bernard Roth, 311–318. Cambridge: Robotics Research/MIT Press.

[109] Marr, David. 1982. *Vision: A Computational Investigation into the Human Representation and Processing of Visual Information*. San Francisco.

[110] Dickmanns, Ernst Dieter, and Hans-Joachim Wünsche. 2009. Advanced Sensing Techniques for Automated System Applications: Perception based on Dynamic Vision. In *Automation Technologies and Application Considerations for Highly Integrated Mission Systems, Technical Report TR-SCI-118*, ed. J. Ramage et al. North Atlantic Treaty Organization, p. 5.

with methods from AI",[111] integrating "data-driven bottom-up and model driven top-down" steps.[112] This can be explained with the help of the lower part of Fig. 4.12. The left side shows the real world, captured by a camera. This is the data-driven bottom-up step. On the right side, an internal representation of the outside world *in space and time* is created by the main computer. This is the model driven top-down step. Also called *imagined world*, this internal world is supported by the camera signals and by background knowledge that is pre-stored in the computer: *Geometrical models* allow classifying the spatial objects that are detected in the image. *Dynamical models* in form of differential equations – containing the laws of physical motion – allow to understand the typical spatial behavior of a vehicle (self-motion) and its interaction with the environment (typical motion behavior of other objects). As the equations for the dynamical models are created with the help of spatial models, they are also called spatiotemporal models.

As shown in the center of the lower part of Fig. 4.12, data-driven and model-driven steps are confronted: The current image from the camera – capturing what happens in the real world – is compared with the internal computer prediction derived from the previous image sequence. As only the current image is analyzed, no previous images need to be stored. Using the specific 4-D models for objects, subjects and motion processes mentioned above, it is then possible to predict where particular features of objects should appear in the next video frame.[113] Instead of only delivering an image of the past taken from the external world, the computer generates an internal spatiotemporal model for the near future. The result: It cannot "see" the future, but it can estimate its probability. For example, if the perception process detects a car overtaking the vehicle from the left rear, the world model together with the laws of perspective projection predict where stable features of this car will be found sideways in the next image. Based on the prediction step the positions of the rectangular windows for image evaluation can be chosen correctly, depending on previous results of image sequence analysis.

This feedback of prediction errors of the previous image sequence was realized with a Kalman Filter, a mathematical method published in 1960 by the Hungarian-American mathematician Rudolph E. Kálmán (1930–2016). It is today considered as one of the "great discoveries of *mathematical engineering*, which uses mathematical modeling to solve engineering problems."[114] The Kalman filter had even

[111] Dickmanns, Ernst Dieter. 1997. Vehicles Capable of Dynamic Vision. *IJCAI'97 Proceedings of the Fifteenth International Joint Conference on Artificial Intelligence*, Vol 2, p. 1577.

[112] Dickmanns, Ernst Dieter, R. Behringer, D. Dickmanns, T. Hildebrandt, M. Maurer, F. Thomanek, and J. Schiehlen. n.d. The seeing passenger car VaMoRs-P. *Proceedings of the Intelligent Vehicles '94 Symposium*, p. 68.

[113] Dickmanns, Ernst Dieter. 2017. Developing the Sense of Vision for Autonomous Road Vehicles at UniBwM. *IEEE Computer 50* (12): 27.

[114] Grewal, Mohinder S., and Angus P. Andrews. 2015. *Kalman Filtering: Theory and Practice with MATLAB*, p. 2. New Jersey.

been qualified as "the greatest achievement in estimation theory of the twentieth century"[115] and laid the foundation of modern control theory.[116]

As the Kalman Filter is crucial for the understanding of the approach of Dickmanns, a short presentation of this method and its historical background will be given here. The term filter is normally used to describe a physical device that separates unwanted components of liquid-solid mixtures. But it is also used in the electronics domain; analog circuits in transistor radios for instance filter electronic signals from mixtures of different frequencies.[117] In the 1930s and 1940s this concept "was extended to the separation of "signals" from "noise" – unwanted disturbances of electrical signals.

And that is exactly what the Kalman filter is able to do: It allows to reduce – to filter – noise in data sequences. The Kalman Filter is an algorithm – a sequence of instructions – that allows to use the incomplete information of inaccurate measurements to *estimate* the state of complex dynamic systems like spacecrafts, aircrafts, ships, mobile robots and autonomous vehicles. The term *state* means in this context a set of quantities, as explained by Timothy Barfoot: The state of a vehicle for example can be described by its position, orientation and velocity. If these quantities are known, the motion of that vehicle over time can be described.[118]

The Kalman Filter works in two stages: prediction and correction.[119] The *prediction step* – based e.g. on wheel odometry – indicates where the vehicle might probably be at the next point in time. But as odometry is prone to errors and inaccuracy, the *correction step* will take observations from other sensors – e.g. camera data – into account. The Kalman Filter then compares predictions and observations with each other to correct the predictions of the vehicles state. It can be resumed that the filter provides a prediction of the *future* system state based on *past* estimations assuming known error statistics. As prediction and correction steps are following each other – updating the previous step while creating the next step – the method is also called *recursive* estimation technique. It can also be called a *probabilistic* state estimation technique as it expresses "ignorance more precisely – using probabilities."[120]

---

[115] Ibid., p. 19.

[116] It should be remembered that the first method aiming at an optimal estimate from noisy data was the *method of least squares*. It is attributed to Carl Friedrich Gauss (1777–1855) who used it 1795 in mathematical astronomy "to minimize the impact of measurement error in the prediction of orbits" (Barfoot, Timothy D., 2017. *State Estimation for Robotics*, Cambridge, p. 2). This method allowed him in 1801 to predict the position of the dwarf planet Ceres after passing behind the sun. Most of today's estimation techniques are related to Gauss' method.

[117] Grewal, Andrews, p. 3.

[118] Barfoot, p. 1.

[119] Cyrill Stachniss, 2020. Kalman Filter – 5 Minutes with Cyrill, uploaded on April 14, online in: https://www.youtube.com/watch?v=o_HW6GnLqvg

[120] Grewal, Andrews, p. 4.

Initially, the method of Kalman had been received with skepticism among his peers.[121] But with the dawn of both the Computer Age and the Space Age, the field of estimation techniques gained importance, especially in the area of space flight. In 1960 Kálmán presented his research to Stanley F. Schmidt of the NASA Ames Research Center in Mountain View. Schmidt recognized the usefulness of the Kalman Filter for the control and trajectory estimation of the Apollo project that had the goal to go to the moon. In what is probably its "first full implementation", the filter was then used by NASA for its Apollo Guidance Computer. Without this method, "the precise and efficient navigation of spacecraft through the solar system could not have been done."[122]

Today, the Kalman filter is used in many domains, especially for the localization, tracking, navigation, trajectory estimation and control of many different types of vehicles, primarily aircrafts. It is also used by radar systems to track targets. In the domain of autonomous driving, it is used to localize a vehicle in the environment, to track the camera position as well as local features in the environment.

With this background knowledge about the Kalman-Filter we can now come back to the scene understanding system based on dynamic machine vision of UniBwM. It should be highlighted that prediction errors are not only used to adjust the features in the image – most importantly they are used to correct the parameters in the imagined 4-D world directly. By exploring the difference between *detected* and *predicted* features the estimation process is continuously improved. The internal spatio-temporal world is continuously updated and adapted according to the observed motion processes in the external world. This means that the features are reduced to intermediate data only. Feature extraction serves the goal to create knowledge about the *state* of the objects that had been detected. It is used to understand the *motion processes* of objects in the real world directly.

Let us now come back to the start of the chapter: We learned that Dickmanns coined the term *dynamic machine vision* for his approach to autonomous driving. He argues that the term *computer vision* is no longer appropriate for the UniBwM-concept as the visual perception process is not only based on input from camera data but from other sensors, e.g. inertial measurement units that detect acceleration and rotation rates of the vehicle.[123] In contrast to Computer Vision-concepts, *Dynamic machine vision* integrates knowledge about the vehicle dynamics in the perception process – all basic movements of the vehicle carrying the cameras are taken into account, including any interference with visual perception. For example, fast angular rates measured directly by inertial sensors lead to smearing of edges in the images of objects in the outside world; it is therefore advantageous to suppress these visual features during these periods. This concept is much more flexible and effi-

[121] Ibid., p. 18.

[122] Ibid., p. 19.

[123] Dickmanns, Ernst Dieter, and Hans-Joachim Wünsche. 2009. Advanced Sensing Techniques for Automated System Applications: Perception based on Dynamic Vision. In *Automation Technologies and Application Considerations for Highly Integrated Mission Systems, Technical Report TR-SCI-118*, ed. J. Ramage, et al. North Atlantic Treaty Organization, p. 2.

cient in perceiving and controlling dynamic motion processes, especially those observed from a moving platform.

The computer only processes camera data and sensor signals. It is the entire vehicle body in its actual state of motion that experiences accelerations and turn rates. It is thus the vehicle at a whole that is able to perceive significant changes in the images and that has to react appropriately with driving maneuvers. For that reason, the vehicle – the machine – *should be recognized as subject of the perception process,* not the computer.

Finally it is important to understand that the 4-D approach does not directly depend on measurement data from previously analyzed images, from precise maps or from the today often used GPS-coordinates. As it needs only vehicle-based sensor sources, it allows driving in completely unknown, unmapped environments. For that reason, this system is of special interest for the military and also for aerospace agencies that develop space rovers for missions on other planets.

4.3.2 The VaMoRs-Van and the First Vision-System (1985–1991)

After a series of positive results in visual road vehicle guidance with the Hardware-In-the-Loop simulation, the ISF acquired in 1984 a 5-ton Mercedes Benz D 508 van (*Kastenwagen*) to perform real-world experiments. Serving as a laboratory on wheels, the test vehicle was dubbed *Versuchsfahrzeug für autonome Mobilität und Rechnersehen* (VaMoRs), or *Test vehicle for autonomous mobility and computer vision.* Over the next two decades it became an important tool for the development of three generations of computer vision systems for autonomous driving.

The purchase of the van was funded by the appointment budget of Dickmanns. The university had to invest about three times the amount of its price to equip the van for tests of autonomous driving.[124] A small automotive shop, the Zinkl-company south of Munich equipped the D 508 with a "stepping motor for activating the throttle",[125] electrohydraulic brakes, an odometer and a torque motor for steering in the left front wheel bay. Wheel revolutions were used to calculate speed and distance covered by the van. As safety feature, automatic control could be switched off when the driver stepped on one of the pedals or manually hit a red button on the dashboard. An additional 220 V-electric power generator and a 19″ industry rack for sensors and parallel processors were also integrated.

Two computers were used in the vehicle: An IBM IC received the measurement data in real-time, controlled steering wheel, brakes and throttle and was used as

[124] Lossau, Norbert. 2017. Das erste autonome Auto kostete 200.000 D-Mark. *Die Welt,* October 13.

[125] Dickmanns, Ernst Dieter, and A. Zapp. 1987. Autonomous high speed road vehicle guidance by computer vision. In *10th International Federation of Automatic Control (IFAC) World Congress,* Munich, July 1987, Preprints, p. 4.

human machine interface.[126] The second unit was a more powerful version of the image processing system for dynamic vision BVV 2. Developed by Volker Graefe of the *Institut für Messtechnik* (IfM / LRT), it used a dozen 16-bit Intel 8086 microprocessors for edge feature extraction with adjacent average grey values in a receptive field and was used with VaMoRs from 1986–1992.[127] The lower left part of Fig. 4.13 shows the BVV2 and a monitor for presenting the results of feature extraction in the back part of VaMoRs where a systems engineer observed the perception process.

The most important element of the van was the vision system: It had to recognize small elements like letters or numbers on traffic signs at reasonable distances (e.g. up to 20 m in range). On the other hand, medium-sized objects like lane markings (at least 12 cm wide) should be recognizable at distances up to about 200 m. Since planar roads of constant width yield cone-like images, road tracking at larger look-ahead distances needed the capability of gaze control.

To meet these requirements, a bifocal camera arrangement with gaze control had been selected for VaMoRs in 1986. The vision system employed two monochrome CCD-TV-cameras (Sony XC 77 CE). As can be seen in Fig. 4.13, video cameras were still rather large at that time. Different focal lengths allowed to simulate the vision capabilities of a biological eye: As in the simulation loop, the bifocal system combined a short focal lens giving a wide field of view (up to ~20 m), with a telescopic lens for a higher resolution to detect obstacles at greater distance.[128]

The cameras have been mounted on a two-axis gaze control platform on the dashboard. This platform allowed turning the cameras in two rotational degrees of freedom – in yaw around a vertical axis, and in pitch around a horizontal axis. The gaze control-platform gave the possibility to focus interesting regions seen in the low-resolution image with the higher resolution of the other camera. Most of the angular perturbations coming from the vehicle driving on a bumpy surface were eliminated using inertial rate sensors on the platform and feeding their signals back to the angle control with a negative sign.

It had been noted that the high performance of the vision system was due to its "system architecture, rather than to the exceptional speed" of its hardware.[129] As already explained in Sect. 4.2.2, it focused all the power of the parallel processors on a few relevant sections of the images – called windows. Only with the capability

---

[126]The IBM IC was the industrial version of the IBM-AT02 with an INTEL 80286 16-bit Microprocessor.

[127]Dickmanns, Ernst Dieter. 2016. Summary of contributions to Vision-based Autonomous vehicles, in: http://www.dyna-vision.de/main/Review2016%20of%20ContribVisBasAutVeh.html

[128]Mysliwetz, Birger. 1990. *Parallelrechner-basierte Bildfolgen-Interpretation zur autonomen Fahrzeugsteuerung, Dissertation, Universität der Bundeswehr München, Fakultät für Luft- und Raumfahrttechnik.* Institut für Systemdynamik und Flugmechanik, p. 2.3.

[129]Graefe, Volker, and Klaus-Dieter Kuhnert. 1992. Vision-Based Autonomous Road Vehicles. In *Vision-Based Vehicle Guidance*, ed. Ichiro Masaki, p. 13. New York.

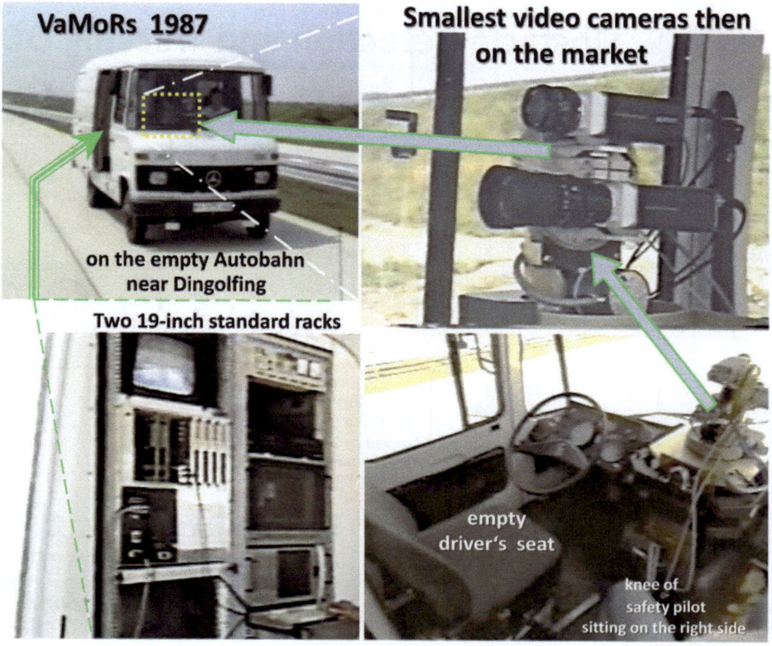

Fig. 4.13 The vision system of the VaMoRs-van with bifocal gaze controllable cameras, here in use during test runs on the highway in 1987. (Credits & Image Source: Dickmanns)

to process image sequences at a high rate (at least 10 images per second), VaMoRs was able to run autonomously at its maximum speed on the highway.[130]

The software architecture of the system is shown in Fig. 4.14. The right side shows the standard lens camera. Its images were used by the BVV to extract edge features. The overall real-time vision system was then able to detect and track the *road and lanes* in front of the VaMoRs van. The same principle was used for obstacle detection: On the left side, a camera with a tele-lens for *obstacle* detection is shown. It focused at vertical features in a horizontal search and horizontal features in a vertical search. When both sets of features encapsulated a rectangle (see lower left part of Figs. 4.8 and 4.14), an obstacle was detected. Vertically in the middle level, the parallel processors (PP) for edge feature extraction (16-bit Intel 8086 microprocessors) are shown, leading to the interpretation step at the GPP-level. Both "eyes" contributed to vehicle guidance, shown at the highest level.

After having explained the vision system and the software architecture, it should be mentioned that autonomous driving had been developed at UniBwM with a special environment in mind. In 1980, Dickmanns described a special use case for the systems his laboratory developed: A "highway autopilot" could be considered as a

[130] Zapp Alfred. 1988. *Automatische Straßenfahrzeugführung durch Rechnersehen*, Dissertation. Universität der Bundeswehr München, Fakultät für Luft- und Raumfahrttechnik, Institut für Systemdynamik und Flugmechanik.

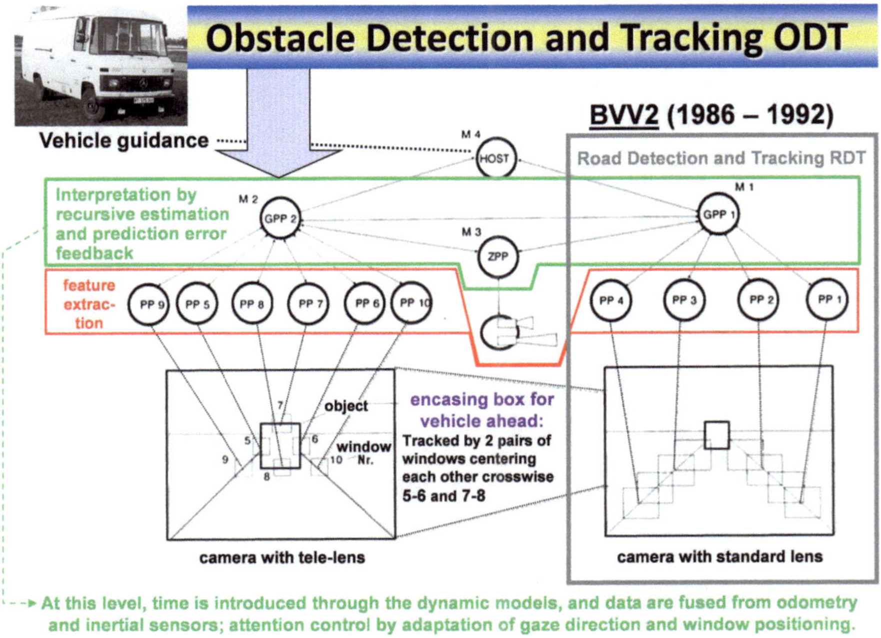

Fig. 4.14 System architecture of the first-generation bifocal perception system (1988) based on the 4-D approach. (Credits & Image Source: Dickmanns)

long-term application goal, relieving the driver „of the tiring control activity under the relatively simple highway conditions".[131] The idea had already come to his mind in 1977 when he had to take the highway each morning to drive to the university in Neubiberg.[132]

Dickmanns emphasizes that the development of autonomous guidance functions should start with the "most structured limited access highway (motorway, Autobahn) since it puts least demand on image processing and understanding."[133] First, it is a highly normed environment with static lane widths and limits to maximum curvatures. Second, it has the advantage that it is restricted to high speed vehicles only – no animals, pedestrians or bicycles will appear on the road. Third, without

[131] "Als langfristiges Anwendungsziel könnte ein "Autobahn-Autopilot" betrachtet werden, der dem Fahrer die ermüdende Steuerungsaktivität unter den relativ einfachen Autobahnbedingungen abnimmt" (Dickmanns, Ernst Dieter. 1980. Untersuchungen und mögliche Arbeitsschritte zum Thema "Künstliche Intelligenz: Rechnersehen und -steuerung dynamischer Systeme", p. 13 [Translation: F.K.]).

[132] Alte Schule - die goldene Ära des Automobils (Youtube-Account), Alte Schule, Folge 95 mit Ernst Dieter Dickmanns (der Podcast) (Title), Uploaded 13.05.2021. https://www.youtube.com/watch?v=YBa0TwJAA2s, min. 28:17 (accessed January 14, 2022).

[133] Dickmanns, Ernst Dieter. 1986. Computer Vision in Road Vehicles – Chances and Problems. *ITCS-Symposion on "Human Factors Technology for Next-Generation Transportation Vehicles"*, p. 2. Amalfi, Italy.

intersections, it is also free from incoming traffic and traffic lights.[134] Due to their reduced complexity, highways seemed to be ideally suited environments to demonstrate experimental machine vision systems.

For these reasons UniBwM decided in 1985 to select freeway driving on the *Autobahn* as "first task domain" for autonomous vehicle guidance. To prepare the first test drive, the D 508 was manually driven on the Autobahn München Garmisch in October 1985 to take video images of the highway. These images of a real road were then used for a real-time analysis in the simulation loop that was presented above. According to Dickmanns and Graefe, this worked without any adjustment and the team started to prepare the van for its first autonomous outdoor-run.[135] The first test drives with this real-time image analysis system for full autonomous driving were then realized in May/June 1986. In the same year, results in visual guidance of road vehicles with the simulation loop were published.[136]

4.3.3 VaMoRs Fully Autonomous High Speed Tests on the Autobahn *(1987)*

Parallel to the *Autonomous Mobile Systems*-project – described in the next section –, UniBwM made under its own responsibility one of its most important test runs: VaMoRs demonstrated in August 1987 the world's first fully autonomous longitudinal and lateral vehicle guidance by computer vision at high speed. The test drive took place at a 24-kilometer-long section of the new Autobahn A 92 between Landshut/Altheim and Dingolfing, not yet opened for public traffic.[137] It was the director of the *Autobahndirektion-Süd* who proposed this stretch of highway to UniBwM after having read about the research in media reports.[138] Using autonomous visual driving, the 5-ton van reached a maximum speed of 96 km/h (60 mph), only limited by its engine power.[139] 20 kilometers were driven without manual inter-

[134] Dickmanns, Ernst Dieter. 1997. Vehicles Capable of Dynamic Vision. *IJCAI'97 Proceedings of the Fifteenth International Joint Conference on Artificial Intelligence*, Vol 2, p. 1579.

[135] Dickmanns, Ernst Dieter, and Volker Graefe. 1988. Sehendes Auto fährt Weltrekord. *Hochschulkurier* 14, p. 18.

[136] Dickmanns, Ernst Dieter, and A. Zapp. 1986. A Curvature-Based Scheme for Improving Road Vehicle Guidance by Computer Vision. In *Mobile Robots, SPIE Proceedings*, vol. 727, 161–168. Cambridge, MA, Dickmanns, Ernst Dieter. 1986. Computer Vision in Road Vehicles – Chances and Problems. *ITCS-Symposion on "Human Factors Technology for Next-Generation Transportation Vehicles"*, Amalfi, Italy.

[137] A video showing VaMoRs on the Autobahn can be found online in: https://dyna-vision.de/wp-content/uploads/2021/03/1987-VaMoRs-AB-Dingolfing.mp4 (accessed January 30, 2022).

[138] Industrial AI Podcast (Youtube-Account), KI und der Erfinder des autonomen Fahrens – KI in der Industrie (Title), Uploaded19.12.2019, in: https://www.youtube.com/watch?v=UYDtI6njaNM min. 26:05 (accessed May 14, 2020).

[139] Dickmanns, Ernst Dieter, and A. Zapp. 1987. Autonomous high speed road vehicle guidance by computer vision. *Proceedings of the International Federation of Automatic Control (IFAC)* 20 (5,

ruption. A safety driver on the driver's seat monitored the operation, but no intervention was necessary.

The test run attracted a lot of media attention. In 1988, a film team from the BBC visited UniBwM and made a short film about VaMoRs at the university campus in Neubiberg and the Autobahn near Dingolfing. The TV-presenter Peter Macann took a ride in the van and explained the 4-D approach to the audience. The clip was then shown in the television series Tomorrow's World.

Especially if compared to international competitors the test-run was an important milestone: For the first time in history, the road was tracked visually – using edge detection and recursive estimation – at higher speeds on the Autobahn. Considering the international research context, the result was about an order of magnitude better performance than demonstrated by any other vehicle around the globe guided by machine vision.

As we have learned in the first part of the book, the test vehicles that were developed at the same time in the United States used much more processor power but reached only walking speed: The Autonomous Land Vehicle (ALV) of Martin Marietta made in August 1987 the first cross-country traverse with a speed of 3 km/h (see Sect. 2.4). The NavLab 1 of Carnegie Mellon University advanced in 1987 with 2.7 mph (4,3 km/h) following a preprogrammed path (see Sect. 2.3.3). Therefore it seems to be accurate to consider VaMoRs at this moment in time as "the world's fastest autonomous road vehicle".[140]

As explained above, the more powerful computer capacity in the US was not effectively used because the machine vision and vehicle control of all these vehicles exclusively analyzed single, whole, static images.[141] Therefore their image extraction rate was very slow (ALV: 2 seconds, Navlab: 20 seconds per image) and the resulting speeds were limited.[142] In contrast, real-time image analysis with the 4D-worldmodel of the ISF only needed 17–33 milliseconds to adapt the internal model of the situation and to calculate steering commands.[143] These results and those of Hans-Joachim Wünsche[144] led in 1987 to the first publication of the new

Part 4); Zapp Alfred. 1988. *Automatische Straßenfahrzeugführung durch Rechnersehen*, Dissertation. Universität der Bundeswehr München, Fakultät für Luft- und Raumfahrttechnik, Institut für Systemdynamik und Flugmechanik.

[140] Graefe, Kuhnert 1992, p. 4.

[141] Dickmanns, Ernst Dieter, V. Graefe, and W. Niegel. Abschlußbericht, Definitionsphase PROMETHEUS, Pro-Art, 1.10.86 bis 14.11.87, Universität der Bundeswehr München, p. 4.

[142] Graefe, Kuhnert 1992, p. 3.

[143] Dickmanns, Ernst Dieter, V. Graefe, and W. Niegel. Abschlußbericht, Definitionsphase PROMETHEUS, Pro-Art, 1.10.86 bis 14.11.87, Universität der Bundeswehr München, p. 2 and Dickmanns, Ernst Dieter, and Volker Graefe. 1988. Sehendes Auto fährt Weltrekord. *Hochschulkurier* 14, p. 21.

[144] Wünsche, Hans-Joachim. 1986. Detection and Control of Mobile Robot Motion by Real-Time Computer Vision. In *Advances in Intelligent Robotics Systems, Proceedings of the Society of Photo-Optical Instrumentation Engineers (SPIE)*, ed. Nelson Marquina, vol. 727, 100–109, Wünsche, Hans-Joachim. 1987. *Bewegungssteuerung durch Rechnersehen*, Dissertation.

image analysis method, called "4-D approach" to real-time computer vision exploiting spatiotemporal (dynamical) models.[145]

4.3.4 The Autonomous Mobile Systems-Project with Daimler (1986–1988)

The real-time demonstrations in the simulation loop and with VaMoRs at UniBwM influenced some players of the automotive industry to join the efforts in developing machine vision systems for ground vehicles.

Daimler-Benz (DBAG) had started in 1985 to work in the area of automated driving, but the company did not exploit machine vision approaches.[146] For that reason, its research division invited UniBwM in 1986 to participate in a public-private partnership called 'Autonomously Mobile Systems' (*Autonom Mobile Systeme*). It was one of the most important precursors of the European PROMETHEUS-project (see Sect. 4.4).

Funded in the years 1987–1988 by the Federal Ministry of Research and Technology (*Bundesministerium für Forschung und Technologie BMFT*), its aim was to transfer the UniBwM-technology for real-time vision onto a *Spur Bus* of DBAG that was available from previous projects. In order to get the project approved by the director of the DBAG-Research Branch, UniBwM had to demonstrate the capabilities of VaMoRs at the skidpan-test-track of DBAG in Untertürkheim near Stuttgart. Built in 1956, this circular test track with a radius of 50 meters is equipped with seven different road surfaces.

After several preparatory test-drives on the taxiways, the runway and parking spaces of the former Airport in Neubiberg near UniBwM, the requested demonstration of VaMoRs was held in December 1986. Accelerating along a straight line up to 10 m/s (36 km/h), the van entered the spiral curve, followed a line on the ground and adapted its speed to the radius of the curvature such that lateral acceleration was about 1 m/s^2. VaMoRs successfully demonstrated autonomous longitudinal and lateral guidance, based on machine vision.

During the first 1½ years of the AMS-project, all software developments and tests have been performed with VaMoRs. For the final demo at the end of the year 1988, the results had been transferred onto a 10-ton electrical *Spur Bus* (O 205) of Daimler-Benz. It was chosen as test platform because Daimler aimed to demonstrate the new technology with a second vehicle. The bus was used to show automated lateral and

---

Universität der Bundeswehr München, Fakultät für Luft- und Raumfahrttechnik, Institut für Systemdynamik und Flugmechanik.

[145] Dickmanns, Ernst Dieter. 1988. 4-D-Dynamic Scene Analysis with Integral Spatio-Temporal Models. In *Proceedings of the Fourth International Symposium on Robotics Research, Santa Cruz 1987*, ed. Robert C. Bolles and Bernard Roth, 311–318. Cambridge: Robotics Research.

[146] Ulmer, Berthold. 1994. VITA II – Active Collision Avoidance in Real Traffic. In *Proceedings of the Intelligent Vehicles Symposium 1994*, p. 1. Paris.

Fig. 4.15 VaMoRs stopping in front of a garbage bin, Rastatt 1988. (Credits & Image Source: Dickmanns)

longitudinal control at speeds of up to 60 km/h on the test track in Rastatt.[147] In addition, the detection of a stationary obstacle – a 0.5 m^2 garbage bin in a curve – had been developed using edge-based 4-D vision. Stopping in front of it was demonstrated with VaMoRs in autumn 1988 up to a driving speed of 40 km/h (Fig. 4.15). The test-drive was documented on video.[148]

The system first searched for features on a horizontal level. After finding a feature, it switched to a vertical search for horizontal features. During all these tests no radar or laser sensors were used, only the 4-D approach with the BVV2 system for the extraction of edge features.

In the years after the AMS-project, VaMoRs learned to react properly to an "increasing number of obstacles and generically known other vehicles in its environment."[149] These research efforts laid the ground for the development of the second generation vision system with transputers (1992) that will be presented in Sect. 4.4.2.

---

[147] Mysliwetz, Birger. 1990. *Parallelrechner-basierte Bildfolgen-Interpretation zur autonomen Fahrzeugsteuerung, Dissertation, Universität der Bundeswehr München, Fakultät für Luft- und Raumfahrttechnik*. Institut für Systemdynamik und Flugmechanik, p. 1.15.

[148] Video of the VaMoRs-test run in Rastatt 1988, online in: https://dyna-vision.de/wp-content/uploads/2021/03/1988-VaMoRs-Rastatt_ODT-AMS_Demo.mp4 (accessed January 30, 2022).

[149] Dickmanns, Ernst Dieter. 1994. The 4D-Approach to Dynamic Machine Vision. *Proceedings of the 33rd Conference on Decision and Control*, p. 3772.

4.4 The PROMETHEUS-Program of the European Union (1986–1994)

Since the 1960s the field of computer and machine vision had been pushed by the electrical engineering and computer science communities. This changed in the 1980s, when the mechanical engineering and automotive industry began to invest in this field in Europe.[150]

Initiated by Daimler-Benz AG[151] and backed by the R&D departments of the European automotive industry, a big European research project centering on the development of vehicle technology was created in the second half of the 1980s. It was named after Prometheus, one of the most tragic figures of Greek mythology. The name of the antique god became an acronym: PROMETHEUS meant *PROgraMme for a European Traffic with Highest Efficiency and Unprecedented Safety*. The adjectives that were chosen reveal the optimist believe in new mechatronic car technology during this time.

Today, PROMETHEUS is often associated with the quest for autonomous driving. But this research direction, part of the sub-project Pro-ART, was only a small fraction of the overall program. As pointed out by Hermann Winner, the main goal of Prometheus was *not* limited to autonomous driving.[152] Its aim was to develop technologies for the second century of motorized ground vehicles and to address several problems of the car based traffic system: The increasing number of vehicles and the limited capacity of the European road system caused a high number of road accident victims, traffic jams and pollution. As the construction of new roads revealed not to be the answer to these problems, new solutions had to be found.

The main idea of Prometheus was to merge the technologies of two important fields, communication and transport, as Gerhard Prätorius emphasizes in his study on the project.[153] As its name reveals, the project had the goal to enhance the safety and efficiency of the transport system, focusing on new guidance and communication technologies. It was part of the fourth period of vehicle electronics that went beyond the limits of car-based systems and aimed at the development of vehicles that communicate with their environment.[154] Autonomous driving was right from the beginning one small element in this technology based approach.

[150] Dickmanns, Ernst Dieter. 2002. The Development of Machine Vision for Road Vehicles in the Last Decade. *Intelligent Vehicle Symposium* 1, p. 269.

[151] Daimler Benz AG (1926–1998), Daimler-Chrysler AG (1998–2007), since 2007 Daimler AG.

[152] Winner, Hermann. 2015. PROMETHEUS – Welche Visionen wurden Realität? In *VDA, Technischer Kongress*, p. 97.

[153] Prätorius, Gerhard. 1993. *Das PROMETHEUS-Projekt, Technikentstehung als sozialer Prozeß*. Wiesbaden, p. 11, 91.

[154] The first generation from the beginning of the 1960s on focused on the substitution of mechanical systems with electronics. During the second period (from the mid-1970s on) electronics was used to realize new functions that were not possible with mechanical means only. The third generation (mid-1980s) integrated both, mechanical and electronical functions in mechatronic devices, but the technology remained based on the vehicle only (Prätorius, p. 85).

The idea of a common European technology development project was discussed in the automotive industry for the first time at the end of 1985. As we have seen in the preceding chapters, Prometheus was a reaction to similar research projects – the FGCS of MITI and 13 industrial players in Japan and the DARPA-Initiative *On Strategic Computing* in the US.[155] Winner and Prätorius highlight that the participants of PROMETHEUS had the common interest to counter what had been perceived as a "technological menace" from Asia, especially Japan.[156] The European automotive industry saw the decline of the European photo- and audio-industry as a warning signal for its own business.

The strategic goal to reinforce its position in the competition with the US and Japan explains why stakeholders in heavy competition rapidly accepted to cooperate: In February 1986 a Memorandum of Understanding was signed by 14 European car manufacturers, documenting their will to work together. PROMETHEUS then became part of EUREKA, a technology development program of several European governments that had been initiated by France in November 1985. The 11 European countries that participated in the EUREKA-program – Austria, Belgium, Finland, France, Germany, Italy, the Netherlands, Norway, Sweden, Switzerland and the United Kingdom – funded PROMETHEUS with 749 Million US-Dollar. Two third of the funding was provided by the industry and one third from the governments of the countries that participated.[157] Funding was only distributed to national research institutions, but full exchange of information with the partners in other countries was mandatory in the EUREKA-funding system.

PROMETHEUS was started in October, 1986 and went through a preparation phase until December 1987. The main research phase lasted from January 1988 to December 1994. During these seven years, participants from 19 countries, including 19 companies of the automotive industry and 70 research institutes worked together. Initially, only European manufacturers were included in the project but then the protectionist approach was discarded and the European counterparts of Ford and GM were allowed to participate. The participants were Alfa Romeo, BMW, Daimler Benz, Fiat, Ford, Iveco, Jaguar, Lancia, MAN, Matra, Opel, Porsche, PSA, Renault, Rolls Royce, Saab Scania, Steyr-Daimler Puch, Volvo and VW.

PROMETHEUS was separated in academic and industrial research fields with different subprojects: First, *basic research* effectuated by academic research institutes was regrouped in the four areas Pro-ART (knowledge-based systems), Pro-CHIP (basic research on hardware), Pro-COM (communication standards) and Pro-GEN (traffic scenarios for traffic management systems). Second, the automotive *industry* centered their research in three areas: Pro-CAR (assistance systems), Pro-NET (vehicle-to-vehicle communication) and Pro-ROAD (communication between vehicle and infrastructure). These three fields were much more application oriented.

---

[155] Winner, p. 95.

[156] Ibid., p. 93–124, 95. Prätorius, p. 84.

[157] Ibid., p. 95.

Academic and industrial researchers both worked together to realize experimental research vehicles – called Common European Demonstrators (CEDs). Often each partner used its own vehicle to demonstrate a new technology to the automobile industry and the press during the three Board Member Meetings of the project (January 1989 in Munich, September 1991 in Turin and October 1994 in Paris).

As the list of participants above shows, PROMETHEUS was a very big project with many different initiatives and over 300 researchers of the automotive industry. For that reason, an overview over all the research projects would go far beyond the size of this chapter. For our purpose, the "academic" sub-domain Pro-ART is the most interesting. Its research efforts centered on sensors and Artificial Intelligence with the aim to realize a vehicle that is able to orient itself in well-structured environments like highways, using machine vision.

The system envisioned had to function as an intelligent partner for the driver or to execute driving tasks independently from the driver, as coined by Dickmanns in 1987.[158] Funding of this part of the project included less than 10% of the total sum of 749 Million US-Dollars.

The role that had been accorded to autonomous driving went through an important transformation process during the project: In the years 1986/1987 PROMETHEUS was dominated by a purely technical vision, centered on individualized automobile traffic.[159] According to Prätorius, the human factor was seen as a risk factor that should be replaced by technology. During this time, automated driving was one of the most important lead images of PROMETHEUS.[160] Initially, Daimler planned to develop automatic lateral guidance using electromagnetic fields generated by cables buried in the center of the lanes. But Dickmanns claimed that this would be a step back as it would put cars on rails without the possibility to make lane changes. He proposed instead a big European effort in machine vision that would not only allow lateral guidance but also the simultaneous detection of obstacles. Cable guidance was then substituted by the cheaper and more flexible concept of machine vision for vehicle guidance.[161]

But after the start of the research phase of PROMETHEUS in 1988, a network-centered vision became more dominant, including also public and freight transport.[162] According to Prätorius, the lead image of autonomous driving lost

[158] "Als fokussierende Aufgabe für die Methodenentwicklung in der Projektphase wurde die Fähigkeit eines Fahrzeugs definiert, sich auf einer Autobahn (gut strukturierte, genormte Umgebung) vollständig selbständig durch Rechnersehen zu orientieren und entweder den Fahrer zur Erhöhung der Verkehrssicherheit im Sinne eines intelligenten, aufmerksamen Beifahrers zu unterstützen oder Teile der Fahrzeugführungsaufgabe selbständig durchführen zu können." (Dickmanns, Ernst Dieter, V. Graefe, and W. Niegel. Abschlußbericht, Definitionsphase PROMETHEUS, Pro-Art, 1.10.86 bis 14.11.87, Universität der Bundeswehr München, p. 2).

[159] Prätorius, p. 91, 95.

[160] Ibid., p. 87.

[161] Dickmanns, Ernst Dieter. 2002. The Development of Machine Vision for Road Vehicles in the Last Decade. *Intelligent Vehicle Symposium* 1, p. 268.

[162] Prätorius, p. 91, 95.

momentum until 1991 when a project report claimed that "PROMETHEUS does not advocate an auto-pilot which replaces the driver".[163] The management of the automotive industry became skeptical about fully autonomous driving and limited the development of computer vision to assistance functions only. In contrast to 1986, human driver and technical systems were now conceived as partners with equal capabilities.[164] The new lead-image was the informed driver, assisted by an intelligent co-pilot. The emphasis on the technical enhancement of individual motorized traffic was reduced and it became a more social concept as confirmed by members of the steering committee, reports Prätorius.[165] The project changed from a basic science oriented approach to an application or goal oriented approach.

Prätorius identified several reasons for this shift: While technical challenges of automated driving were not seen as main hindrances, automated driving collided with conceptions of driver agency that were dominant in the automotive industry. The missing societal demand, critical discourses about the relation between human and technology, the lack of a juridical framework and liability concerns of the industry were other important factors for a reclassification of automated driving for a more distant future.[166] Changes in the social environment are a decisive element of technical development, argues Prätorius. Depending on the question if comfort, safety or traffic efficiency is set as main goal, an automated system can be designed in very different ways.[167] If social demands change, technology development also has to change.[168]

After this introduction the next sections will present the research vehicles that were developed in a public-private partnership between UniBwM and Daimler. This is followed by a section about the research efforts of PSA in France, Lucas in the UK and Volkswagen in Germany.

4.4.1 The VITA Research Van of Daimler-Benz AG with UniBwM-Technology (1991)

Encouraged by the high-speed demonstration of VaMoRs in Dingolfing 1987, Daimler-Benz established during the PROMETHEUS-project its own research vehicle, using a modified 7,5-ton Mercedes van (Kastenwagen T2) named VITA

[163] PROMETHEUS, The PROMETHEUS Common European Demonstrators, 1991 (cit. in: Prätorius, p. 101).

[164] Prätorius, p. 122.

[165] Ibid., p. 92.

[166] Ibid., p. 94, 123.

[167] Ibid., p. 123.

[168] Ibid., p. 110.

Fig. 4.16 The experimental VITA-I research van of Daimler-Benz, 1991. (Credits: Daimler AG, Image Source: https://media.daimler.com/marsMediaSite/en/instance/picture/Prometheus-VITA-1.xhtml?oid=13794091)

(Vision Information Technology Application) (Fig. 4.16). The primary goal of VITA was "automatic collision prevention".[169]

For this purpose, the van was equipped with the vision technology of UniBwM – a bifocal camera system and a copy of the BVV2 that had already been used in the VaMoRs-van. The complete software development regarding vision for VITA had also been done by UniBwM.[170]

In 1991, the half time-demo of PROMETHEUS was held at the test-track of Fiat in Torino (Italy). The group DBAG/UniBwM demonstrated with VITA lane following and convoy driving down to a full stop: By visual means and up to a speed of 60 km/h, the intelligent cruise control maintained automatically a safe distance to the vehicle in front.[171] This was probably the first time that a vehicle demonstrated "purely vision-based 'Stop&Go'-capabilities" publicly, estimates Dickmanns.[172] As shown in a video of the demo, VITA also stopped behind a parking car, using no

[169] Daimler 2016a, b.

[170] A subproject of VITA was the intelligent cruise control that automatically maintains a safe distance to the vehicle in front. The system was later commercialized by Daimler as Distronic and Distronic plus (Daimler 2016a). These systems use a stereo camera *and* radar.

[171] Video Source: Prometheus, Turin 1991, DB-VITA, online in: https://dyna-vision.de/wp-content/uploads/2021/03/1991-Turin-VisionBasedStop-BehindCar.mp4 (accessed January 30, 2022).

[172] Dickmanns, Ernst Dieter. 2002. The Development of Machine Vision for Road Vehicles in the Last Decade. *Intelligent Vehicle Symposium* 1, p. 273.

radar or laser range finder, just monocular motion stereo based on 4-D vision.[173] It even showed road-running at night, including lane changing and forking into another road.

4.4.2 The Second Generation Vision-System Based on Transputers (1991–1997)

Parallel to the PROMETHEUS-project, UniBwM developed with VaMoRs and in the simulation-loop the second generation of its vision system, used from 1991–1997. The main difference to the first system was more powerful hardware. First, a transputer system with up to 60 processors was used. Each of these processors had four direct links to its neighbors, accelerating the communication between the components. Second, a 32-bit Personal Computer was used.[174] The van was also equipped with a GPS-antenna on the roof and a corresponding receiver: Together with map information, this technology allowed easier detection and estimation of the intersection parameters of cross-roads while approaching it. An electric generator (220 V, 10 kW) with an external cooling unit on the top of the van had also been integrated.[175]

Equipped with the new vision system, VaMoRs was in 1994 able to keep lanes while adapting its speed automatically in curves, to transition from free-lane driving to convoy-driving behind another vehicle, to perform stop-and-go maneuvers in a traffic jam and to change lanes after being triggered by a human operator who had to check if the lane is free.[176]

The second generation vision system hanging from the roof included a new gaze control platform. Compared to the previous camera, the size of the „vehicle eye" was reduced. Its installation under the roof (Fig. 4.17) improved visibility. This allowed VaMoRs to turn-off autonomously into a two-lane crossroad with lane markings, as a video from 1996 shows.[177] These capabilities were first demonstrated in the aforementioned Hardware-In-the-Loop simulation, where the system interacted with a mathematical representation of its future environment.[178] Real-world

---

[173] Video Source: Prometheus, Turin 1991, DB-VITA, online in: https://dyna-vision.de/wp-content/uploads/2021/03/1991-Turin-VisionBasedStop-BehindCar.mp4 (accessed January 30, 2022).

[174] Two years later, this transputer architecture was integrated in a new research vehicle (VaMP) that will be presented in the next paragraph.

[175] Dickmanns, Ernst Dieter., H.2.3 Van VaMoRs, online in: https://dyna-vision.de/hardware-used/plants-controlled-autonomously/van-vamors-1985-2004 (accessed January 30, 2022).

[176] Dickmanns, Ernst Dieter. 1994. The 4D-Approach to Dynamic Machine Vision. *Proceedings of the 33rd Conference on Decision and Control*, p. 3772, see also: Brüdigam, Claus. 1994. Intelligente Fahrmanöver sehender autonomer Fahrzeuge in autobahnähnlicher Umgebung, Dissertation. Universität der Bundeswehr München, Fakultät für Luft- und Raumfahrttechnik.

[177] https://dyna-vision.de/wp-content/uploads/2021/03/1996-turnoff-right-taxiway-Neubibberg.mp4 (accessed January 30, 2022).

[178] A.3.7 Turning off onto an unknown crossroad, online in: https://dyna-vision.de/application-areas-and-tasks/road-vehicle-guidance/turning-off-onto-an-unknown-crossroads (accessed January 30, 2022).

Fig. 4.17 VaMoRs of phase-b) with a new "eye" with 3 cameras under the roof, a GPS- antenna and a powerful transputer system for perception, 1991. (Credits & Image Source: Dickmanns)

experiments were conducted on the test track of the former NATO-airport in Neubiberg and the test site Pfullendorf.

4.4.3 VaMP and VITA-II at the Final Demo of PROMETHEUS in Paris (1994)

According to Dickmanns, the successful demonstrations of UniBwM in Dingolfing 1987 and in Turin 1991 slowly changed the mind of car industry executives. Daimler-Benz asked the Dickmanns-team if they could demonstrate vision based autonomous driving to guests on a three-lane public highway for the final event of the PROMETHEUS-project in Paris 1994. This idea first seemed "absurd" to Dickmanns,[179] but later he estimated that with the power of transputers and up to 20 research employees his team should be capable of reaching this goal.

Daimler and UniBwM then decided to acquire two Mercedes 500 S-class sedans (W140) as new research vehicles. While the basic vehicles had been procured by the research groups, their equipment with the additional hard- and software necessary has been funded by the EUREKA-Prometheus program. Daimler was in charge of installing the hardware components for basic sensing and acting in both vehicles while the ISF of UniBwM developed the architecture and software for cognitive capabilities on the new transputer hardware.

[179] Delcker, Janosch. 2018. The man who invented the self-driving car. *Politico*, July 19, 2018, online in: https://www.politico.eu/article/delf-driving-car-born-1986-ernst-dickmanns-mercedes/

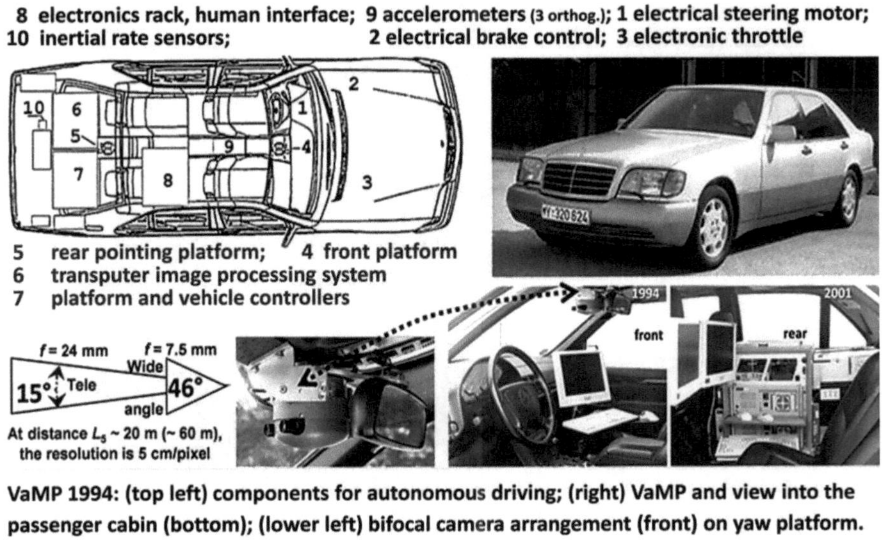

Fig. 4.18 Hardware components for autonomous driving in the test vehicle VaMP, 1994. (Credits & Image Source: Dickmanns)

UniBwM purchased the new research vehicle in 1993 and used it until 2004. It was called VaMoRs-PKW, in short VaMP *(Versuchsfahrzeug für autonome Mobilität PKW*). To prepare the Mercedes 500 for autonomous driving test-runs, it was equipped with a modified, electrically controlled hydraulic brake, a torque motor for the actuation of the steering system and inertial sensors for acceleration and angular rates (see Fig. 4.18).

The new generation of computers that had been tested with VaMoRs – as described in the last section – was now also used in the VaMP-Mercedes. Its trunk was filled with up to 60 transputers – parallel processors with four links to neighbors – 16-bit T-222 for image processing and communication and 32-bit T-800 for number crunching and knowledge processing.[180] A PC was used as human-machine interface. As before, a safety driver took always place on the left front seat and a computer system operator had its work station in the left back seat near a PC to the right (see Fig. 4.18, lower right corner).

But the most important modification concerned the vision system. For the first time "forward and rearward perception of the environment"[181] became possible: Two miniaturized "finger"-CCD-cameras with different focal lengths (8 mm and 24 mm) each looked to the front and to the rear, covering 100 m in both directions.

180 Ibid., Dickmanns, Ernst Dieter. 2002. The Development of Machine Vision for Road Vehicles in the Last Decade. *Intelligent Vehicle Symposium* 1, p. 272.

181 Dickmanns, Ernst Dieter, R. Behringer, D. Dickmanns, T. Hildebrandt, M. Maurer, F. Thomanek, and J. Schiehlen. n.d. The seeing passenger car VaMoRs-P. *Proceedings of the Intelligent Vehicles '94 Symposium*, p. 69.

With the visual detection of three lanes nearby and far away, in front and in the back of the vehicle, up to 12 cars – two in the own lane and two cars in the directly neighboring lanes in both the front and rear hemispheres – could be detected and tracked and their movements could be estimated.[182] If a truck appeared in front of a detected vehicle, it was also detected by the bi-focal system. The tracking range was up to about 100-meter distance.[183]

The second Mercedes 500 – called VITA-II – of the Daimler-Benz AG was equipped with the same forward-backward vision system but in addition to the four finger cameras Daimler used "7 pairs of cameras for stereo vision. One pair looked to the front and three each to the sides".[184] Instead of transputers, it used "different types of microprocessors".

The three-level software structure of the second-generation vision system of ISF-UniBwM is shown in Fig. 4.19. The *feature level* at the bottom includes the recognition of features of objects (road lanes, obstacles and signal lights). 3-D shape recognition and the detection of moving humans were not available due to missing computer power. The center of the image shows the *object-level* where features are merged to hypotheses of objects and their dynamic states (3 D positions and motion); the relative 4-D-states of these objects are thus estimated. Compared to image data, the transition to objects and their 4-D states reduces the amount of data by two to three orders of magnitude. The *situation-level* on top is charged with interpreting the situation regarding several objects together. This includes, beside situation assessment also behavior-, vehicle- and viewing direction control with regard to the mission to be performed. Some information is also given to the human operator (host manager).

In October 1994 the final demo of PROMETHEUS took place on Autoroute 1 near the Airport Charles de Gaulle Paris, France. In total 88 vehicles were shown at the final event, covering all domains of the whole project. Ten vehicles belonged to CED 3 (collision avoidance).[185]

UniBwM participated under cover of the industrial company Daimler with VaMP – also called Common European Demonstrator (CED) 303 – in the demo. The two S-class Mercedes are shown on the left side of Fig. 4.20. Before heading to France, Daimler substituted the Y-license plate of VaMP – specific for vehicles of

[182] Dickmanns, Ernst Dieter. 2002. The Development of Machine Vision for Road Vehicles in the Last Decade. *Intelligent Vehicle Symposium* 1, p. 273, Dickmanns, Ernst Dieter. 1994. The 4D-Approach to Dynamic Machine Vision. *Proceedings of the 33rd Conference on Decision and Control*, p. 3772.

[183] Dickmanns, Ernst Dieter, R. Behringer, D. Dickmanns, T. Hildebrandt, M. Maurer, F. Thomanek, and J. Schiehlen. n.d. The seeing passenger car VaMoRs-P. *Proceedings of the Intelligent Vehicles '94 Symposium*, p. 68.

[184] Dickmanns, Ernst Dieter. 2002. The Development of Machine Vision for Road Vehicles in the Last Decade. *Intelligent Vehicle Symposium* 1, p. 272 (referring to Ulmer, Berthold. 1994. VITA II – Active Collision Avoidance in Real Traffic. In *Proceedings of the Intelligent Vehicles Symposium 1994*, Paris, pp. 1–6, Brauckmann, M.E., C. Goerick, J. Groß, and T. Zielke. 1994. *Towards All Around Automatic Visual Obstacle Sensing for Cars, Proceedings of International Symposion on Intelligent Vehicles*, pp. 79–84. Paris).

[185] Prometheus – Proart (ed.), 1994. Rapport de fin de contrat de recherche, December, p. 177.

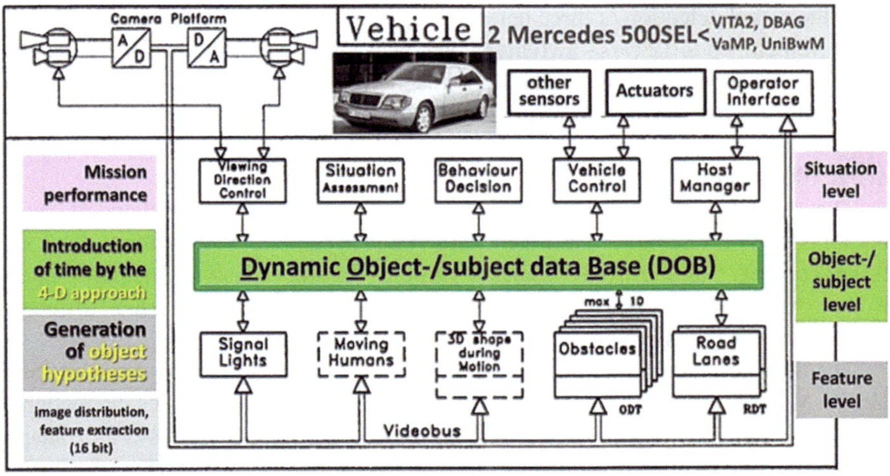

Fig. 4.19 The software structure of the transputer system with three knowledge levels (features, objects/subjects, situations). (Credits & Image Source: Dickmanns)

Fig. 4.20 The three autonomous road vehicles at the PROMETHEUS demonstration in Paris, October 1994. From left to right: UniBwM VaMP (CED 303), Daimler VITA-II (CED 302) and Daimler VITA-I. (Credits: Reinhold Behringer. Image Source: Delcker 2018)

the German Federal Armed Forces – by a license plate of Stuttgart. This was probably decided because the vehicle had to cross the border to France but it also reduced the recognition of UniBwM in the final event. VITA-II (CED 302) of Daimler-Benz AG can be seen in the center of the image. The third vehicle on the photo is the VITA-I-van of Daimler Benz that was only used to transport equipment.

At the International Symposium on Intelligent Vehicles (IV'94), taking place the following week in Paris, the contributions of UniBwM to the vision systems of VITA-2 and VaMP have been demonstrated to an international audience. The capabilities of the two autonomous research vehicles were outstanding at that time. It was the first time worldwide that autonomous driving by machine vision was demonstrated on a public *Autobahn* with three lanes and in dense standard traffic. Equipped with the new machine vision system described above, the vehicles were now able to judge by themselves if it would be safe to change lanes.

During the final demo, the two Mercedes 500 were aware of their position in the lane and recognized road markings and the relative position of other vehicles in the lane. In total they demonstrated about a thousand kilometers of autonomous free-lane driving, autonomous transition into convoy driving and even *autonomous lane changes* at speeds up to 130 km/h in normal three-lane traffic on the Autoroute near the Airport Charles de Gaulle. A human safety driver who remained always in the left front seat – the hands mostly near, not on the steering wheel – only had to validate the decision.[186] Some high-ranking guests had the opportunity to take a ride in both research vehicles.

The performance of VaMP on the Autoroute near Paris has been documented in a video film.[187] The clip shows how the vehicle took the autonomous decision to pass a bus in front. But the system then noticed that the bus driver also decided to do a lane change – VaMP stopped its acceleration and switched to convoy driving in the second lane. It then decided to make a second lane change (Fig. 4.21).

The safety driver set the blinker – giving the car the signal to perform the autonomously proposed passing maneuver. After passing the bus in the left lane, VaMP decided to change back to the medium lane in front of the bus. Again, the driver did set the blinker to allow the vehicle to do so. This capability to decide *autonomously* if a lane change should be made was a significant difference to the systems of the past that were only *automated*.

Despite this success it should also be mentioned that the cameras had some difficulties with shadows. The vision system tracked other cars by mapping horizontal edges in the lower parts of the vehicles as the shadow zone under the vehicles was one of the most stable features. But the reliability of the vehicle recognition system "under various lighting and weather conditions still was not yet satisfactory for

[186] Dickmanns, Ernst Dieter. 1997. Vehicles Capable of Dynamic Vision. *IJCAI'97 Proceedings of the Fifteenth International Joint Conference on Artificial Intelligence*, Vol 2, p. 1589.

[187] 1994 Twofold LaneChange Paris: https://dyna-vision.de/wp-content/uploads/2021/03/1994-Twofold-LaneChange-Paris-VaMP.mp4

Fig. 4.21 VaMP demonstrating an autonomous lane changing maneuver on Autoroute-1 near Paris, 1994. (Credits: Dickmanns. Image Source: https://dyna-vision.de/wp-content/uploads/2021/03/1994-Twofold-LaneChange-Paris-VaMP.mp4)

market introduction" reports Dickmanns.[188] For example, when the light of the deep-standing sun in the evening appeared under big fuel trucks, the area of shadows was no longer homogenous.[189] On the contrary, radar working with reflected electromagnetic radiation actively sent out is not affected by light in the visual range. This was one of the reasons why most of the automotive industry decided to privilege radar instead of vision systems for obstacle detection.[190] Radar is also able to directly measure speed differences by Doppler shift. That is even more important for safety.

4.4.4 Prolab (France), Lucas (UK) and Volkswagen (Germany)

At the end of this chapter it should be mentioned that during PROMETHEUS a multitude of research vehicles were developed by the different partners in each country. They all followed very different goals and methods. We decided to limit our presentation to VaMP and the two VITA vehicles as they represent the biggest progress with regard to the goal of autonomous driving.

---

[188] Dickmanns, Ernst Dieter. 2002. The Development of Machine Vision for Road Vehicles in the Last Decade. *Intelligent Vehicle Symposium* 1, p. 272.

[189] Knop, Carsten. 2018. Ein Deutscher war Tesla 30 Jahre voraus. *Frankfurter Allgemeine Zeitung*, February 1, 2018, online in: https://www.faz.net/aktuell/wirtschaft/autonomes-fahren-ein-deutscher-war-tesla-30-jahre-voraus-15424550.html

[190] In the year 2000, the S-class was the first car that could be ordered with a radar system called Distronic – a system that maintains the distance to the car in front while the driver has to keep the lane.

It would extend the frame of this book too much to give a complete survey on all European activities till the end of Prometheus 1994. But it should be mentioned that most European countries did not fund own research projects on autonomous driving. Their projects were instead mostly centered on assistance systems only, as a short overview on projects in France and the UK shows:

The vehicles that were developed in France as part of the Common European Demonstrator (CED) 3 (anti-collision systems) were focused on the development of an assistance system that informs the driver in case of danger – "the driver remaining in control of actions on the vehicle".[191] In the frame of the French Pro-ART project, two vehicles were equipped and demonstrated under the direction of the laboratory Heudyasic in Compiègne.[192] Prolab1 (1988–1991), a Renault 21 Nevada, was "not an autonomous vehicle" emphasizes the research report of Pro-ART France.[193] The project had the goal to inform the driver about the danger of leaving a lane or approaching an obstacle, but "without acting on the vehicle". Prolab1 was presented in June 1991 in Compiègne and in September 1991 at the test circuit of Fiat in Turin where it was the only French research vehicle.

The Prolab 2 vehicle (1991–1994), a Peugeot 605, demonstrated a driver information system only. Again, it had "no automatic control of the vehicle".[194] PSA equipped the vehicle with front and rear cameras, an infrared responder, a scanning laser range finder and linear cameras at the front. Cameras in the mirrors on the left and the right looked to both sides.[195] ProLab2 had been realized by nine French research laboratories cooperating with Renault and PSA and was presented at the final demo of Prometheus in October 1994 in Mortefontaine (CERAM).

Another example is from the UK: The English automobile companies Lucas Industries and Jaguar developed in 1994 together with the UK Department of Trade and Industry a semi-automated vehicle in the frame of PROMETHEUS. According to Raglan Tribe, leader of the Lucas research program at this time, the research vehicle, a Jaguar XJ6, traveled 17,000 miles on public highways in the UK, France, Germany and Italy.[196] Similar to the ProLab vehicles of PSA, the main focus was on warning systems. The approach of Lucas was based on vision enhancement technologies used by military tanks and fighter jets to pinpoint targets in low-visibility

---

[191] "L'objectif était de développer un système d'aide à la conduite, le conducteur restant maître des actions sur le véhicule." (Prometheus – Proart (ed.), 1994. Rapport de fin de contrat de recherche, December, p. 163 [Translation: F.K.]).

[192] Ibid., p. 9.

[193] "Prolab1 n'est pas un véhicule autonome" (Ibid., p. 163f.).

[194] Chatila, Raja. 1996. Autonomous Mobile Robots in France: Some Projects and Scientific Approaches. In *Autonome Mobile Systeme 1996*, ed. Günther Schmidt and Franz Freyberger, p. 8. München.

[195] Laugier, Christian. 2002. Towards "Automated Road", Some Projects and New Technologies, Keynote ITSC 2002 Singapore, online in: http://emotion.inrialpes.fr/laugier/Research/itsc02.pdf, p. 8 (accessed October 12, 2020).

[196] Raglan tribe (Youtube-Account), Robot Car | Raglan Tribe (Title), Uploaded 23.2.2013, https://www.youtube.com/watch?v=AlLZhcnpXYI

conditions.[197] An infrared camera was fixed behind the windshield, detecting heat emissions of objects in a range of 100 meters. A screen beneath the gloves compartment of the Jaguar test-vehicle showed the image to the driver. Later, a head up display developed by Pilkington PLC for military airplanes was used to project the infrared image to the windshield.

Lucas developed also a more advanced system that went beyond warning systems. An *adaptive* version of the cruise control had been developed, meaning a speed regulator that not only maintains a certain speed but also a safe distance to the vehicle in front. Therefore throttle and brakes had to be controlled by the system. Collision avoidance was realized with microwave radar sensors that measured distances between vehicles. Obstacle detection was capable of "tracking simultaneously 12 vehicles" within a range of 120 meters ahead.[198] Radar and vision were then combined to realize a lane guard system that warned the driver when he unintentionally left his lane (rumble strip warning). It could keep the vehicle automatically in the lane, controlling the steering wheel, but it could also activate an automatic emergency braking.

A project that would be interesting to be presented further was conducted by the automotive industry in Germany: Based on the studies of Wolfgang Zimdahl with the Opto-Pilot in 1982 (see Sect. 4.2.3), Volkswagen continued its experiments on automated driving during the PROMETHEUS-program. In 1989, Volkswagen tested a multiple-vehicle platoon named "convoy driving" that operated under fully automatic steering and longitudinal control. A video source shows five cars – a Volkswagen T3 bus and four VW Passat B3 (Fig. 4.22) – following each other at speeds "up to 100 km/h" with a gap between two vehicles of "less than two meters"on a test circuit.[199] But according to Ioannou, the project was discontinued after the test run and no further details are known about the experimentation as the technology remains proprietary.[200] For this reason an appreciation of this experiment remains difficult.

It can be resumed that PROMETHEUS made an important contribution to the communication and network-based assistance systems of today. According to Daimler, its findings "have long since been integrated into numerous everyday technologies", for instance the adaptive cruise control in modern cars.[201] From a historical perspective the research vehicles that were presented are important because most of today's driver assistance systems have their roots in these technology testbeds.

[197] Yazigi, Monique P. 1992. Promethean Plans for the Next Generation of Cars. *New York Times*, September 13, 1992.

[198] Raglan tribe (Youtube-Account), Robot Car | Raglan Tribe (Title), Uploaded 23.2.2013. https://www.youtube.com/watch?v=AlLZhcnpXYI 1, 40 min.

[199] Automatedhighway (Youtube-account), Automated Highway History (video title), uploaded on April 24, 2011, online in: https://www.youtube.com/watch?v=dyqU85D0lkA, 4:11–5:10 (accessed on April 21, 2020).

[200] Ioannou, p. 16.

[201] Daimler 2016a, b.

Fig. 4.22 Volkswagen Convoy driving, 1989. (Credits: Volkswagen AG. Image Source: Automatedhighway (Youtube-account), Automated Highway History (video title), uploaded on April 24, 2011, online in: https://www.youtube.com/watch?v=dyqU85D0lkA, 4:11–5:10 [accessed on April 21, 2020])

Hermann Winner evaluated which of the 27 systems developed during the Prometheus-project were commercialized and whether the chosen path of technology development finally led to a product available on the market. He identified four categories of systems, (1) driver information systems, (2) active driver support, (3) cooperative driving and (4) traffic information and guidance systems. In the second domain, he mentions "autonomous driving", developed in the CED 3: For instance, the Pro-ART-subprojects on Lane Keeping (CED 2.2), Collision Avoidance (CED 3) and Autonomous Intelligent Cruise Control (CED 5) contributed to the market introduction of ACC in 1999 and to automated emergency braking systems in 2003. Since 2013, systems with automatic longitudinal and lateral control in one lane had been commercialized depending on a human operator in a position of surveillance.[202]

4.5 Research After Prometheus

As shown in the preceding chapters, the most important European projects on autonomous driving during the 1980s had been undertaken in Germany, followed by the UK. Influenced by the PROMETHEUS-project this situation changed in the 1990s. Other European countries started their own research projects. The following chapter will first focus on France, an interesting object of study because researchers followed another philosophy than many other countries. An important part of the French research efforts of the 1990s focused on *fully autonomous vehicles* for cities,

[202] Winner, p. 102.

circulating in *controlled environments at low speeds*. The goal was to develop a hybrid mobility concept between public transport and individual automobiles. Second, the chapter will present the development efforts in Italy during the 1990s.

4.5.1 Praxitèle (1993–1999) and CyCab (1997) in France

One of the pioneers of automated driving in France is the computer scientist Michel Parent. From 1991–2010 he was director of the research team *Informatics, Mathematics and Automation for La Route Automatisée* (IMARA) at the *Institut National de Recherche en Informatique et en Automatique* (INRIA), the French National Institute for Research in Computer Science and Control.[203] INRIA was founded in 1967 with the aim to create knowledge in the domain of information and communication technologies. An important part of the research on Intelligent Transport Systems in France during the 1990s was effectuated by the IMARA team. It developed the concept of the Cybercars, small automated urban vehicles conceived to complete public transport systems.

During a meeting with the author of this book Michel Parent remembered that a sequence in the film *Total Recall* (US 1990) with a robot-taxi called JohnnyCab inspired him to start a research project on an automated taxi service at INRIA.[204]

The preparation of the project was started in December 1991 under the preliminary name *Développement d'un engin de déplacement automatique en libre service (DEDALS)*.[205] The initial idea was that a taxi user could call a vehicle with his phone, enter a destination and the taxi would then drive him to the location of his choice. The long-term objective was the realization of an automated vehicle.[206]

In 1993, several academic and industrial partners joined the project, now called Praxitèle. The name was chosen by Evelyne Bénéjam-François, R&D director of the Compagnie générale française des transports et entreprises (CGFTE) with the intention to make a reference to Greek culture. Praxiteles was one of the most renowned fourth century Greek sculptors.

[203] Today, IMARA has become RITS (Robotics and Intelligent Transportation Systems).

[204] Conversation of the author with Michel Parent at INRIA Roquencourt, October 9, 2020.

[205] Plet, Cécile. 2001. Michel Parent, "Les voitures entièrement automatiques? Vous en verrez dans 5 ans." In Internetactu.net, February 22, 2001, online in: http://www.internetactu.net/2001/02/22/michel-parent-les-voitures-entirement-automatiques-vous-en-verrez-dans-5-ans/ (accessed October 8, 2020).

[206] Beltran, Alain, and Pascal Griset. 2007. Des véhicules intelligents pour assurer la sécurité routière? *Code Source, L'hebdomadaire des 40 ans de l'INRIA*, No 36, November 5, 2007, p. 1, online in: https://www.inria.fr/sites/default/files/2019-11/Code_Source_1998-2007.pdf (accessed October 8, 2020).

The project had the goal to demonstrate a "new mode of public transport"[207] for urban centers based on shared individual cars with electric propulsion. Seven teams of INRIA under the direction of Michel Parent worked together with industrial partners as Électricité de France (EDF), Renault, Dassault Électronique and the urban transport and waste company Compagnie Générale d'Entreprises Automobiles (CGEA) – today Veolia propreté.[208] The research center INRETS, the Ecole des Mines Paris and the University of La Rochelle also participated. Praxitèle was funded from 1993–1997 by INRIA-INRETS.[209]

An Optima Sun Ligier electrical vehicle from Aleph Technologies was used as test vehicle. It was equipped with a linear CCD-camera (2049 pixels, operating at a frequency of 1000 Hz) with a cylindrical lens and an infrared and polarized filter. A belt of ultrasonic range sensors (Polaroid 9000) and an infrared target made of three sets of LEDs organized along vertical lines completed the sensor side. The motion controller did run on a SUN Workstation and the code was "transmitted via Ethernet to the vehicle control unit."[210] The automation of the driving task was realized with a "control unit based on a Motorola VME162-CPU board and a transputer net."[211]

Three automated driving behaviors were demonstrated with the Ligier vehicle. First, the immaterial coupling or *platooning* of two vehicles was developed, as can be seen in an INRIA-video from 1995 (Fig. 4.23).[212] The lead vehicle – driven by a human driver – was followed by a second vehicle without driver. The already mentioned linear camera in the vehicle observed infrared signals coming from the trunk of the lead vehicle. The information captured by the camera was treated by an on-board computer system that gave steering and brake commands. The development of a platooning capability followed an important international trend of this decade, as we have seen in the preceding chapters. But the French concept was very different as it did not focus on highways only: The idea behind this system was to couple

[207] INRIA, Rapport d'activité 1997, Action Praxitele, 3. Présentation générale et objectifs, in: https://raweb.inria.fr/rapportsactivite/Ra-Dev97/PRAXITELE/praxitele3.html (accessed October 11, 2020).

[208] INRIA, Rapport d'activité 1997, Action Praxitele, 1. Le partenariat, in: https://raweb.inria.fr/rapportsactivite/Ra-Dev97/PRAXITELE/praxitele1.html (accessed October 11, 2020).

[209] Laugier, Christian, S. Sekhavat, F. Large, J. Hermosillo, and Z. Shiller. 2001. Some steps towards autonomous cars. In *IFAC Symposium*, p. 17. Japan.

[210] Ibid., p. 12.

[211] Paromtchik, Igor E., and Christian Laugier. 1997. Automatic Parallel Parking and Returning to Traffic Maneuvers. *Proceedings of the 1997 IEEE/RSJ International Conference on Intelligent Robot and Systems, Innovative Robotics for Real-World Applications, IROS '97, Grenoble*, France, Vol. 3, pp. V21–V23, in: http://emotion.inrialpes.fr/~paromt/infos/papers/paromtchik:laugier:video:iros:1997.pdf, p. 3, see also: Paromtchik, Igor E., and Christian Laugier. 1996b. Autonomous Parallel Parking of a Nonholonomic Vehicle, 1996. In *Proceedings of the IEEE Intelligent Vehicles Symposium*, Tokyo, Japan, September 19–20, 1996, pp. 13–18, online in: http://emotion.inrialpes.fr/~paromt/infos/papers/paromtchik:laugier:iv:1996.pdf

[212] Parent, Michel. 1995. Le future de l'automobile, Présentation INRIA-IRISA, December 3, 1995, in: https://www.irisa.fr/manifestations/seminaires-1999/3dec99/parent/sld007.htm (accessed October 12, 2020).

Fig. 4.23 Platooning of two Ligier vehicles, first with a driver, second following without driver, realized by INRIA in 1994. (Credits: INRIA, Image Source: Screenshot of a Video: laraimara, Parking automatique à l'aide d'ultrasons – automated parking using ultrasounds, 1994, Uploaded on February 2, 2010, in: https://www.youtube.com/watch?v=BkHHS4uAuOk [accessed October 12, 2020])

several vehicles of the envisioned shared car service automatically into a train and to tow them to a station with high vehicle demand.

Second, "one of the world's first experimental prototypes" of *automatic parallel parking* was developed by INRIA in 1995, emphasizes Christian Laugier.[213] With this capability, the vehicles, once arrived at their destination, should park themselves automatically. Third, *lane changing and obstacle avoidance* were also demonstrated.

Four years later, Praxitèle became an experimental mobility service in the real-world: From October 1997 until July 1999[214] a car-sharing service with 50 manually driven electrical Renault Clio had been made available to the public in Saint-Quentin-en-Yvelines near Paris.[215] The main goal of this experiment was to demonstrate how public transport and individual vehicles could be merged.

[213] Laugier, Christian. 2019. A Journey into the History of Automated Driving, Invited pioneer's talk. *IROS 2019 – IEEE/RSJ International Conference on Intelligent Robots and Systems*, Nov 2019, Macau, China, p. 2, in: https://hal.inria.fr/hal-02428196/document (accessed November 11, 2020). The research results were published in: Paromtchik, Igor, and Christian Laugier. 1996a. Motion Generation and Control for Parking an Autonomous Vehicle. In *Proceedings of the IEEE International Conference on Robotics and Automation*, 3117–3122. Minneapolis. and Paromtchik, Igor E., and Christian Laugier. 1996b. Autonomous Parallel Parking of a Nonholonomic Vehicle, 1996. In *Proceedings of the IEEE Intelligent Vehicles Symposium,* Tokyo, Japan, September 19–20, 1996, pp. 13–18, online in: http://emotion.inrialpes.fr/~paromt/infos/papers/paromtchik:laugier:iv:1996.pdf

[214] Massot, Marie-Hélène. 2000. Praxitèle: un concept, un service et une expérimentation, Bilan d'un prototype, in: Transports Environnement Circulation N° 159, May–June, p. 25–32.

[215] INRIA, Rapport d'activité. 1997. https://raweb.inria.fr/rapportsactivite/Ra-Dev97/PRAXITELE/praxitele3.html (accessed October 8, 2020).

Fig. 4.24 The Praxitèle electric car-sharing service, Saint-Quentin-en-Yvelines, 1997. (Credits: Michel Parent. Image Source: Parent, Michel, Yang, Ming, Cybercars: A Complement to Mass Transport, Powerpoint-Presentation (without date), p. 10, online in: https://slideplayer.com/slide/3354708/)

But due to liability concerns, low performance and a low reliability of the technology none of the automated driving functions developed with the Ligier test-vehicle were integrated in the Renault Clio fleet. As the vehicles had no platooning function, human employees had to steer them manually back to their delivery station. Instead, technology development was focused on the real-time localization of the vehicles, a video-surveillance system, the communication between the Clio-fleet and battery recharging by induction (Fig. 4.24).[216]

But INRIAs efforts on vehicle automation continued with another vehicle, the CyCab, presented to the public in June 1997 (Fig. 4.25). It was realized in cooperation with the company ANDRUET,[217] who had delivered two prototypes in mid-1996.[218] The CyCab was a completely new designed electrical two-seater vehicle without steering wheel and without pedals. The dashboard was only equipped with a joystick and a flatscreen as human-machine interface. Four driving modes were available – manual steering by joystick, teleoperation, platooning and fully automated driving.[219] The CyCab used laser and infrared technologies and a new camera for platooning,

[216] Parent, Michel. 1995. Le future de l'automobile, Présentation INRIA-IRISA, December 3, 1995, p. 6, in: https://www.irisa.fr/manifestations/seminaires-1999/3dec99/parent/sld006.htm (accessed October 12, 2020).

[217] INRIA, Action PRAXITELE, Programme de travail et état d'avancement, Rapport d'activité 1997, in: https://raweb.inria.fr/rapportsactivite/Ra-Dev97/PRAXITELE/praxitele4.html (accessed November 10, 2020).

[218] INRIA, Action PRAXITELE, Résultats et transferts industriels, Rapport d'activité 1997, in: https://raweb.inria.fr/rapportsactivite/Ra-Dev97/PRAXITELE/praxitele5.html (accessed November 10, 2020).

[219] Parent, Michel, Le future de l'automobile, Présentation INRIA-IRISA, December 3, 1995, p. 8, in: https://www.irisa.fr/manifestations/seminaires-1999/3dec99/parent/sld008.htm

Fig. 4.25 The CyCab (1996) and the CyberCab (2000) (Credits: Michel Parent. Image Source: Parent, Michel, Yang, Ming, Cybercars: A Complement to Mass Transport, Powerpoint-Presentation (without date), p. 13 and 14, online in: https://slideplayer.com/slide/3354708/)

developed with the company COSE.[220] Automated parking and the detection of obstacles nearby was realized with ultrasound-sensors. Its maximal speed was 20 km/h.[221]

Around 15 years later, Google X presented in 2014 a prototype of an autonomous vehicle that resembled the CyCab in many ways, but the technology was radically different. Called Firefly, it was a low speed vehicle driving with a speed up to 40 km/h, it had no steering wheel and a friendly exterior shape with a front that looked like a koala.[222] But finally it served only as platform to develop the self-driving technology and got abandoned in 2017.

From the Praxitèle and CyCab-prototypes emerged a concept called Cybercar, meaning a small automated electrical road vehicle for urban centers that should fill the gap between individually owned automobiles and public mass transport systems.

Christian Laugier from INRIA Rhône-Alpes explains the Cybercar-concept in a very precise definition: While the automotive industry at this time aimed to achieve full automation *in the future* with the successive development of assistance systems, the Cybercars-approach *started* with full automation in very controlled environments with low speeds, experimenting its use in pedestrian zones and aiming at intercity tracks as final goal.[223]

---

[220] INRIA, Action PRAXITELE, Résultats et transferts industriels, Rapport d'activité 1997, in: https://raweb.inria.fr/rapportsactivite/Ra-Dev97/PRAXITELE/praxitele5.html (accessed November 10, 2020).

[221] A more detailed technical report can be found in: Baille, Gérard, Garnier, Philippe, Mathieu, Hervé, Pissard-Gibollet, Roger, Le cycab de l'INRIA Rhône-Alpes, Rapport de recherche, INRIA 1999.

[222] Davies, p. 177f.

[223] Laugier, Christian. 2002. Towards "Automated Road", Some Projects and New Technologies, Keynote ITSC 2002 Singapore, online in: http://emotion.inrialpes.fr/laugier/Research/itsc02.pdf, p. 2 (accessed October 12, 2020).

Fig. 4.26 The RobuCAB vehicle of Robosoft. (Credits: Michel Parent. Image Source: http://www.cybercars.fr/wp-content/uploads/2013/10/robucab4-1.jpg)

Based on the CyCab experiments, the Automated People Mover RobuCAB (Fig. 4.26) was created by Robosoft, an INRIA start-up. Fifty vehicles were commercialized for use in fenced environments like amusements parks or airport parking spaces.

Following the first model of the CyCab (1996), INRIA and Yamaha developed in the year 2000 another vehicle called CyberCab (right side of Fig. 4.25). Its shape resembled a golf cart with big windows. The CyberCab was guided by a wire in the ground, a system that is called in French "filo-guidé".[224] This is remarkable as it reminds us that technology development often comes back to old ideas: While Japan and the US abandoned the concept of cable-based vehicle guidance completely, researchers in France renewed this concept after 1997. At the end of the year 2000 five automated electric vehicles were shown at a demonstration of INRIA Rocquencourt.[225]

CyberCab led to the European Project CyberCar (2001–2005) with 10 industrial partners (Fiat Yamaha, Frog) and 7 research institutes (INRIA, INRETS, ENSMP), working on automated vehicles for urban spaces. 12 Cities were involved (Rome,

[224] The ParkShuttle from Frog in the Netherlands and driverless electrical vehicle Serpentine conceived by Bernard Saugy in Switzerland are other examples of Cybercars.

[225] Blosseville, J.M., and Saïd Mammar. 2000. LaRA Update, the French Program La Route Automatisée, Powerpoint-Presentation. http://aramis.iup.univ-evry.fr:8080/~smam/publications/publis2000/lara_2000.pdf (accessed October 12, 2020).

Rotterdam, Lausanne, Antibes…).[226] In 2011 INRIA presented the Cybus, an autonomous shuttle (navette), demonstrated with 25 vehicles during the Exposition Floriade in Schiphol, Amsterdam.

4.5.2 *La Route Automatisée (LaRA) in France (1997–2003)*

In 1997 the informal consortium *La Route Automatisée (LaRA)* – the Automated Road – was created in France, exploring the scenario of an automated highway lane and the automatization of light vehicles in semi-urban environments at low speed.[227]

LaRA was a Joint Research Unit (JRU) that associated several academic and public partners. The team *Informatique, Mathématiques et Automatique pour la Route Automatisée (IMARA)* of the *Institut National de Recherche en Informatique et en Automatique (INRIA)* participated with expertise on Intelligent Transport Systems (ITS), the *Centre de Robotique (CAOR)* – created in 1989 by Claude Laurgeau – of the *Ecole Centrale des Mines de Paris (ENSMP)* developed the vehicles,[228] the *Laboratoire Central des Ponts et Chaussées (LCPC)* was in charge of the infrastructure, the *Laboratoire sur les interactions véhicules-infrastructure-conducteurs (LIVIC)* de INRETS[229] worked on the assistance systems and the *Télécom Paris tech (École nationale supérieure des télécommunications)* was in charge of the telecommunication systems.[230] LaRA was financed in 1997 and 1998 by the second *Programme interministériel de recherche et d'innovation dans les transports terrestres* (PREDIT) that covered the period 1996–2000. It also received European funding in 1999.[231] From 1999–2003 it was funded by INRIA.[232]

---

226 Laugier, Christian. 2002. *Towards "Automated Road", Some Projects and New Technologies, Keynote ITSC 2002 Singapore*, online in: http://emotion.inrialpes.fr/laugier/Research/itsc02.pdf, p. 21 (accessed October 12, 2020).

227 Laboratoire Central des Ponts et Chaussées (LCPC), LA ROUTE AUTOMATISÉE, Un scenario périurbain, Brochure, May 2001, p. 13, 15.

228 In 2009 Claude Laurgeau emphasized that "its most essential activities were centered on an intelligent vehicle" (Laurgeau, Claude, 2009. *Le siècle de la voiture intelligente*, p. 7. Paris).

229 On January 1, 2011, INRETS and the Laboratoire central des ponts et chaussées merged to create the Institut Français des Sciences et Technologies des Transports, de l'Aménagement et des Réseaux (IFFSTAR) which became part of Université Gustave-Eiffel on January 1, 2020.

230 Plet, Cécile. 2001. Michel Parent: Les voitures entièrement automatiques? Vous en verrez dans 5 ans. Internetactu.net, February 22, 2001, online in: http://www.internetactu.net/2001/02/22/michel-parent-les-voitures-entirement-automatiques-vous-en-verrez-dans-5-ans/ (accessed October 12, 2020). See also https://www.irisa.fr/manifestations/seminaires-1999/3dec99/parent/sld009.htm

231 Parent, Michel. 1995. Le future de l'automobile, Présentation INRIA-IRISA, December 3, 1995, p. 9, in: https://www.irisa.fr/manifestations/seminaires-1999/3dec99/parent/sld009.htm (accessed October 12, 2020).

232 Laugier et al. 2001, p. 17.

A research report from 2001 reveals the two main goals of LaRA. First, the project explored the potential of automated vehicles for suburban contexts and city centers as it had been developed by INRIA with the CyCab and Cybercar-concepts. Second, LaRA had the aim "to develop assistance systems that would lead eventually to an Automated Highway System"[233] like the one proposed by Caltrans/PATH during Demo '97 in the United States. For that reason, the possible design of the entry and exit points of automated lanes was an important research topic.

The main goal of LaRA was to increase both, the road capacity and the safety.[234] With manual driven vehicles both goals are normally in conflict – increased road capacity decreases safety for several reasons, for example because the distances between the number of vehicles are too short for adequate human reactions.[235] While manual driven cars only allow 2.200 vehicles per hour to pass a highway,[236] an automated system, using anti-collision systems and vehicle-to-vehicle communication would allow a capacity of 5.000 vehicles per hour.[237]

But in contrast to the San Diego-Demo, the approach of LaRA was not centered on technology development only. It took into consideration possible user reactions. The 20 researchers of LaRA explored the social acceptability of automated systems and also the question how to finance such an automated transport infrastructure. The main hindrances for an automated transport network were identified in the socio-economic and political domain, not in the technical challenges.[238] The creation of a low speed automated system (until 60 km/h) during a period of 10–20 years was proposed to introduce the system to the public and change cultural mentalities.[239] The report concluded that an automated road network could then be developed in a time span of 30 years – the same time that the TGV needed from first draft to realization.[240]

233 Laboratoire Central des Ponts et Chaussées (LCPC), LA ROUTE AUTOMATISÉE, Un scenario périurbain, May 2001 (Brochure), p. 100 (Translation from French: F.K.).

234 Rousselot, Michel. 2001. Préface. In *La route automatisée, Un scenario périurbain*, ed. LCPC, p. 9. Paris.

235 Laboratoire Central des Ponts et Chaussées (LCPC), LA ROUTE AUTOMATISÉE, Un scenario périurbain, May 2001 (Brochure), p. 26.

236 Ibid., p. 25.

237 Ibid., p. 33.

238 Ibid., p. 15.

239 Ibid., p. 16, 55.

240 Ibid., p. 16.

4.5.3 VaMPs Long-Distance Drive from Munich to Odense (1995)

While researchers in France explored low-speed automated vehicles and automated lanes for highways during the 1990s, UniBwM continued to develop its concept of fully autonomous driving by machine vision on highways.

From 1983 to 1998, the European Union organized five programs on information technology and technology transfer, named ESPRIT (European Strategic Program for Research and Development in Information Technologies). One of the projects (No 8329) supported by ESPRIT II was the technology transfer program CLEOPATRA (Cluster of Embedded Parallel Time Critical Applications). It started on March 1, 1994 and had a formal duration of 24 months (April 1, 1994 – March 31, 1996). Seven industry partners from Germany, three from France and one from Denmark worked with two public research institutions from Germany on "High Performance Computing and Networking" (HPC) technologies.

The project included two subprojects, covering a Smart Vehicle and a Smart Helicopter. For the Smart Helicopter-project UniBwM cooperated with Eurocopter Deutschland and the DLR Brunswick. In the sub-project Smart Vehicle that is of primary interest here, UniBwM acted as contributor to Daimler Benz AG and received all funding from this automotive company. The goal of the Smart Vehicle project was to replace the transputers for image sequence processing in the VITA-II and VaMP-vehicles by faster Motorola PowerPC 601 processors. With the more powerful hardware, the image evaluation time was cut in half (from 80 to 40 ms) and the number of processors needed was only one fifth (20%).

Within the framework of CLEOPATRA "several thousand kilometers of automatic driving have been performed"[241] with the VaMP research vehicle. In November 1995, the Mercedes was used for a fully autonomous long distance drive – with lateral and longitudinal control – over more than 1600 km on the German Autobahn from Munich to Odense (Denmark) near Kopenhagen, where a meeting of the above mentioned smart vehicle partners took place. UniBwM organized the trip in response to the growing concurrence in the research field: Three months before, Carnegie Mellon University organized a semi-autonomous 2800-mile trip from Pittsburgh to San Diego (No Hands Across America). Their vehicle, called Navlab 5, was steered laterally by vision while the human operator handled throttle and brake (see Sect. 2.3.6). After this demonstration, other research groups were under pressure to show their own performance capabilities.

During the long-distance drive to Odense, the VaMP-Mercedes performed autonomous lane keeping, lane change, longitudinal control and collision avoidance. It also supervised its own functionality including a stop of the autonomous vehicle control and a hand over to the human driver.

---

[241] Maurer, M., R. Behringer, S. Fürst, F. Thomanek, E.D. Dickmanns. 1996. A Compact Vision System for Road Vehicle Guidance. *Proceedings of 13th International Conference on Pattern Recognition*, p. 316.

The engineering team resumes the event as follows: The mission planned by the driver "started at the Munich city 'Autobahn' A 99, followed the 'Autobahn' A9 to Nuremberg, the A3 to Würzburg and the A7 via Kassel, Hannover and Hamburg to Flensburg; in Denmark VaMP drove the E45 to Kolding and the E20 to Odense; on the way back it took the same route. Altogether the system was switched on for more than 1600 km; 95% were driven automatically, which corresponds to more than 1500 km. More than 400 lane change maneuvers had been performed automatically"[242] after being started manually by the safety pilot. He had to supervise the rear and side hemispheres because in the framework of the CLEOPATRA-project no backward looking cameras were available, "only the front vision system was used". The safety driver also had to choose the desired travel speed and had to adapt this speed to traffic signs and environmental conditions. A speed of up to 175 km/h was reached in fully autonomous driving on the Autobahn.

One essential goal of the trip was to learn about the limits in dynamic scene understanding under real-world conditions of this final second-generation system, and then to decide which parts should be improved for the third-generation perception system planned. According to the engineering team, the system had problems first due to the limited operating range of the cameras, e.g. at construction sites with ambiguous lane markings: As an intensity-based edge detection system was used, the recognition of "yellow lane markings overriding the still existing normal white ones" at construction sites was not possible.[243] Another difficulty occurred when the sun looked directly in the camera at dusk or dawn. The darkness of tunnels also was a challenge for standard cameras at that time, a problem that also was noted during the Mille Miglia-Demo of ARGO (University of Parma) in Italy two years later (see next Sect. 4.5.4). Under bright sunshine with shadows, "the dynamic range of the intensity signal" was too large for normal cameras.[244] The second category of problems was caused by hypothesis generation of the obstacle detection program (ODT). These were "not only met by vehicles but sometimes by other objects like certain patterns on the road" or by shadows of bridges.[245] The knowledge of these flaws was used to improve the design of the third-generation trinocular vision system of UniBwM that used "two divergently oriented wide-angle cameras with a central region of overlap" and a high-resolution color camera.[246]

After the Prometheus-project with the successful Paris-Demo of VITA-II and VaMP in 1994, the test-drive of VaMP from Munich to Odense in 1995 and the PATH-Demo in 1997, the automotive industry confirmed its preference for radar with respect to obstacle recognition. Sufficiently powerful color vision systems

[242] Ibid. Note: Interestingly, the engineering-team used the term automatic, not autonomous. But as this system took decisions, it should be described as autonomous.

[243] Ibid, p. 317, Dickmanns, Ernst Dieter. 2002. The Development of Machine Vision for Road Vehicles in the Last Decade. *Intelligent Vehicle Symposium* 1, p. 276.

[244] Dickmanns, Ernst Dieter. 2002. The Development of Machine Vision for Road Vehicles in the Last Decade. *Intelligent Vehicle Symposium* 1, p. 276.

[245] Maurer et al. 1996, p. 317.

[246] Dickmanns, Ernst Dieter. 1997. Vehicles Capable of Dynamic Vision. *IJCAI'97 Proceedings of the Fifteenth International Joint Conference on Artificial Intelligence*, Vol 2, p. 1589.

were still too expensive and complex, as confirmed by Dickmanns.[247] Daimler for instance focused its research efforts on lane keeping and convoy-driving for trucks. In 1996, UniBwM decided to end the cooperation with Daimler and continued to develop autonomous driving with funding from defense-related projects. This will be detailed in Sect. 4.6.

The new paradigm that appeared during this period was based on knowledge of the environment, delivered by precise maps and precise positioning. GPS-technology in combination with dead reckoning, cameras and lidars instead of only visual approaches became the dominant sensor choice in the automotive field for road following. Some years later, this shift became fully visible in the DARPA-challenges. The participants received a CD with GPS-waypoints to be followed by their vehicles. For the use case that was chosen – a desert scenario in Iraq – a relatively high number of waypoints in curves were sufficient so that vision systems were not necessary. As all vehicles of the DARPA Challenges followed a trace of GPS-waypoints, Sebastian Thrun argued that these robots were blind.[248] Guided through a sequence of waypoints, these vehicles are not self-sufficient, not really autonomous, they depend on the broader infrastructure of a satellite system. This approach can be called *confirmation-type-vision*, in contrast to the *pathfinder- or scout-type-vision* preferred by UniBwM that allows driving in unknown terrain with the help of a more demanding vision system.

4.5.4 ARGO and **Mille Miglia in Automatico,** *VisLab, University of Parma (1998)*

Like Carnegie Mellon University in the United States and UniBwM in Germany, the Laboratory for Artificial Vision and Intelligent Systems (VisLab) at the Dipartimento di Ingegneria dell'Informazione at Università degli Studi di Parma in Italy also developed machine vision solutions for autonomous driving.

In the framework of the PROMETHEUS-program University of Parma developed and tested a real-time computer vision system, using a Fiat Ducato 18 Maxi as research vehicle. The van was called MOB-LAB (MOBile LABoratory) and its main goal was to demonstrate vision-based navigation with an emphasis on lane and obstacle detection. This was realized with a system of analog B/W and color cameras in the front and the back of the vehicle. Only grey level images were processed because the technology for data processing of color images was unaffordable at this time.[249] The MOB-Lab vehicle was not automated but "manually driven" – its goal

---

[247] Lossau 2017 and Knop 2018.

[248] Davies, p. 99.

[249] Bertozzi, Massimo, Alberto Broggi, and Alessandra Fascioli. 2020. *The Evolution of Vision-based Unmanned Ground Vehicles: the VisLab Perspective*, (undated, without publication source), in: http://citeseerx.ist.psu.edu/viewdoc/download?doi=10.1.1.73.7759&rep=rep1&type=pdf (accessed July 20, 2020), p. 3.

Fig. 4.27 The ARGO vehicle (Credits: Alberto Broggi. Image Source: http://www.vislab.it/argo/ARGO/images/ufficiali/31.jpg)

was to display warning signals to the driver.[250] It was presented during the final PROMETHEUS meeting in Mortefontaine near Paris in October 1994.

In the second half of the 1990s, VisLab carried two pioneering projects, ARGO and RAS (Surface Antarctic Robot).[251] ARGO, that is of primary interest here, was the follow-up program of PROMETHEUS at the University of Parma. The name ARGO was chosen as a reference to the ship of the mythical Greek hero Jason and the many-eyed Greek god Argus.[252]

ARGO used the results of the MOB-LAB program to equip a new research vehicle, a Lancia Thema 2000 passenger car (Fig. 4.27).

Initially, the system was not developed to realize an automatic vehicle. Its goal was an active safety system that supervises the driver's behavior, issues acoustical or optical warning signals and can take control of the vehicle in dangerous situations.[253] But ARGO was then extended to develop an automatic vehicle that could be tested on the highway.

---

[250] Bertozzi, Massimo, Alberto Broggi, Gianni Conte, and Alessandra Fascioli. 1997. Obstacle and Lane Detection on ARGO, *Proceedings of IEEE Intelligent Transportation Systems conference 1997,* p. 1, online in: http://www.ce.unipr.it/people/broggi/publications/iv97.pdf

[251] The Surface Antarctic Robot (RAS)-project aimed at autonomous driving of a platoon of snow-cats in the Antarctica region. Vision was used to follow a lead vehicle with high precision. It was successfully tested in the Italian Alps and in Antartica.

[252] Broggi, Alberto, Massimo Bertozzi, and Alessandra Fascioli. 1999b. ARGO and the MilleMiglia in Automatico Tour. *IEEE Intelligent Systems and Their Applications* 14 (1), p. 55.

[253] Broggi, Alberto, Masssimo Bertozzi, Alessandra Fascioli, Lo Bianco, Corrado Guarino, and Aurelio Piazzi. 1999a. The ARGO autonomous vehicle's vision and control systems. *International Journal of Intelligent Control and Systems* 3 (4), p. 411.

Fig. 4.28 The interior of the ARGO vehicle with cameras, control panel, steering motor, on-board monitor, and two emergency buttons. (Credits: Alberto Broggi. Image Source: http://www.vislab.it/argo/ARGO/images/ufficiali/61.jpg)

For two reasons, the engineering team at VisLab decided to use only passive sensors (cameras) in the Lancia. First, active sensors like radar may be confronted with interferences from other sensors, especially when a multitude of vehicles share the same space at the same time.[254] Vision systems in contrast can acquire data without altering the environment. Second, the costs of the system had to be kept low to "ease its integration into a large number of vehicles".[255] For that reason, commercial PCs with 450 MHz Pentium II processors were chosen to extract image features and control the vehicle trajectory in real-time (50 images per second).[256] To sense the environment, two synchronized low-cost B/W-cameras and a speedometer that detected the velocity of the vehicle were used. This stereoscopic vision system of ARGO was mounted inside the car at both top corners of the windscreen (Fig. 4.28). Beneath the pair of stereo cameras for the detection of obstacles on the road a monocular image was used to "extract the road geometry".[257] Both were based on the

[254] Broggi, Alberto, Masssimo Bertozzi, Alessandra Fascioli, Lo Bianco, Corrado Guarino, and Aurelio Piazzi. 1999a. The ARGO autonomous vehicle's vision and control systems. *International Journal of Intelligent Control and Systems* 3 (4), p. 410.

[255] Broggi, Alberto, Bertozzi, Massimo, Fascioli, Aleassandra, ARGO and the MilleMiglia in Automatico Tour, 1999b, p. 55.

[256] Broggi, Alberto, Masssimo Bertozzi, Alessandra Fascioli, Lo Bianco, Corrado Guarino, and Aurelio Piazzi. 1999a. The ARGO autonomous vehicle's vision and control systems. *International Journal of Intelligent Control and Systems* 3 (4), p. 412.

[257] Bertozzi, Massimo, Broggi, Alberto, Fascioli, Alessandra, 2020. The Evolution of Vision-based Unmanned Ground Vehicles: the VisLab Perspective, (undated, without publication source), in: http://citeseerx.ist.psu.edu/viewdoc/download?doi=10.1.1.73.7759&rep=rep1&type=pdf (accessed July 20, 2020), p. 3.

inverse-perspective mapping (IPM) approach that removed the perspective effect from the images.

ARGO was driven by a system called Generic Obstacle and Lane Detection (GOLD). Its main task was "road following".[258] The turning of the steering wheel was derived from visual perception. An "electric stepping motor was coupled to the steering column by means of a belt."[259] Three driving modes were available – *manual* driving with acoustical and visual warning signals to the driver, *supervised* driving with the system taking control of the vehicle in dangerous situations and *automatic* driving "following lanes, localizing obstacles, and performing lane changes."[260] In the automatic mode, road following based on lane detection or platooning could be chosen. With this system, the Lancia Thema was able to "determine its position with respect to the lane, to compute the road geometry, to detect generic obstacles on the path, and to localize a leading vehicle."[261] It was also able to "overtake slower traffic."[262]

ARGO is one of only four research vehicles that performed tests on public roads during this period: Since 1992 VaMoRs from UniBwM regularly performed fully autonomous driving tests on public roads of any kind around Munich; VaMP joined in 1994. In July 1995 the NavLab 5 from Carnegie Mellon University showed semi-autonomous driving from Pittsburgh to San Diego *(Hands Off Across America-*Demo). In November 1995, VaMP drove autonomously on the highway from Munich to Odense (Denmark).

Three years later, from June 1–6, 1998 University of Parma followed with a long-distance test drive of ARGO on Italian highways. The tour was called *Mille Miglia in Automatico* – named after the famous car race that takes place in Italy each year – and broadcasted live on the internet. Like NavLab 5, ARGO was a semi-autonomous vehicle – only lateral control was automated, longitudinal control (acceleration, brake) was assured by a human driver. The experiment was centered on road following (lane and obstacle detection).

The 2000 km long test run was carried out on the Italian highways (Parma-Turin-Pavia-Ferrara-Ancona-Rome-Florence-Parma), including various landscapes, weather conditions, road works, absent road signs, diversions and heavy traffic.

---

[258] Broggi, Alberto, Bertozzi, Massimo, Fascioli, Alessandra, 1999b. ARGO and the MilleMiglia in Automatico Tour, Jan/Feb, p. 55, online in: http://www.ce.unipr.it/people/broggi/publications/expert.pdf

[259] Broggi, Alberto, Masssimo Bertozzi, Alessandra Fascioli, Lo Bianco, Corrado Guarino, and Aurelio Piazzi. 1999a. The ARGO autonomous vehicle's vision and control systems. *International Journal of Intelligent Control and Systems* 3 (4), p. 412.

[260] Bertozzi, Masssimo, Alberto Broggi, and Alessandra Fascioli. 2006. VisLab and the Evolution of Vision-Based UGVs. *Computer* 39 (12), p. 33.

[261] Broggi, Alberto, Masssimo Bertozzi, Alessandra Fascioli, Lo Bianco, Corrado Guarino, and Aurelio Piazzi. 1999a. The ARGO autonomous vehicle's vision and control systems. *International Journal of Intelligent Control and Systems* 3 (4), p. 411.

[262] Broggi, Alberto. 2008. History of AHS in Italy and future issues. *Proceedings of the 2008 IEEE International Conference on Vehicular Electronics and Safety*, Columbus, OH, USA, September 22–24, 2008, p. 11.

An on-board low-cost 200 MHz Intel Pentium MMX computer with multimedia extensions was used as processing system for the real-time vision-based driving functionalities.[263] The system processed "up to 25 pairs of stereo frames per second" and the control signal for autonomous steering was provided every 40 ms. During speeds of 90–95 km/h "there was no noticeable difference in comparison to a human driver", claims the engineering team. But at higher speeds up to 123 km/h "the vehicle tended to oscillate inside the lane."[264]

Around 90% of the tour was driven automatically. Road signs, guard-rails, junctions and exits were detected under various lightning conditions and during high temperatures. The maximum distance driven automatically without interruption was 54,3 km.[265] This can be explained by the fact that the system used the right lane marking as orientation for navigation. But at highway exits two lanes appeared – a continuous line for the exit and a dashed line that separated the exit lane and the lane where the car was driving on the highway. In this situation the system had to ask for a user decision which line it should choose. After user intervention, the system quickly resumed automated steering.[266]

As many other outdoor tests worldwide, MilleMiglia showed that reflections, rain, road works, interferences, temperature variations and human-machine interfaces remain challenging fields open to improvement.[267] Especially the quick and extreme changes in illumination during the entry and exit of tunnels were difficult for the vision system.

In 2004, University of Parma equipped together with Oshkosh and the Ohio State University a big truck called TerraMax and participated in the first DARPA-Challenge, an off-road course through the desert. But after 1.4 miles, the truck stopped due to a software failure.[268]

In 2005, VisLab participated again in the DARPA-Grand Challenge. VisLab had the goal to "apply the results of basic and advanced research in the development of UGVs for unmanned missions in real environments."[269] Its aim was to improve "soldier survivability in the battlefield by removing soldiers from harms way, especially

[263] Broggi, Alberto, Bertozzi, Massimo, Fascioli, Alessandra, 1999b. ARGO and the MilleMiglia in Automatico Tour, Jan/Feb, p. 60.

[264] Broggi, Alberto, Masssimo Bertozzi, Alessandra Fascioli, Lo Bianco, Corrado Guarino, and Aurelio Piazzi. 1999a. The ARGO autonomous vehicle's vision and control systems. *International Journal of Intelligent Control and Systems* 3 (4), p. 438.

[265] Broggi, Alberto, Bertozzi, Massimo, Fascioli, Alessandra, 1999b. ARGO and the MilleMiglia in Automatico Tour, Jan/Feb, p. 59.

[266] Ibid., p. 61.

[267] Bertozzi, Masssimo, Alberto Broggi, and Alessandra Fascioli. 2006. VisLab and the Evolution of Vision-Based UGVs. *Computer* 39 (12), p. 34.

[268] Özgüner, Umit, Keith A. Redmill, and Alberto Broggi. 2004. Team TerraMax and the DARPA Grand Challenge: A General Overview. In *Intelligent Vehicles Symposium*. University of Parma, p. 237.

[269] Broggi, Alberto, Paolo Grisleri, Chris Yakes, and Chris Hubert. 2007. *The Oshkosh-VisLab Joint Efforts on UGVs: Architecture, Integration, and Results*. Platform Innovations and System Integration for Unmanned Air, Land and Sea Vehicles, 146th NATO Workshop on Applied Vehicle

during convoy operations."[270] Collaborating with Oshkosh Trucks and Rockwell Collins, VisLab developed the vision system of the TerraMax truck. It was one of only five participants that completed the 132-mile (215 km) desert course of the Grand Challenge, defined by a list of waypoints.[271] The truck was based on Oshkosh's Medium Tactical Vehicle Replacement (MTVR) MK23 truck, developed for the Marine Corps. It used a fix-mounted trinocular vision system with different distances between the cameras. The decision to use different vision streams removed many mechanical problems of other vision systems, claim the engineers.[272] TerraMax reached "a maximum speed of 68 km/h"[273] or 42 mph,[274] avoiding obstacles, recognizing paths and planning trajectories. It was the only team that arrived at the finishing line using vision as primary sensor source combined with laser-scanners (LIDAR), GPS / Inertial Navigation System (INS) for geo-localization and basic information about speed and acceleration.

The vehicle demonstrated "obstacle avoidance, negotiating tunnels, narrow roads and cliffs, GPS waypoint following and 28 h of nonstop continuous operation".[275] Steering, throttle, brake and transmission were controlled by the computer system. The vision system was now also able to compensate different lighting situations, allowing to sense the environment even during situations with direct sunlight. It was later installed in a 10x10 Oshkosh Palletized Loading System (PLS) military truck and demonstrated in 2006 at Yuma Test Center. One year later, Oshkosh and University of Parma participated in the Urban Challenge in 2007, demonstrating a 60-mile (97 km) supply mission in a city environment. It had to be completed in 6 hours, respecting road rules and negotiating through traffic. Partners were Teledyne, Auburn University and IBEO Automobile Sensor GmbH.[276]

Technology Panel, Florence 2007, p. 1. http://www.ce.unipr.it/people/broggi/publications/natoavt146.pdf

[270] Braid, Deborah, Alberto Broggi, and Gary Schmiedel. 2006. The TerraMax Autonomous Vehicle, Field Report. *Journal of Field Robotics* 23 (9), p. 707.

[271] Bertozzi, M. Broggi, A. et al., 2021. Vision Technologies for Intelligent Vehicles p. 6 (online in: http://www.ce.unipr.it/people/broggi/publications/ivia2007.pdf, (accessed May, 5, 2021).

[272] Braid, Deborah, Alberto Broggi, and Gary Schmiedel. 2006. The TerraMax Autonomous Vehicle, Field Report. *Journal of Field Robotics* 23 (9), p. 707.

[273] Bertozzi, Masssimo, Alberto Broggi, and Alessandra Fascioli. 2006. VisLab and the Evolution of Vision-Based UGVs. *Computer* 39 (12) , p. 38.

[274] Braid, Deborah, Alberto Broggi, and Gary Schmiedel. 2006. The TerraMax Autonomous Vehicle, Field Report. *Journal of Field Robotics* 23 (9), p. 706.

[275] Ibid., p. 694.

[276] Broggi, Alberto, Paolo Grisleri, Chris Yakes, and Chris Hubert. 2007. *The Oshkosh-VisLab Joint Efforts on UGVs: Architecture, Integration, and Results*. Platform Innovations and System Integration for Unmanned Air, Land and Sea Vehicles, 146th NATO Workshop on Applied Vehicle Technology Panel, Florence 2007, p. 3.

4.6 Defense-Funded Research on Autonomous Driving in Europe

The last chapter has reminded us that beside civilian traffic military scenarios are another possible use case for autonomous vehicles. Not only in the US but also in Europe military actors became interested in the potential of unmanned ground vehicles. With the aim to give insights into these activities, the following chapter will go back in chronology to present defense funded research projects on autonomous vehicles in France and Germany.

First, a short presentation of the French military research project DARDS is given. The following sections detail the implication of UniBwM in several defense-funded programs in Germany. Parallel to the PROMETHEUS project, UniBwM participated in the robotics program PRIMUS (1989–2003). After 1996, UniBwM developed its third-generation vision system EMS in the context of the AUTONAV-agreement (1997–2003) with US-American partners.

4.6.1 The Military Research Project DARDS in France (1984–1996)

When DARPA started 1984 the Strategic Computing Initiative in the US, the Autonomous Land Vehicle (ALV) was an important element (see Sect. 2.4). In the same year, the French military started a project called *Démonstrateur autonome à rapidité de déplacement pour la surveillance (DARDS)* – Rapid Autonomous Surveillance Demonstrator. Funded by the *Direction des Recherches, Études et Techniques de l'armement (DRET)* – the Bureau of Research of the French Ministry of Defense – the project aimed at a four-wheeled semi-autonomous vehicle for mine clearance, reconnaissance and urban combat missions. It was one of the first industrial demonstrators with control-command performances. To preserve its manufacturing secrets, DARDS had been classified as confidential and remains until today unknown to a broader public.

In contrast to the ALV, which consisted of a completely new designed prototype, DARDS was based on a Mercedes G-class jeep for military purposes (Fig. 4.29). Most of its robotization was realized by Dassault Électronique under Jean-Philippe Quin as the company's product owner. The 3D path detection capability had been done by the company *Industry and Technology for Machine Intelligence* (ITMI), founded by the roboticist Jean-Claude Latombe from Grenoble University who later held a chair in robotics at Stanford University. ITMI was one of the first companies in the world able to do 3D trajectory path planning *in real-time* with vision algorithms.

According to Dr.-Ing. Catherine Fargeon, who was at the time project manager of DARDS for the French MoD, the first test drives started in the years 1987–1988. They took place at the *Établissement Technique d'Angers (ETAS)*, a test center of

Fig. 4.29 Mercedes G-class Jeep used for DARDS, seen here at the beginning without full autonomy. (Image Credits: Ministry of Armed Forces, France. Image use authorized by the program manager of the research program)

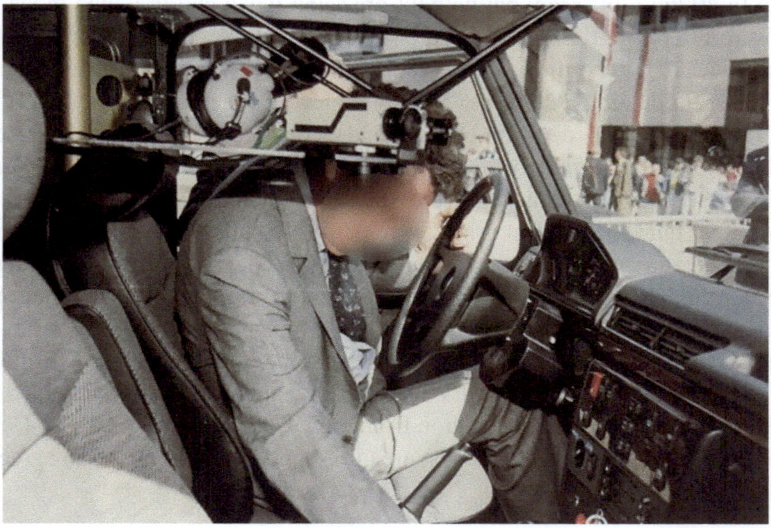

Fig. 4.30 The 3D vision system of DARDS, fixed at the roof, Cité des Sciences, Paris, 1991/1992. (Credits & Image Source: Figs. 4.30 and 4.31 provided by Catherine Fargeon)

Fig. 4.31 This photo shows the big radome that protects the radar antenna on the roof of DARDS, Cité des Sciences, Paris, 1991/1992

the *Direction générale de l'Armement (DGA)*.[277] These first autonomous driving tests started with free outdoor runs. Equipped with video cameras and a 40 m-laser range finder, the Mercedes-Jeep was able to realize "basic motion tasks under human supervision" reports the Armed Forces Journal.[278] All together 10.000 kilometers of trials had been performed "in different semi-autonomous modes in all kinds of environments" during the project. Fargeon confirms that speeds of over 100 km/h were reached on straight runways, up to 40 km/h with dummy obstacles and up to 40 km/h off-road, but without precising when and in which mode these tests were effectuated. She also reports that Dassault électronique embedded the driving styles of two Formula 1-World champion drivers in the control-command of the wheels, for dealing with turns. The lowest level concerned the controls of each axis. The higher level decision-making and behavior intelligence was based on research from the *Office National d'Études et de Recherches Aérospatiales (ONERA)*, the French national aerospace research center.

For safety purposes, all test drives were performed under distant supervision. At any moment, autonomous driving could be neutralized by remote control if a problem occurred. The technical principle implemented was that of aircraft autopilots that can be disengaged.

The successful experiments in France attired the attention of the US military. After the discontinuation of the ALV in 1987, new approaches for unmanned vehicles were needed. In 1989 a delegation from the US-American Ministry of Defense and certain universities visited the test area in Angers in the frame of an agreement

[277] I am grateful to Catherine Fargeon for providing me with information about the DARDS project.
[278] Armed Forces Journal International, Vol. 139, 2001, p. 36.

of cooperation. Only one year later, the US military started the DEMO 1-program, aiming at teleoperation and path-retrace capability of UGVs (see Sect. 2.5.1), abilities that resembled very much those of DARDS.

After the success of the first tests with DARDS its capabilities were transferred on a Wiesel-tank, an operation led by the *Direction des Armements Terrestres* and not by DRET. The tank was not only used to test semi-autonomous driving. A remote operation of the weapons system was added, respecting the Ottawa convention which prohibits firing without human supervision. Tests took again place at the *Etablissement Technique d'Angers*. Following a reform of the DGA, the *Direction des Recherches, Études et Techniques de l'armement* was closed in 1996. This was also the end of DARDS. The Wiesel-tests continued a few years after the closure of DRET, centering on operational and not on autonomous driving functions.

From a historical perspective, DARDS is foremost remarkable by its timing. It was one of the earliest European military projects in the field of Unmanned Ground Vehicles. It is also remarkable that US-American partners came to visit the French group as reported by Fargeon. Finally it shows that defense funded research on semi-autonomous robotics technologies started earlier than projects of the automotive industry. But as the description of DARDS oscillates between supervision, teleoperation and semi-autonomous driving, it remains unclear what kind of task the jeep was able to accomplish autonomously.

4.6.2 The Experimental Robotics Program PRIMUS (1989–2003)

Some years after France began experimenting unmanned ground vehicles, the German industry also launched its first defense funded research programs. Since the mid 1980s several studies came to the conclusion that unmanned systems should become an important element of the German Federal Armed Forces in the twenty-first century.[279] Following these recommendations, a first meeting with academic and industrial partners about military robotics was held in January 1985 at the German Federal Ministry of Defense (BMVg) in Bonn. In November of the same year the *Heeresamt* organized a conference about robots on the battlefield (*Roboter auf dem Gefechtsfeld*). After these preparatory activities the Federal Ministry of Defense initiated a project called *Autonomes Landfahrzeug* 1986–1992), the same term that DARPA used in the US (Autonomous Land Vehicle).[280] The University of

[279] Schaub, Günter, Alfred Pfändner, and Christoph Schäfer. 2004. PRIMUS: Autonomous Navigation in Open Terrain with a Tracked Vehicle. In *Unmanned Ground Vehicle Technology VI*, ed. Grant R. Gerhart, Chuck M. Shoemaker, and Douglas W. Gage, vol. 5422. Proceedings of SPIE, p. 156.

[280] Phase I of *Autonomes Landfahrzeug* lasted from September 15, 1986 to December 31,1989, Phase II from January 1, 1990 to December 31, 1992.

Fig. 4.32 The Mercedes GD-300 test vehicle, used in the PRIMUS A program from 1989–1992. (Image Credits: Dornier. Image Source: Schaub, Pfändner, Schäfer, p. 157)

the Bundeswehr Munich (UniBwM) participated in this project, cooperating with Dornier GmbH Friedrichshafen (later part of EADS, now Airbus).

After the first phase of the project the German Federal Ministry of Defense started in 1989 a long term research and development program in experimental robotics called PRIMUS (*Program of Intelligent Mobile Unmanned Systems*), focusing on "autonomous driving in unknown open terrain".[281] Its task was to develop algorithms for autonomous navigation and to integrate them into military vehicles. PRIMUS was carried out in four phases: A (1989–1992), B (1993–1995), C (1996–1999) and D (2001–2003).[282]

Dornier was chosen as the prime contractor of PRIMUS A. As part of the experimental 'battlefield robotics'-program of Dornier (January 29 – July 31, 1992), the first generation vision system developed at UniBwM had to be ported to a Mercedes GD-300 all-terrain test vehicle (Fig. 4.32). During the final demo of PRIMUS A in summer 1992 on the Bundeswehr training area in Pfullendorf, the Jeep was able to "drive autonomously (not teleoperated) in open terrain with a speed of up to 5 km/h, detecting and avoiding obstacles in the driving path."[283]

[281] Schaub, Pfändner, Schäfer, p. 156.

[282] Thanks to Günter Schaub for help with the exact dates of the PRIMUS-programs (E-Mail to the author on July 1, 2020).

[283] Schaub, Pfändner, Schäfer, p. 157.

With an improved LADAR from Dornier, a speed "of up 25 km/h in open terrain" and "of up to 60 km/h" on a country road was achieved in autonomous driving mode, detecting obstacles.

The next major milestone was reached during PRIMUS C (1996–1999). A tracked vehicle was chosen for the experiments. Pursuing the goal of autonomous obstacle detection and avoidance in an unknown environment, a method for the autonomous detection of paths (contours) developed by UniBwM was transferred on a Wiesel 2 tank.[284] The Wiesel 2 is a bigger version of the first Wiesel (1990) from Rheinmetall Landsysteme GmbH, a lightly armored air-loadable weapons carrier capable of night combat that was primarily developed for reconnaissance missions of paratroopers. It was used by the German Federal Armed Forces in many international missions (e.g. Somalia, Afghanistan).

Two digitized drive-by-wire Wiesel 2-tanks were equipped for experimental purposes, one as a robot and the other one as a Command and Control station (CCS). The third generation vision system from UniBwM – that will be presented in Sect. 4.6.4 – was transferred to the robot vehicle for road following. The tank was equipped with a color camera on a two-axis-platform and a computer system with Motorola PowerPC processors. The university developed also the camera pointing platform of the tank. Dornier integrated a 1 Hz 3-D LADAR system for distance measurement, called *Entfernungsbildkamera* (EBK).

4.6.3 The AUTONAV-Project (1997–2003)

At the end of PRIMUS B in 1995, the efforts to develop technologies for unmanned ground vehicles led to a transnational cooperation between the United States and Germany. In March 1995 a Memorandum of Understanding (MoU) was signed between the US-American and the German Ministries of Defense. It led to the *Autonomous Vehicle Navigation Control System* (AUTONAV)-project that lasted from 1997–2003.

As already mentioned in the US-chapter (see Sect. 2.5.3), the AUTONAV-project codified a technology transfer between the Demo III program (1998–2002) – managed by the US-American Army Research Lab – and similar unmanned ground vehicle programs in Germany.[285] According to Siedersberger, AUTONAV was born out of the two German programs PRIMUS and IFF and the results achieved under both programs contributed to AUTONAV.[286] After the separation from Daimler in 1996, UniBwM received funding from the German Federal Ministry of Defense and

[284] Dickmanns, Ernst Dieter, Forschungsbericht 1.10.1998–30.9.2002. https://dokumente.unibw.de/pub/bscw.cgi/999528 (accessed May 21, 2020).

[285] Albus, James S., and Alexander M. Meystel. 2001. *Engineering of Mind, An Introduction to the Science of Intelligent Systems*, p. 297. New York.

[286] Siedersberger, Karl-Heinz. 2003. *Komponenten zur automatischen Fahrzeugführung in sehenden (semi-)autonomen Fahrzeugen*, Dissertation. Universität der Bundeswehr München, p. 13.

the *Bundesamt für Wehrtechnik und Beschaffung* (BWB).[287] BWB is a civil administration charged with the development, testing and acquisition of military equipment. It is separated from the armed forces but part of Bundeswehr.[288] Funding was used for an already existing project with the internal name *Intelligente Fahrzeug-Funktionen* (IFF).[289] From 1996 on, IFF was bound to the obligation to participate in what was called in the outside communication the AUTONAV-project. IFF became the German academic part of AUTONAV with the third generation vision system of VaMoRs. The Wiesel 2 tank – mentioned above – that Dornier had to be developed into an autonomous test vehicle in the context of the PRIMUS-C program became the industrial part of AUTONAV.

The AUTONAV-project aimed at the development of an autonomous vehicle for both on- and off-road navigation and was made between several industrial, academic and military partners in both countries: The German side was represented by UniBwM and Dornier GmbH Friedrichshafen – since 1989 part of Deutsche Aerospace AG (DASA).[290]

On the American side, the National Institute of Standards and Technology (NIST) in Washington (James Albus) and the Sarnoff Corporation in Princeton (Peter Burt) participated in the AUTONAV agreement.

To understand the cooperation between the two countries, a comparison of the research activities of both laboratories is helpful: In the decade before the 1997 AUTONAV-agreement, UniBwM and Sarnoff had developed complementary technologies for machine vision:

On the one hand, the 4-D approach developed by UniBwM relied on spatiotemporal perception of static and moving objects after the "extraction and analysis of simple spatial features such as edges, lines and corners".[291] This approach proved effective on "paved roads and man-made environments" with distinguishable spatial features.

---

[287] Dickmanns, Ernst Dieter, Forschungsbericht 1.10.1998–30.9.2002. https://dokumente.unibw.de/pub/bscw.cgi/999528 (accessed May 21, 2020).

[288] The separation between administration and armed forces is prescribed in Art. 87b of the constitution of the Federal Republic of Germany. Both are forming the *Bundeswehr*.

[289] An ongoing third-party-funded research project from the BMVg was continued from 1993 under the new name IFF. The first stage of IFF began on February 1, 1993. The second stage of IFF lasted from January 1996 to December 1998. From November 1996 on it became part of the AUTONAV-project. The third stage of IFF lasted from January 1999 to December 2002 (Dickmanns, Ernst Dieter, Forschungsbericht 1.10.1998–30.9.2002, in: https://dokumente.unibw.de/pub/bscw.cgi/999528 (accessed May 21, 2020)).

[290] The Deutsche Aerospace AG (DASA) was founded on May 19, 1989 through the merger of Dornier GmbH - part of the Daimler-Benz group -, the Motoren- und Turbinen-Union (MTU) and two parts of AEG AG. DASA was the largest arms exporter in Germany. In 1995, DASA was renamed Daimler-Benz Aerospace AG. In 1998 it became DaimlerChrysler Aerospace AG. In the year 2000 the company merged with the French Aérospatiale-Matra and the Spanish CASA and became the European Aeronautic Defense and Space Company (EADS). Since 2014 it is called Airbus Group.

[291] Baten, Stefan, Michael Lützeler, Ernst Dieter Dickmanns, Robert Mandelbaum, and Peter J. Burt. 1998. Techniques for Autonomous, Off-Road Navigation. *IEEE-Intelligent Systems and Their Applications* 13 (6), p. 57.

On the other hand, Sarnoff developed area-based techniques for stereo-image analysis that did not require "distinguishable spatial features in the image".[292] For that reason, the system performed robustly in natural and off-road environments and worked also in "unfavorable lighting conditions, where edge-based image processing is prone to failure." One aim of the AUTONAV-agreement was to combine the 4-D perception and control architecture of UniBwM with the area-based stereo-vision techniques of Sarnoff.

4.6.4 The Third-Generation Vision System (EMS) of UniBwM

The following section will present in more detail the third-generation vision system of UniBwM that was developed with VaMoRs from 1997–2004 during the AUTONAV-project. Together with a German industrial partner it had been applied in reduced form to the Hybrid Adaptive Cruise Control (HACC)-project with VaMP.[293] The main goal of EMS-vision was to include all capabilities that UniBwM had developed in the past while adding the detection of negative obstacles (like ditches) during cross-country or off-road navigation.

The third generation vision system was given the name *Expectation-based, Multi-focal Saccadic Vision* (EMS). The term *expectation-based* is a reference to the 4D-approach, *multifocal* means that the system uses two or more focal lengths of the video cameras in parallel. And the term *saccade* had been coined in the 1880s by the French ophthalmologist Émile Javal, describing a rapid movement of both eyes fixing visually different spatial locations. As our eyes do not capture all of a large scene at the same time, they focus with quick movements some details to map interesting parts of the whole image. EMS-vision followed this biological role model. It allowed VaMoRs to "look over the shoulder" with fast eye movements and to identify incoming crossroads in advance, preparing a turning movement (Fig. 4.33).[294]

This was a first difference to the previous vision systems of UniBwM. Second, the advancement of microprocessor development in the mid-1990s allowed for the first time full-frame-rate image sequence processing with *commercial-off-the-shelf* (COTS) hardware.[295] Instead of the very expensive custom-adapted hard-

---

[292] Ibid., p. 58.

[293] Hofmann, Ulrich, Zur visuellen Umfeldwahrnehmung autonomer Fahrzeuge, Dissertation, 2004, UniBwM., Dickmanns, Ernst Dieter. 2002. The Development of Machine Vision for Road Vehicles in the Last Decade. *Intelligent Vehicle Symposium* 1, p. 277.

[294] Dickmanns, Ernst Dieter, 2022. Turning off onto an unknown crossroad. https://dyna-vision.de/application-areas-and-tasks/road-vehicle-guidance/turning-off-onto-an-unknown-crossroads (accessed January 30, 2022).

[295] Gregor, R., M. Lützeler, M. Pellkofer, K.H. Siedersberger, and Ernst Dieter Dickmanns. 2002. EMS-Vision: A Perceptual System for Autonomous Vehicles. *IEEE Transactions on Intelligent Transportation Systems* 3 (1), p. 48 and Dickmanns, Ernst Dieter, H.1.3 EMS vision system on COTS – PC hardware (3rd generation), online in: https://dyna-vision.de/hardware-used/vision-systems/ems-vision-system-on-cots-pc-hardware-3rd-generation (accessed January 30, 2022).

Fig. 4.33 Intersection, seen by the MarvEye-camera system, wide-angle images on the left and the right, tele-image in the center. (Credits: Dickmanns. Image Source: Lützeler, M., Dickmanns, Ernst Dieter, EMS-Vision: Recognition of Intersections on Unmarked Road Networks, in: Proceedings of the IEEE Intelligent Vehicles Symposium 2000 (Cat. No.00TH8511), 2000, pp. 302–307, cit. version found in: https://dyna-vision.de/wp-content/uploads/2020/10/2000-EMS-Vision-RecogIntersectionsOnUnmarkedRoadNetworks.pdf, p. 5)

ware and custom-made software that had been used in the decades before, it was now possible to use computers available on the consumer market and to adapt them to specific needs. This allowed a reduction of costs and time for software development. UniBwM decided in 1996 to use a network of four Intel Pentium dual-processor PC's with 0.3 to 1 GHz clock rate for the third-generation of its vision system.[296] Three of these PC's were used for image processing and one for behavior generation.

Third, it employed more cameras in its "eye". In the two decades before, the Dickmanns-team had always used one (monocular approach) or two (bifocal approach) cameras in their experiments. For the saccadic vision system designed in 1995, called *Multi-focal, active/reactive Vehicle Eye* (MarVEye), the field of view was enlarged on both sides – employing now four cameras (Fig. 4.34).

Two black and white wide angle cameras were used to cover the immediate spatial environment of the robot vehicle (30 m). A color medium tele-camera was centered between these two cameras and assured large look-ahead distances (100 m) "to detect obstacles sufficiently early for collision avoidance".[297] The fourth camera was a highly sensitive b/w camera with a strong tele-lens (300 m) that was part of a longer term planning for early detection of obstacles and intersections. All four cameras were mounted on a two-axis pan-tilt camera platform.

After this short presentation of the EMS vision system we can now come back to the digitized Wiesel 2 robot that was developed during PRIMUS-C and AUTONAV. Its autonomous driving and navigation skills were demonstrated in

[296] Two types of PC's were used: Dual Pentium II with 333 MHz and Dual Pentium III with 450 MHz (Gregor, R., M. Lützeler, M. Pellkofer, K.H. Siedersberger, and Ernst Dieter Dickmanns. 2002. EMS-Vision: A Perceptual System for Autonomous Vehicles. *IEEE Transactions on Intelligent Transportation Systems* 3 (1), p. 4).

[297] Gregor, R., M. Lützeler, M. Pellkofer, K.H. Siedersberger, and Ernst Dieter Dickmanns. 2002. EMS-Vision: A Perceptual System for Autonomous Vehicles. *IEEE Transactions on Intelligent Transportation Systems* 3 (1), p. 1.

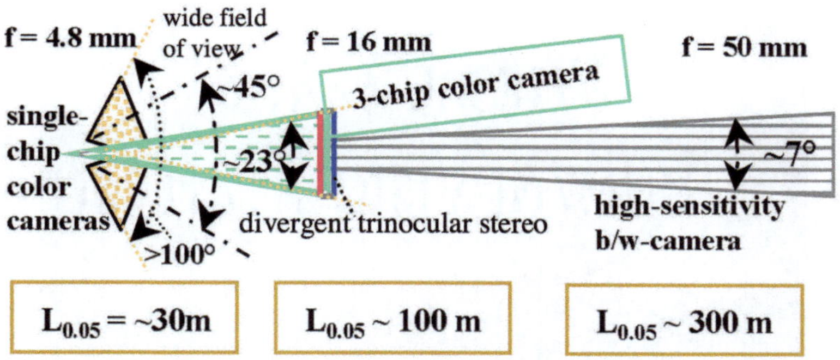

Fig. 4.34 The specification of the four cameras in the MarVEye-Vision system. (Credits: Dickmanns. Image Source: Lützeler, M., Dickmanns, Ernst Dieter, EMS-Vision: Recognition of Intersections on Unmarked Road Networks, in: Proceedings of the IEEE Intelligent Vehicles Symposium 2000, pp. 302–307, cit. version found in: https://dyna-vision.de/wp-content/uploads/2020/10/2000-EMS-Vision-RecogIntersectionsOnUnmarkedRoadNetworks.pdf, p. 4)

October 2000 to an international audience in Pfullendorf (Fig. 4.35).[298] With obstacle avoidance and global navigation, the Wiesel 2 drove autonomously in open terrain with a speed of 25 km/h as reported by Ingo Schwartz. One year before, the tank drove at a speed of 50 km/h on dirt roads, using a contour tracker with a TV-camera.[299]

Also in autumn 2000, a camera-based stereo system (Visual Front End) from the American AUTONAV-partners Sarnoff Corporation and Pyramid Vision Technology (PVT) was integrated into the EMS Vision system. Pursuing the goal to create the world's first *real-time system with full images at full frame rate*, it was tested with VaMoRs of UniBwM, now in its third development stage. VaMoRs was able to stop autonomously in front of a 60 cm wide ditch from speeds up to 16 km/h on a grass surface, a scene that had been captured on video (Fig. 4.36).[300]

During PRIMUS D (2001–2003), this "detection and classification of ditches and trenches" was developed further.[301] As result of the AUTONAV-agreement the system was now compatible with the 4-D/RCS approach of the Demo III program. The architecture was developed by James Albus while UniBwM delivered the theoretical background.

[298] Schaub, Pfändner, Schäfer, p. 157.

[299] Dickmanns, Ernst Dieter, X.4 Table of „Firsts" in 'Dynamic machine vision / Visual guidance of autonomous vehicles', online in: https://dyna-vision.de/survey-of-experimental-results/table-of-firsts-in-dynamic-machine-vision-visual-guidance-of-autonomous-vehicles (accessed on January 30, 2022) and Schwartz, Ingo, Primus: an autonomous driving robot, Proceedings of the SPIE, Vol. 3693, AeroSense Session on Unmanned Ground Vehicle Technology, Orlando (Florida), April 1999.

[300] Detect ditch, Stop VaMoRs 2000 in: https://dyna-vision.de/wp-content/uploads/2020/10/26-VaMoRs-NegObstacles-Series2000EMS-Vision.mp4

[301] Schaub, Pfändner, Schäfer, p. 158.

Fig. 4.35 Wiesel 2 Demo in Pfullendorf, 2000. (Credits: Dickmanns. Image Source: http://dyna-vision.de/main/WebQualVideo/25%20TrackedVehicle%20Wiesel2%20UnsealedRoad%202000.wmv [accessed on June 26, 2020])

Fig. 4.36 VaMoRs stopping autonomously in front of a ditch, September 2000. (Credits: Dickmanns. Image Source: Screenshot, Detect ditch, Stop VaMoRs 2000 in: https://dyna-vision.de/wp-content/uploads/2020/10/26-VaMoRs-NegObstacles-Series2000EMS-Vision.mp4)

During PRIMUS D the Wiesel tank presented above was again used as test platform. The tank had five modes of operation: (1) autonomous driving with waypoint navigation, (2) semi-autonomous driving with an operator who had to give initial commands before the robot started to drive autonomously, (3) remote control with an operator who drove the robot with a joystick and throttle, (4) autonomous road following with the EMS-vision-software of UniBwM and (5) autonomous vehicle following without radio communication to the vehicle in front which is only detected by sensors.[302] For the mode last mentioned "a heated wire formed as a square was fitted at the rear" of the leading vehicle and the following vehicle tracked it with an IR-camera for teleoperation that was adapted to the visual perception module of the EMS-vision software.

It is important to understand that the system was "designed to use only terrain information from its own sensors" – meaning that obstacles were only recognized when they appeared in the range of the LADAR. No further knowledge from GPS or maps was used because in military scenarios these sources are not always available or not trustworthy. But the operator had the possibility to designate forbidden areas, representing for instance mine-fields that are not detectable by LADAR.

In 2002 and 2003 the results of PRIMUS D were shown at a proving ground of the German Armed Forces. At the final presentation in October 2003, the Wiesel "detected, classified and crossed ditches reliably", "drove a distance of more than 1,3 km on an unpaved road passing meadows and a forest" and was able to identify "shadows and additional contours on the road, turnoffs and junctions."[303]

The third phase of the project *Intelligente Fahrzeug Funktionen* (IFF), part of AUTONAV, ended in October 2001[304] with the final demo of the AUTONAV project at the former military airport Neubiberg. VaMoRs demonstrated lane keeping, turning into a crossroad while estimating its angle of intersection, the distance to it and its road width, as well as the transition from road driving to cross-country driving, following a sequence of GPS-way points and visual landmarks (way point navigation).[305] Most importantly VaMoRs was now able to identify and avoid positive and negative obstacles (ditches), stop in front of them or go around. After this demo in October 2001, the AUTONAV-project was prolonged with the aim to reduce the size of the US-American stereo vision system (30 liter volume) to a standard board that could be inserted into one of the computers of the EMS-vision system. This was the main progress from 2001–2003.

Two additional AUTONAV-demos followed in 2003, one in Germany (Neubiberg), the other one in the US (Fort Knox). Again, the detection of positive and negative obstacles was demonstrated on a navigable lawn with a trench of

---

302 Ibid., p. 160.

303 Ibid., p. 164.

304 Hofmann, p. 15.

305 Siedersberger, Karl-Heinz. 2003. *Komponenten zur automatischen Fahrzeugführung in sehenden (semi-)autonomen Fahrzeugen*, Dissertation. Universität der Bundeswehr München, p. 19.

unknown position and direction. A very impressive video of the test run in Neubiberg shows how the MarVEye-system – now with five cameras – turned its gaze on the two-axes platform to prepare a right turn of the vehicle.[306] Then VaMoRs followed a sequence of GPS-waypoints and autonomously stopped in front of a ditch in the grassland. Since the entire ditch could not be covered from a single gaze direction, the system performed saccadic gaze control like the human eye – the stereo vision system turned to both sides forth and back. The vehicle then decided to bypass the ditch at the right-hand side.

The experiment showed that VaMoRs was able to detect, perceive and drive around a negative obstacle. Most importantly, the demonstration showed that the system was not only automated but *autonomous* – it made decisions that were not codified in advance as it is the case in *automated* routines. It had the capability of performing an avoidance maneuver around a negative obstacle at a safe distance.

4.7 Conclusion

As in Japan, the first European experiments on automated driving in the 1960s were influenced by the US-American concepts of RCA and GM. The first country in Europe that started tests with guidance-cables (1960) was the UK. Germany followed with initiatives of Continental and Siemens (1968) and MAN and Bosch (1971). Depending on control towers, these experiments were reminiscent of concepts from the 1950s. But they also experimented with sensor technologies (radar, infrared) that became now available for the automotive sector. In contrast to the US-robotics research of these years, nobody in Europe experimented with camera-based vehicle guidance at this stage.

During the 1970s the guidance cable approach got abandoned in Europe and was substituted by research on machine vision for vehicle control. The mobile robot HILARE from LAAS, inspired by Shakey and the Stanford Cart, was the first French mobile robot able to navigate autonomously through an unknown environment. Ernst Dieter Dickmanns from the University of Bundeswehr Munich (UniBwM) selected machine vision as long-term research goal for vehicles on the ground, in the air and in space-flight scenarios. He started in 1979 with laboratory experiments on visual motion control, using a small pole-balancing cart, an air-cushion vehicle for satellite- docking, and a simulation loop for road vehicle guidance.

In contrast to the US, military use cases of autonomous systems had not been a main topic in Europe until the mid-1980s. This was also the moment when the automotive industry became interested in the field: Volkswagen tested in 1985 the

[306] Videofilm "On-Off-road mission performance". https://dyna-vision.de/wp-content/uploads/2021/03/2003%20EMS-On-OffroadDriving%20VaMoRs.mp4 (accessed January 30, 2022).

experimental vision system Opto-Pilot for lateral control on a highway and Daimler started to work on automated driving. They joined the much more advanced efforts of UniBwM, an institution that realized already in 1986 its first test drives with a real-time system for the analysis of image sequences aiming at fully autonomous longitudinal and lateral driving on the Autobahn by computer vision. This motivated Daimler in 1986 to start the public private partnership *Autonom Mobile Systeme* with UniBwM. Equipped with technology from UniBwM, a Spurbus of Daimler demonstrated 1988 in Rastatt fully autonomous driving with speeds of up to 60 km/h and obstacle detection as well as stopping in front of it at a speed of 40 km/h. One year before, UniBwM performed the first fully autonomous high-speed demo with the VaMoRs-van on a free real highway with speeds up to 96 km/h.

In 1986 Daimler initiated the European Prometheus-project that pushed during its research phase (1988–1994) the development of communication technologies in the mobility sector. During the first phase of the project, Daimler used the hard- and software of UniBwM for the intelligent cruise control of its VITA research vehicle (1991) that was as VaMoRs exclusively based on vision, without radar. In parallel, Daimler insisted in switching to the newly appeared "Transputers" for the next step, since they allowed four direct parallel connections to neighboring processors as well as a reduction in size of the overall system. This development had been done by ISF-UniBwM with support from the company Parsitec that delivered the transputers in Germany. This yielded the second-generation vision system of UniBwM that was used exclusively in the final phase of PROMETHEUS.

In 1993 Daimler and UniBwM each acquired a Mercedes S-Class vehicle (VaMP and VITA- 2) and demonstrated at the final demo of PROMETHEUS 1994 in Paris fully autonomous driving up to the maximum speed allowed in France of 130 km/h, with lane changes and passing in actual public three-lane highway traffic.

In parallel and independent from Prometheus the German Federal Ministry of Defense (BMVg) started in 1989 the experimental robotics program PRIMUS that focused on autonomous driving in open terrain. UniBwM cooperated with Dornier and delivered the vision system for a military Jeep that demonstrated its capabilities in 1992. Early in the 1990s, VaMoRs has also been equipped with the new transputer vision system. It has been used for developing the capability of recognizing a crossroad and turning off onto it.

Throughout the 1990s, military interest in autonomous vehicle technology was high, not only in Europe: In 1997 the AUTONAV-agreement between the German and the US-American Ministries of Defense, focusing on off-road navigation, was born out of contacts over several years between Jim Albus (the initiator from NIST) and Ernst D. Dickmanns in connection with the PRIMUS-program. UniBwM, Dornier and DASA cooperated with NIST and Sarnoff to work on the detection of negative obstacles like ditches and trenches. In this context UniBwM developed the EMS-vision system that employed technical "eyes" with four cameras able to perform saccadic gaze changes. With a pointable field of view, VaMoRs was now able to identify crossroads on both sides at longer distances. UniBwM provided the EMS-vision system for a Wiesel-2 tank; it was capable of autonomous road following as well as obstacle detection and avoidance. In the early 2000s the Wiesel tank successfully followed the road autonomously and detected ditches and trenches.

In these years, many other research groups had entered the field, but their efforts were focused on driver information (France), warning and cruise control systems (UK) and convoy driving (Volkswagen in Germany). From the mid-1990s on, France began to focus its research efforts on low-speed autonomous vehicles for cities, aiming at a hybrid between public transport and individual automobiles. Platooning, lane changing, obstacle detection and automatic parking functions had been developed by the public research institution INRIA. In 1997 a prototype called CyCab, a 20 km/h-vehicle without steering interface was presented. In 2000, the concept of cable guidance was reactivated with another vehicle called CyberCab. Inspired by the US-American highway platooning-experiments, France started in 1997 its own project *La Route Automatisée*, combining research on low-speed automation for suburban areas and highway automation.

In reaction to the semi-autonomous long-distance drive of the NavLab 5 in the United States, UniBwM realized in 1995 with the VaMP-Mercedes a fully autonomous trip of about 1.600 kilometers from Munich to Odense, demonstrating 400 autonomous lane changing maneuvers. The goal was acquiring statistical data from many real-world situations when the system decided to quit autonomous driving; this was to be used for improvements for the third-generation vision system planned. In 1998 the University of Parma followed with a long-distance drive of 2.000 km through Italy, using machine vision for autonomous lateral control.

Bibliography

Albus, James S., and Alexander M. Meystel. 2001. *Engineering of Mind, An Introduction to the Science of Intelligent Systems*. New York.

Barbaza, Pierre. 1984. Profession Pilote Essayeur. *Echappement*, 90.

Baten, Stefan, Michael Lützeler, Ernst Dieter Dickmanns, Robert Mandelbaum, and Peter J. Burt. 1998. Techniques for Autonomous, Off-Road Navigation. *IEEE-Intelligent Systems and Their Applications* 13 (6): 57–65.

Beltran, Alain, and Pascal Griset. 2007. Des véhicules intelligents pour assurer la sécurité routière? *Code Source, L'hebdomadaire des 40 ans de l'INRIA*, No 36, November 5, 2007, p. 1, online in: https://www.inria.fr/sites/default/files/2019-11/Code_Source_1998-2007.pdf. Accessed October 8, 2020.

Bertozzi, Massimo, Alberto Broggi, Gianni Conte, and Alessandra Fascioli. 1997. Obstacle and Lane Detection on ARGO, *Proceedings of IEEE Intelligent Transportation Systems conference 1997*, online in: http://www.ce.unipr.it/people/broggi/publications/iv97.pdf.

Bertozzi, Massimo, Alberto Broggi, and Alessandra Fascioli. undated. The Evolution of Vision-based Unmanned Ground Vehicles: the VisLab Perspective (undated, without publication source), in: http://citeseerx.ist.psu.edu/viewdoc/download?doi=10.1.1.73.7759&rep=rep1&type=pdf. Accessed July 20, 2020.

Bertozzi, Masssimo, Alberto Broggi, and Alessandra Fascioli. 2006. VisLab and the Evolution of Vision-Based UGVs. *Computer* 39 (12): 31–38.

Blosseville, J.M., and Saïd Mammar. 2000. LaRA Update, the French Program La Route Automatisée, Powerpoint-Presentation, 2000. http://aramis.iup.univ-evry.fr:8080/~smam/publications/publis2000/lara_2000.pdf. Accessed October 12, 2020.

Braid, Deborah, Alberto Broggi, and Gary Schmiedel. 2006. The TerraMax Autonomous Vehicle, Field Report. *Journal of Field Robotics* 23 (9): 693–708.

Brauckmann, M.E., C. Goerick, J. Groß, and T. Zielke. 1994. Towards All Around Automatic Visual Obstacle Sensing for Cars, *Proceedings of International Symposion on Intelligent Vehicles*, 79–84. Paris.

Broggi, Alberto, Masssimo Bertozzi, Alessandra Fascioli, Lo Bianco, Corrado Guarino, and Aurelio Piazzi. 1999a. The ARGO autonomous vehicle's vision and control systems. *International Journal of Intelligent Control and Systems* 3 (4): 409–441.

Broggi, Alberto, Massimo Bertozzi, and Alessandra Fascioli. 1999b. ARGO and the MilleMiglia in Automatico Tour. *IEEE Intelligent Systems and Their Applications* 14 (1): 55–64.

Broggi, Alberto, Paolo Grisleri, Chris Yakes, and Chris Hubert. 2007. The Oshkosh-VisLab Joint Efforts on UGVs: Architecture, Integration, and Results. Platform Innovations and System Integration for Unmanned Air, Land and Sea Vehicles, 146th NATO Workshop on Applied Vehicle Technology Panel, Florence 2007, pp. 16/1–14, online in: http://www.ce.unipr.it/people/broggi/publications/natoavt146.pdf.

Broggi, Alberto. 2008. History of AHS in Italy and future issues. *Proceedings of the 2008 IEEE International Conference on Vehicular Electronics and Safety*, Columbus, OH, USA, September 22–24, 2008, pp. 11–12.

Brüdigam, Claus. 1994. *Intelligente Fahrmanöver sehender autonomer Fahrzeuge in autobahnähnlicher Umgebung*, Dissertation. Universität der Bundeswehr München, Fakultät für Luft- und Raumfahrttechnik.

Bundeswehr, Die Laufbahn der Offiziere (M/W/D). 2020. https://www.bundeswehr.de/de/ueber-die-bundeswehr/dienstgrade-laufbahnen-bundeswehr/laufbahn-offiziere. Accessed Mai 8, 2020.

Cardew, K.H.F. 1970. The Automatic Steering of Vehicles: An Experimental System Fitted to a DS 19 Citroen Car, Driver Aids and Abilities Section, Road Research Laboratory, Crowthorne, Berkshire, RRL report LR 340 (1970), online in: https://trl.co.uk/sites/default/files/LR340.pdf. Accessed June 22, 2019.

Chatila, Raja. 1996. Autonomous Mobile Robots in France: Some Projects and Scientific Approaches. In *Autonome Mobile Systeme 1996*, ed. Günther Schmidt and Franz Freyberger, 2–7. München.

CNRS, La science de la machine intelligente, Interview with Georges Giralt, interviewed by Patrick Danès, *La lettre du LAAS*, N° 37, October 2008, p. 58, online in: https://www.laas.fr/LettreDuLaasStatic/content/download/793/4835/version/3/file/La-Lettre-n37–5.pdf.

Continental AG, Controlled by Electronics: Continental Launched Its First Driverless Vehicle 50 Years Ago, Press Release September 11, 2018, online in: https://www.continental.com/en/press/press-releases/controlled-by-electronics-144932. Accessed September 25, 2020.

Continental AG, Controlled by Electronics: Continental Launched Its First Driverless Vehicle 50 Years Ago, Press Release September 11, 2018, 50 Jahre fahrerloses Fahren: Pressematerial, online in: https://publicarea.admiralcloud.com/p/a49d3d8ba92f3cdbfa864f. Accessed September 25, 2020a.

Continental AG, Controlled by Electronics: Continental Launched Its First Driverless Vehicle 50 Years Ago, Press Release September 11, 2018, The "Ghost Car", online in: https://www.continental.com/en/company/history/the-ghost-car-145316. Accessed September 25, 2020b.

Daimler, The PROMETHEUS project launched in 1986: Pioneering autonomous driving, September 20, 2016a, online in: https://media.daimler.com/marsMediaSite/ko/en/13744534.

Daimler, Das Projekt PROMETHEUS ab 1986: Vorreiter des autonomen Fahrens, September 20, 2016b, online in: https://media.daimler.com/marsMediaSite/de/instance/ko.xhtml?oid=13744534.

Delcker, Janosch. 2018. The man who invented the self-driving car. *Politico*, July 19, 2018, online in: https://www.politico.eu/article/delf-driving-car-born-1986-ernst-dickmanns-mercedes/.

Dickmanns, Ernst Dieter. 1980. Untersuchungen und mögliche Arbeitsschritte zum Thema "Künstliche Intelligenz: Rechnersehen und -steuerung dynamischer Systeme".

———. 1986. Computer Vision in Road Vehicles – Chances and Problems. *ITCS-Symposion on "Human Factors Technology for Next-Generation Transportation Vehicles"*, Amalfi, Italy.

———. 1987. PROMETHEUS, 11 170 Integrated Approaches, State of the art review, March 1987.

———. 1988. An Integrated Approach to Feature Based Dynamic Vision. *Proceedings of the Computer Society Conference on Computer Vision and Pattern Recognition*, Ann Arbor, pp. 820–825.

———. 1993. Interview, Der Lotse geht an Bord. *Hochschulkurier*, pp. 6–7.

———. 1994. The 4D-Approach to Dynamic Machine Vision. *Proceedings of the 33rd Conference on Decision and Control*, pp. 3770–3775.

———. 2002. The Development of Machine Vision for Road Vehicles in the Last Decade. *Intelligent Vehicle Symposium* 1: 268–281.

———. Contributions to Visual Autonomous Driving, Part I, 2015, online in: http://dyna-vision.de/main/Text_PDF/2015%20ContribPart%20I%20to%20Visual%20Auton.%20Driving%20EDD.pdf. Accessed October 6, 2020.

———. 2016. Summary of contributions to Vision-based Autonomous Vehicles. http://www.dyna-vision.de/main/Review2016%20of%20ContribVisBasAutVeh.html.

———. May a Pair of 'Eyes' Be Optimal for Vehicles Too ?, *Electronics*, 2020, 9, 759, published on May 5, 2020, online in: https://www.mdpi.com/2079-9292/9/5/759.

Dickmanns, Ernst Dieter, and A. Zapp. 1985. Guiding Land Vehicles Along Roadways by Computer Vision. *Proceedings Congrès Automatique 1985*, AFCET, Toulouse, pp. 233–244.

———. 1986. A Curvature-Based Scheme for Improving Road Vehicle Guidance by Computer Vision. *Mobile Robots, SPIE Proceedings*, vol. 727, 161–168. Cambridge, MA.

———. 1987. Autonomous high speed road vehicle guidance by computer vision. *Proceedings of the International Federation of Automatic Control (IFAC)* 20 (5, Part 4): 221–226.

Dickmanns, Ernst Dieter, and Volker Graefe. 1988. Sehendes Auto fährt Weltrekord. *Hochschulkurier* 14: 17–23.

Dunne, Jim. 1966. This 140-m.p.h. Test Car is Driven by Remote Control. *Popular Science*, pp. 82–83.

Ellwein, Thomas, Achatz von Müller, and Harro Plander. 1974. *Hochschule der Bundeswehr zwischen Ausbildungs- und Hochschulreform*, Opladen.

Firgau, Walter. Ohne Fahrer auf richtigem Kurs. *Süddeutsche Zeitung*, September 6, 1973.

Giles, C.G., and J.A. Martin. 1961. Cable Installation for Vehicle Guidance Investigations in the New Research Track at Crowthorne, Report RN/40 57/CGG. Crowthorne: Road Research Lab.

Giralt, Georges, R.P. Sobek, and Raja Chatila. 1979. A multi-level planning and navigation system for a mobile robot; a first approach to HILARE. *Proc. 6th IJCAI*, Tokyo, Japan.

Giralt, Georges. 1990. Robots et intelligence des machines. *Les Cahiers du MURS* n° 22, 4ème trimestre, pp. 27–47.

Graefe, Volker, and Klaus-Dieter Kuhnert. 1992. Vision-Based Autonomous Road Vehicles. In *Vision-Based Vehicle Guidance*, ed. Ichiro Masaki, 1–29. New York.

Gregor, R., M. Lützeler, M. Pellkofer, K.H. Siedersberger, and Ernst Dieter Dickmanns. 2002. EMS-Vision: A Perceptual System for Autonomous Vehicles. *IEEE Transactions on Intelligent Transportation Systems* 3 (1): 48–59.

Hagen, Hans. 1974, April 19. *Lastwagen ohne Fahrer*. VDI-Nachrichten.

Hartwich, Erwin, and Walter Zimdahl. 2012. Regelung von Straßenfahrzeugen: Ein Zeitzeugenbericht aus dem Institut von Prof. W. Oppelt, TH Darmstadt. *at-automatisierungstechnik* 60 (6): 338–343.

INRIA, Rapport d'activité 1997, Action Praxitele. https://raweb.inria.fr/rapportsactivite/RA-Dev97/PRAXITELE/. Accessed October 11, 2020.

Knop, Carsten. 2018. Ein Deutscher war Tesla 30 Jahre voraus. *Frankfurter Allgemeine Zeitung*, February 1, 2018, online in: https://www.faz.net/aktuell/wirtschaft/autonomes-fahren-ein-deutscher-war-tesla-30-jahre-voraus-15424550.html.

Laboratoire Central des Ponts et Chaussées (LCPC), *LA ROUTE AUTOMATISÉE, Un scenario périurbain*, Brochure, May 2001.

Laugier, Christian, S. Sekhavat, F. Large, J. Hermosillo, and Z. Shiller. 2001. Some steps towards autonomous cars. *IFAC Symposium*, 9–17. Japan.

Laugier, Christian. 2002. Towards "Automated Road", Some Projects and New Technologies, Keynote ITSC 2002 Singapore, online in: http://emotion.inrialpes.fr/laugier/Research/itsc02.pdf. Accessed October 12, 2020.

———. 2019. A Journey into the History of Automated Driving, Invited pioneer's talk. *IROS 2019 – IEEE/RSJ International Conference on Intelligent Robots and Systems.*, Nov 2019, Macau, China, in: https://hal.inria.fr/hal-02428196/document. Accessed November 11, 2020.

Laurgeau, Claude. 2009. *Le siècle de la voiture intelligente*. Paris.

Lossau, Norbert. Das erste autonome Auto kostete 200.000 D-Mark. *Die Welt*, October 13, 2017.

MAN Bus & Trucks, Controlled by an invisible hand, January 28, 2020, online in: https://www.mantruckandbus.com/en/moments/1971.html. Accessed September 25, 2020.

Massot, Marie-Hélène. 2000. Praxitèle: un concept, un service et une expérimentation, Bilan d'un prototype. *Transports Environment Circulation* 59: 25–32.

Maurer, M., R. Behringer, S. Fürst, F. Thomanek, E.D. Dickmanns. 1996. A Compact Vision System for Road Vehicle Guidance. *Proceedings of 13th International Conference on Pattern Recognition*, pp. 313–317.

Meissner, Hans-Georg. 1982. *Steuerung dynamischer Systeme aufgrund bildhafter Informationen*, Dissertation UniBwM/LRT.

Meyer, Wolfgang. 1968. Ein Auto ohne Fahrer rast mit 120 Sachen über die Beton-Piste. *Bild*, September 13, 1968.

Mysliwetz, Birger. 1990. *Parallelrechner-basierte Bildfolgen-Interpretation zur autonomen Fahrzeugsteuerung,* Dissertation, Universität der Bundeswehr München, Fakultät für Luft- und Raumfahrttechnik. Institut für Systemdynamik und Flugmechanik.

Özgüner, Umit, Keith A. Redmill, and Alberto Broggi. 2004. Team TerraMax and the DARPA Grand Challenge: A General Overview. *Intelligent Vehicles Symposium*, 232–237. University of Parma.

Parent, Michel. 1995. Le future de l'automobile, Présentation INRIA-IRISA, December 3, 1995. https://www.irisa.fr/manifestations/seminaires-1999/3dec99/parent/sld007.htm. Accessed October 12, 2020.

Paromtchik, Igor E., and Christian Laugier. 1997. Automatic Parallel Parking and Returning to Traffic Maneuvers. *Proceedings of the 1997 IEEE/RSJ International Conference on Intelligent Robot and Systems, Innovative Robotics for Real-World Applications, IROS '97, Grenoble*, France, Vol. 3, pp. V21–V23, in: http://emotion.inrialpes.fr/~paromt/infos/papers/paromtchik:laugier:video:iros:1997.pdf.

Paromtchik, Igor, and Christian Laugier. 1996a. Motion Generation and Control for Parking an Autonomous Vehicle. *Proceedings of the IEEE International Conference on Robotics and Automation*, 3117–3122. Minneapolis.

———. 1996b. Autonomous Parallel Parking of a Nonholonomic Vehicle, 1996. *Proceedings of the IEEE Intelligent Vehicles Symposium*, Tokyo, Japan, September 19–20, 1996, pp. 13–18, online in: http://emotion.inrialpes.fr/~paromt/infos/papers/paromtchik:laugier:iv:1996.pdf.

Plet, Cécile. 2001. Michel Parent, “Les voitures entièrement automatiques ? Vous en verrez dans 5 ans.” *Internetactu.net*, February 22, 2001, online in: http://www.internetactu.net/2001/02/22/michel-parent-les-voitures-entirement-automatiques-vous-en-verrez-dans-5-ans/. Accessed October 8, 2020.

Prätorius, Gerhard. 1993. *Das PROMETHEUS-Projekt, Technikentstehung als sozialer Prozeß*. Wiesbaden.

Prometheus – Proart (ed.), Rapport de fin de contrat de recherche, 1994, December.

Reuters, U.K.. 1960. New Road Research Track Opened at Crowthorne, October 25, 1960, in: https://reuters.screenocean.com/record/547949. Accessed June, 25 2019.

Reynolds, John. 2001. Cruising into the Future. *Telegraph.*, May 26, 2001, online in: https://www.telegraph.co.uk/motoring/4750544/Cruising-into-the-future.html. Accessed June 12, 2019.

Rives, P., and L. Marce. 1985, October. Use of Moving Vision Sensors to Control Robots in an Unknown Universe. *Proceedings of the 5th International Conference on Robot Vision and Sensory Controls*, 165–176. Amsterdam.

Rousselot, Michel. 2001. Préface. In *La route automatisée, Un scenario périurbain*, ed. LCPC. Paris.

Schaub, Günter, Alfred Pfändner, and Christoph Schäfer. 2004. PRIMUS: Autonomous Navigation in Open Terrain with a Tracked Vehicle. *Unmanned Ground Vehicle Technology VI*, ed. Grant R. Gerhart, Chuck M. Shoemaker, and Douglas W. Gage, vol. 5422, 156–165. Proceedings of SPIE.

Siedersberger, Karl-Heinz. 2003. *Komponenten zur automatischen Fahrzeugführung in sehenden (semi-)autonomen Fahrzeugen*, Dissertation. Universität der Bundeswehr München.

Ulmer, Berthold. 1994. VITA II – Active Collision Avoidance in Real Traffic. *Proceedings of the Intelligent Vehicles Symposium 1994*, Paris, pp. 1–6.

Universität der Bundeswehr München, Präsidialabteilung, Struktur- und Entwicklungsplan der Universität der Bundeswehr München 2015–2020, München 2015, online in: https://www.unibw.de/universitaet/universitaet/leitung/struktur-und-entwicklungsplan-unibw-m-2015-2020.pdf. Accessed May 7, 2020.

Vieweg, Christof. 2015, July 31. Wer hat das Roboterauto erfunden ? Die Bundeswehr! *DIE ZEIT*.

Volkswagen Newsroom, Automatically bound for the future, The history of autonomous driving at Volkswagen, August 24, 2018, online in: https://www.volkswagen-newsroom.com/en/stories/automatically-bound-for-the-future-4118. Accessed March 17, 2020.

Waugh, Rob. 2013. How the first “driverless car” was invented in Britain, *Yahoo News UK*, July 17, 2013, online in: https://uk.news.yahoo.com/how-the-first%2D%2Ddriverless-car%2D%2Dwas-invented-in-britain-in-1960-093127757.html. Accessed September 29, 2020.

Winner, Hermann. 2015. PROMETHEUS – Welche Visionen wurden Realität? *VDA, Technischer Kongress*, pp. 93–124.

Wünsche, Hans-Joachim. 1986. Detection and Control of Mobile Robot Motion by Real-Time Computer Vision. In *Advances in Intelligent Robotics Systems, Proceedings of the Society of Photo-Optical Instrumentation Engineers (SPIE)*, ed. Nelson Marquina, Vol. 727, 100–109.

———. 1987. *Bewegungssteuerung durch Rechnersehen*, Dissertation. Universität der Bundeswehr München, Fakultät für Luft- und Raumfahrttechnik, Institut für Systemdynamik und Flugmechanik.

Yazigi, Monique P. 1992. Promethean Plans for the Next Generation of Cars. *New York Times*, September 13, 1992.

Zapp Alfred. 1988. *Automatische Straßenfahrzeugführung durch Rechnersehen*, Dissertation. Universität der Bundeswehr München, Fakultät für Luft- und Raumfahrttechnik, Institut für Systemdynamik und Flugmechanik.

Zimdahl, Walter. 1963. *Die Bewegung des Straßenfahrzeugs auf vorgeschriebener Bahn*, Diploma Thesis, 1963, Institut für Regelungstechnik, Technische Hochschule Darmstadt (supervised by Fritz Frank).

———. 1965. Führungsverhalten des vierrädrigen Straßenfahrzeugs bei Regelung des Kurses auf festgelegter Bahn. *Regelungstechnik, Zeitschrift für Steuern, Regeln und Automatisieren* 13 (5): 221–268. online in: https://doi.org/10.1524/auto.1965.13.112.221. Accessed October 1, 2020.

———. 1969. *Regelungstechnische Probleme bei dem Leitkabelgeführten Kraftfahrzeug*, Dissertation. Fakultät für Elektrotechnik, Technische Hochschule Darmstadt.

Zimdahl, W., I. Rackow, and T. Wilm. 1986. OPTOPILOT – ein Forschungsansatz zur Spurerkennung und Spurführung bei Straßenfahrzeugen. *VDI Berichte Nr.* 162: 49–60.

Chapter 5
Findings of the Inquiry

Our inquiry was started with the aim to identify the main *actors* who participated in the invention of automated and autonomous driving. Peter Bowler emphasizes in his book *The history of the future* that we should forget the idea that "each new technology has a single inventor".[1] According to him, this idea is a product of the patent system. Scientists and inventors are not "isolated individuals". New technologies are "products of large research and development teams" and therefore they do not have "easily identifiable origins". Even when "single individuals were recognized as the key innovators", new systems needed almost "a sequence of inventions" before they were commercialized.

On the one hand, this seems to be an accurate statement for our topic as well: The three case studies have shown that automated and autonomous driving have no single inventor but multiple origins. They are based on a large number of older inventions. If one aims to designate an inventor of vehicle automation at all, the groups and researchers of all institutes and companies in the three regions mentioned in this book should be considered as the inventors. Automated driving was not invented by one person, at one time and place, but by generations of engineering teams in different countries who all participated in the development of very different concepts to realize this dream. The definition of automated driving was never static, it shifted between different expectations and strategies, ideas and concepts, contexts and interests. Constituted of *histories* in the plural, automated driving oscillated throughout its 50 years old history between approaches of electrical engineering, of artificial intelligence and robotics as well as control engineering.

But on the other hand we should also admit that the different concepts of the different actors led to very different results on behalf of full autonomy: As discussed in the introduction, it makes a big difference if a research vehicle has automated or autonomous capabilities. Vehicles with *autonomous* decision-making are much

---

[1] Bowler, Peter. 2017. *A history of the future, Prophets of Progress from H.G. Wells to Isaac Asimov*, p. 15. Cambridge.

F. Kröger, *From Automated to Autonomous Driving*, Archimedes 70,
https://doi.org/10.1007/978-3-031-49881-7_5

more complex than vehicles with *automated* driving functions. For that reason, their invention is a much more significant event. It should also be recognized that certain concepts – for instance the 4D-approach – can be attributed to a single inventor. Ernst Dieter Dickmanns convinced institutions and persons to fund the research that was necessary to develop this concept over several years.

An important finding of our study is the fact that throughout history, two oppositional concepts of vehicle automation were explored:

The first approach was based more on the infrastructure, the second more on the vehicle. The guidance cable approach belongs to the infrastructure-based approach – even if a certain modification of the vehicle was also necessary, as it needed sensors to capture the signal of the wire in the road. The dynamic machine vision-approach of Dickmanns belongs to the second category, aiming at a vehicle that is able to navigate on common roads without any modifications.

The inquiry has shown that several *shifts* between both concepts can be observed in history. In the 1950s – due to missing computer power and missing in-vehicle sensors – research was more concentrated on concepts relying on infrastructure and corresponding antennas in the vehicle. In the 1970s and 1980s the idea of an automated vehicle that approaches *autarchy* became more important. Of course, both approaches were often combined to achieve the desired goals – GPS-guidance for example relies on infrastructure (satellites) outside the vehicle, but its information is always combined with vehicle-based information from radar and vision for obstacle avoidance.

Until today, the relation between infrastructure and vehicle remains an important question: Most of the actual approaches of autonomous driving are based on highly accurate digital maps of the environment. Data delivered by the sensors of each vehicle is compared with those from external databases. These data-driven vehicles cannot navigate in areas without cartography. As they depend on external systems they will never be completely autonomous.

The historian David Edgerton analyzed in his book *The Shock of the old* how old technologies continue to influence the development of new technologies. This can also be observed in the history of automated driving: It is striking that guidance cables, magnetic markers (PATH) and GPS-waypoints have in common to guide a vehicle along a physical or virtual leash, as Dickmanns claimed: "Most of today's autonomous cars are being pulled along the streets by the GPS satellite navigation system on a virtual leash".[2] In this perspective the concept of an infrastructure guided vehicle had never been abandoned – physical cables were only replaced by virtual cords. This is an important difference to the UniBwM-vehicles that Dickmanns conceptualized as "individuals" who should be able to take decisions on their own, only based on camera data, without network-based data. As we have seen, they were able to recognize relevant landmarks in their environment, using different sensors and knowledge bases and imitating human perception principles.

[2] Lossau, Norbert. 2017. Das erste autonome Auto kostete 200.000 D-Mark. *Die Welt*, October 13 (Translation German to English: FK).

Following a pure definition of autonomy, they were able to navigate in unknown terrain, without data from the global GPS-system and without precise maps. Even if all communication networks would go offline, these vehicles would continue to operate. This is an important argument for considering them as the first really autonomous vehicles.

Another goal of our inquiry was to identify who started and who followed. We have seen that the identification of firsts is not always as easy as it may appear. Sometimes research institutions claim that they did things for the first time, either for marketing reasons or because they simply do not know about other inventors.

Ann Johnson emphasized that the answer to the question of the first depends on a precise definition. Sometimes, these definitions are difficult to give.[3] In other cases, it is easier to answer the question of the first, for example while claiming that VaMoRs was the first high-speed autonomous vehicle in history.

5.1 Comparison of the Research Efforts in the Three Regions

The semiotician Ferdinand de Saussure distinguished two analytical perspectives on language: A *diachronic* perspective observes the historical change of language in longer periods while a *synchronic* perspective analyzes language at a certain point in time.[4] These terms are also useful for historical research. In the three parts of the book we followed a diachronic perspective, focusing on the chronological succession of events in the *longue durée* in each region. The following paragraph will resume the history of automated and autonomous driving in a *synchronic* perspective, allowing a closer *comparison* of the transnational technology development in the US, Japan and Europe at the same time in history. This comparative perspective allows to highlight the contacts, the *differences and common elements* of the research efforts in the US, Japan and Europe. It also allows paying attention to time, focusing on the *simultaneity* or *non-simultaneity* of the research efforts: Often several regions were engaged in the same research directions but not at the same time. Therefore the research process can be compared to a ping-pong-game: On the one hand, when experiments undertaken in one region were promising, researchers in other regions also started to work in the same field – for instance when Japan and the UK started cable-guidance experiments following US-concepts. In this case, it was the

[3] For instance, asking who did the *first AHS-demo worldwide* raises the question how to define such a demo: Can a platooning experiment with two cars already count as an AHS-demo? While *experiments* suggest an unstable knowledge status, a *demo* promotes the nearly successful development of a technology – it has the aim to show that a system "is working". But is a demo only a demo when several laboratories participated? Must a demo necessarily include an audience, perhaps even an international audience? Whether these questions are answered in the one or the other direction, the first platooning demo must be attributed to Japan or the United States.

[4] Berger, Arthur Asa. 2005. *Media Analysis Techniques*, p. 18f. Thousand Oaks, London, New Delhi.

non-simultaneity of research that created a pull dynamic. On the other hand, effects of repercussion pushed several regions to reinforce investments in the same direction at the same time, e.g. when the US and Germany invested both in machine vision at the beginning of the 1980s. In this case, a dynamic situation was the result of *simultaneous* research efforts.

The synchronic perspective allows to identify transnational historical *paradigms*. Therefore we will come back to the *periods* coined by Tsugawa with the aim to refine them as they do not always follow each other exactly but overlap at some occasions. Between these paradigms there are *shifts and turning points* that can be dated. Finally, a synchronic perspective has the advantage to reveal those transnational *long historical trends* that Möser suggested to identify.

5.1.1 1950–1960: Guidance Cables and Mobile Robots

The inquiry has shown that the research on automated driving was first dominated by concepts of electrical engineering. The exploration of the guide-wire was the dominant approach to vehicle automation worldwide during the 1950s and 1960s. In all three regions the growing problems of the mainly automobile-based traffic system (pollution, flux capacity, accidents) constituted a favorable background for these studies on automation. Experiments with this infrastructure-based system were first started at the end of the 1950s in the United States – but Japan and the United Kingdom followed quickly at the beginning of the 1960s. In Germany, guidance cables were only used at the end of the 1960s to test tires, not to develop automated driving.

But during the 1960s with the increase in computing power another approach independent from modifications of infrastructure gained momentum: The idea of a visually guided vehicle – finding orientation with the help of cameras – was first explored in the US robotics field, using computer science and artificial intelligence-approaches.

5.1.2 1970s: Shift to Machine Vision Approaches

It took more than one decade before this idea was used for road vehicles: In the 1970s a decline of infrastructure based concepts (guidance-cable) and an international shift to machine vision can be observed. This new path or trajectory of technology development for autonomous driving marginalized the guidewire paradigm that lost confidence. Instead, the research on technical images, cameras and image processing technologies that had been started during the 1960s in the US became an important field of experimentation for a growing number of research groups. However, sufficiently powerful *digital* (micro-) processors were available only in laboratory environments, not for use in a vehicle.

In the late 1970s the *idea of a visually guided vehicle* became more valued by the engineering community that worked on road vehicle automation and planetary rovers. It became a new technical *Leitbild* beyond the robotics domain. Hardware development was an accelerating factor in this shift – the progress of microprocessor development successively widened the space of possibilities for the research efforts in all three regions. With the miniaturization of electronic components computer units could be integrated into cars. New camera and radar systems made it possible to think about an automobile that can "see". This was a big difference to the early car automation concepts that struggled with missing sensors and high costs for the infrastructure needed. Japan was one of the forerunners of this shift, abandoning the guidance-cable approach already at the end of the 1960s and working instead on vehicle-based machine vision exploiting the presence of guide rails at the side of the road and above the ground. In Europe, cable-guidance continued to be explored at the beginning of the 1970s. England abandoned it already in the mid-1970s.

The year 1977 can be called a global *Scharnierjahr*, a German word for a pivotal year representing a *turning point* in history. An important shift in automation concepts can be observed: In this year, research on cable guidance had been abandoned in the US. Early successes in machine vision-concepts were achieved: Japanese researchers fixed for the first time two cameras on a full-scale automobile, developing the first vision-guided autonomous automobile. On-board image processing computers were not yet available, only low performance onboard computers with 8-bit processors. In the same year, Europe entered this field that had been dominated so far by the US and Japan: The mobile robot HILARE was created in France and Ernst Dieter Dickmanns selected machine vision for the guidance of various types of vehicles as long-term research field. He started the development of a real time vision-system aiming at a vehicle that perceives road- and lane boundaries, avoiding obstacles on the road.

5.1.3 1980s: Shift to Digital On-Board Computing and Renaissance of Infrastructure-Concepts

In the 1980s a common socio-economic factor was the economic concurrence between the three regions. Important funding for big computing initiatives – the *Fifth Generation Computing Project* 1981 in Japan, the *Strategic Computing Initiative* 1983 in the United States and the *Prometheus*-program 1986 in Europe – was made available in the regions of the Triade and pushed the research efforts. In all three regions, important financial funding from government or private institutions became foremost available when international competition peaked.

In this context, another international shift can be observed in the mid-1980s: Self-contained road vehicles with all sensors and computer hardware *onboard* – now with digital microprocessors – became a common trend in the US. High performance image processing computers became available, but they were huge and had to be integrated in small vans in the US (ALV 1984 and NavLab1 1986), Germany (VaMoRs 1984) and Japan (PVS 1987). This was an important difference in comparison to the decades before.

The 1980s also were a decade of important progress in machine vision. While the early concepts in the US and Japan used machine vision only in a quasi-static form for simple feed forward control, not only automated, but even autonomous vehicles *with decision making capabilities* became conceivable for the first time in history. The first fully autonomous longitudinal and lateral vehicle guidance by machine vision on a highway was realized with VaMoRs in 1987, exploiting the 4D-approach of UniBwM. The automotive industry in Germany also started own projects on vision guided vehicles and platooning.

During the 1980s not only the vehicles but also the cameras became more mobile. While the camera of the Stanford Cart was only able to move in a static position horizontally, the camera of the CMU Rover was able to pan, tilt and slide. With the vision system of UniBwM the vehicle was now able to control the gaze direction of the cameras.

An important finding that goes beyond our initial hypothesis is the fact that infrastructure based concepts aiming at an Automated Highway System *regained momentum* in the second half of the 1980s, especially in the US: Most ITS-projects focused on the cooperation of vehicle and infrastructure. At UC Berkeley the guidance cable had been substituted by magnetic markers in the roadbed to realize road following. This shows that old concepts are rarely completely abandoned.

Another important development of the 1980s was the growing implication of military actors in the development of autonomous vehicles technology – the first military research programs on Unmanned Ground Vehicles were started in Europe and the US – but not in Japan.

5.1.4 1990s: Big Demos, Shift to Assistance Systems, Military Projects

The big investments of the 1980s were followed by a decade of fierce international competition and large-scale demonstrations of autonomous vehicles. The most important were the mid-term demo of PROMETHEUS 1991 in Torino, the AHS-demo by PATH 1992, the final demo of PROMETHEUS with VITA-II and VaMP 1994, the first and second Japanese AHS-demo 1995, ‘No Hands Across America’ by CMU and the long-distance drive of VaMP from Munich to Odense, both 1995, the San-Diego demo 1997 and Mille Miglia in Automatico, University of Parma 1998. The demos were dominated by Automated Highway and platooning concepts on the one hand and vision-based concepts on the other hand. No big project on highway automation was held in Germany as the research efforts in this country focused predominantly on visual guidance.

The years 1994–1997 can again be identified as *Scharnierjahre*, this time at the transition point from infrastructure-based (AHS) to vehicle-based driver assistance systems. In 1996 the cooperation between Daimler and UniBwM ended – the automotive industry privileged radar instead of camera-based obstacle detection. The historical climax of infrastructure-based approaches was reached with the PATH-Demo in San Diego 1997 that combined on-board systems and magnetic markers. In the next years, a shift from autonomous driving to assistance systems can be observed both in Japan and in the US. From then on, the incremental implementation of vehicle automation was privileged. Earlier than other countries, Japan had already focused on assistance systems in 1991 when it started the Advanced Safety Vehicle.

Another trend of the 1990s is the growing implication of the defense industry and military actors in the field of autonomous driving. Research programs with a specific military focus were started in the US (Demo I-III) and Germany (PRIMUS). Both countries cooperated in this domain. The academic research on vehicle automation in the United States overlapped at some occasions with military projects, for instance when actors like CMU contributed with their NavLab-vehicles to all three fields of highway automation, robotics research and UGV-development. In fact, civil and military research were never completely separated in the US – already in the 1960s most of the robotics research had been funded by DARPA. Without this important funding from military sources, autonomous driving would not exist in its actual form.

5.2 How to Compare the Different Robots?

As outlined in the introduction one goal of our inquiry was to compare the capabilities of the different driving robots and the different concepts of vehicle automation. This comparison can be done with the aim of a hierarchical classification of the different artefacts and approaches. But for several reasons this conclusion has not the aim to give a prize for the "best" or most successful automated vehicle or mobile robot. The inquiry has indeed shown that a direct comparison of all vehicles and technical concepts is only possible within very limited boundaries. Leading CMU-researchers argue in a paper on autonomous driving that "all papers on road following claim success. Most are missing crucial details that would let others evaluate competing algorithms. Even where all the details of the software are spelled out, crucial differences in hardware (processing rates, camera capabilities, vehicle and camera control, and so on) make head-to-head comparisons difficult."[5]

To evaluate an automated vehicle, at least its *speed, the distance and the environment* traversed must be known. If these three parameters would be available for all

[5] Thorpe, Charles, Martial Hebert, Takeo Kanade, and Steven Shafer. 1991. Toward Autonomous Driving: The CMU Navlab, Part I, Perception. *IEEE Expert* 6 (4), p. 51.

mobile robots, automated and autonomous vehicles presented in this book, a direct comparison of the different concepts would have been possible. But as we have seen in the three chapters, this information is often only partially available.

First, a general comparative overview shows that the different vehicles increased their speed. But we have seen that speed indications are difficult to evaluate. Speed is only one factor that can be used to measure progress: Some robots were slower than their predecessors, but this should not be considered as a failure when new technological architectures had been experimented, as it was the case with the Intelligent Vehicle in Japan for instance.

Second, speed indications are often given without precising the *distance* driven and – most important – without indicating the *environment* that was traversed: a concrete road or an off-road-terrain, a street with or without curves, with lane markings or without lane markings, in the flat land or the mountains, with or without obstacles, only following the road or also detecting obstacles, not only detecting obstacles but also avoiding them, by night or day, by rain, fog, even snow? For these reasons, a detailed head-on qualitative comparison of different automated vehicles is not possible.

Instead of a hierarchical *separation* of different robots, it is more productive to use the comparative frame for the identification of some *common* long-term historical trends: Our investigation has shown that the research vehicles of all countries together share an evolutionary path from indoor runs with off-board computers to outdoor runs with on-board computing. Starting as laboratory vehicles connected via cable or radio to large remote computers, the robots became more and more self-contained. They were more and more able to drive autonomously outside the laboratory and even in all-terrain environment – they extended their range. This development can be seen as a direct effect of hardware progress: On-board computing was only possible with the shrinking size and the increasing computing power of microprocessors.

Another important finding is that all the research concepts had to find a way to reduce the complexity of the *environment*. This was the key to success. UniBwM in Germany selected one-way highways without oncoming traffic, crossings, pedestrians and bicycles as ideal environment for autonomous driving at *higher* speeds. INRIA in France instead focused on automated vehicles with *slow* speeds for urban centers. DARPA did choose the desert for its first two robotic challenges, an environment that is less normed but contains mainly rocks and cactus that do not move. However, the vehicles did not have to find their way since this was prescribed by numbers of GPS-waypoints. These examples remind us that the real challenge for the perception systems of autonomous vehicles is the dense traffic in very different types of global cities.

Researchers also reduced complexity by working on the automation of one degree of freedom only: Throughout the 50 years of research efforts that had been presented, lateral control – steering – of a vehicle was more often automated (e.g. RCA and GM 1957, Conti and Siemens 1968, Opto-Pilot of Volkswagen 1982, No Hands Across America test-drive of CMU 1995, Mille Miglia in Automatico of University of Parma 1998) than longitudinal control because lane keeping is easier to automate. Accelerating and braking need obstacle detection, a capability that remains a challenging task even today – as for instance the accident with Elaine Herzberg shows, the first pedestrian killed by a self-driving vehicle.[6] We can state here that automated driving remained for a long time automated steering only – until the final demo of PROMETHEUS in 1994 with VITA2 and VaMP demonstrating longitudinal and lateral autonomous driving on public three-lane highways with normal traffic near Paris.

5.3 History as Linear Timeline

One outcome of our inquiry is the creation of a historical timeline with different research vehicles and different concepts of vehicle automation (see appendix). Timelines with milestones represent the way many members of the engineering community and the industry think about their own creations and the advancements in their field. Timelines are very interesting phenomena because they often seem to be the only way to represent the chronological succession of time: Timelines are by nature linear, a quality that is not a problem *per se*. They allow getting an overview over a multitude of complex and heterogenous events. But what is indeed a problem is the fact that the linear and static perception of time influences our perception of technical progress.

A famous example is the microprocessor development: The American physicist Gordon Moore, one of the founders of Intel, described in 1964 a rule that is known as Moore's law. He observed that since the invention of integrated circuits in 1958 the number of transistors on an integrated circuit doubled every year.[7] In 1970, the production costs of the cheapest chips would only be 10% of the costs from 1965.[8] One of the key factors of this development was miniaturization – small components are cheaper and operate faster. The interval of chip density "soon stretched from 12 to 18 months", and later from 18 to 20 months but Moore's prediction "has held true" since over 50 years acknowledged the historian of technology Paul Ceruzzi in 2005.[9] Two examples may illustrate this development: In 2016, a smartphone had 120 million times the computing power of the control computer for NASA's Apollo lunar program. If the iPad 2 had been commercialized in 1994, it would have been on the list of the fastest supercomputers of the world.[10] For that reason, Ceruzzi

[6] Herzberg was killed on March 18, 2018 after a collision with a self-driving test-vehicle from Uber in Tempe, US.

[7] Moravec, Hans. 1988. *MIND Children, The Future of Robot and Human Intelligence*, p. 68. Cambridge, London.

[8] Gugerli, David. 2018. *Wie die Welt in den Computer kam, Zur Entstehung digitaler Wirklichkeit*, p. 107. Frankfurt a.M.

[9] Ceruzzi, Paul E. 2005. Moore's Law and Technological Determinism: Reflections on the History of Technology. *Technology and Culture* 46 (3), 584–593, here: p. 585.

[10] Gruber, Angela, 2016. Kleiner geht's nicht, in Spiegel Online, March 26, online in: https://www.spiegel.de/netzwelt/web/moore-s-law-die-goldene-regel-der-chiphersteller-broeckelt-a-1083468.html

suggested to “step back from a social constructionist view of technology and consider that, in at least one instance, raw technological determinism is at work”.[11]

It is indeed tempting to tell the history of automation as linear progress as the big picture seems to confirm this vision: Contemporary robots have more capabilities – higher ranges, better perception, more precise localization, faster reactions – than those of the 1960s. But does that mean that “the Waymo vehicle can be thought of as something like a tenth to twelfth generation descendant of the original Stanford Cart”, as Lester Earnest claims?[12] Is this not an example of the linear stairway perception of technology progress that historians like Mahoney, Möser, Mom and Johnson criticized (see Chapt. 1 Introduction)? Is hardware development the only factor of progress?

The imagination of progress as a stairway of technology-generations that successively become more powerful is a seductive, but too simplifying scenario. Instead of reducing history to hardware progress we should recognize that hardware progress is only *one* factor: Hardware indeed “constituted a bottleneck until the 1970s” as Marvin Minsky highlights.[13] And neural networks suffered a setback in the 1960s because of insufficient computer power. But Daniel Crevier has criticized that the history of AI was too often written like a function of hardware power only.[14] If the interest in AI was declining, it was explained by the missing computer power. And if it increased, it was the increasing computer power that promised better results. He reminds us that “hardware progress will not solve all of AI’s problems”.[15] Hans Moravec emphasizes that even with computers that are more powerful, they do not become forcibly much smarter.[16] In the 1980s, when hardware became more powerful, the main problem was how to program the machines. Crevier claimed in 1993 that “in the old days, machine memories were too small to hold the knowledge researchers wanted to pour into them. Now it’s the other way around: you aren’t ever going to fill the new machines with hand code.”[17] That shows that hardware capacity is not the only factor that influences advances in AI-research.

An important result of this study is that processor speed and hardware power alone will not offer the solution for the difficult task of autonomous driving. The first Navlab, equipped with massive processing power, but hurting itself in the image evaluation process (1 to 2 seconds per image) is only one example.[18]

---

[11] Ceruzzi 2005, p. 593.

[12] Earnest 2018.

[13] Crevier 1993, p. 310.

[14] Re-examining the history of Artificial Intelligence “in the light of unrealized hardware constraints can lead to interesting revisions of accepted explanations for why the field took certain orientations” says Crevier (Crevier 1993, p. 309).

[15] Crevier 1993, p. 310.

[16] Moravec 1988, p. 9.

[17] Crevier 1993, p. 310.

[18] Mysliwetz 1990, p. 1.16.

The historians of technology Martina Heßler and Heike Weber recall that the determinist view suggested by Moore's law was not only based on the development of integrated circuits but also on the development of a mass market with users asking for new devices.[19] Some engineers confirm this perspective. Dickmanns emphasizes that the progress of machine vision in the 1990s "is only partially due to increased microprocessor performance". The number of research groups active in the field also "increased strongly".[20] Cyrus Mody called this *the long arm of Moore's Law*.[21] The hardware industry used it as orientation for their development efforts.[22] Only with enormous investments of capital, people, and institutions in the semiconductor industry the prediction of ever increasing computer speed works like a self-fulfilling prophecy.

If history is not a straight path of successive steps to better solutions, if a linear conception of history is misleading – how can we imagine history? What could be a more appropriate image than the timeline? While analyzing the very diverse research projects of automated and autonomous driving, another image than the timeline came into our mind: It is the puzzle. History can be thought of as a *dynamic landscape of puzzle pieces* that is always in the state of becoming and is never finished. Pieces of the puzzle can appear and disappear. They can also hold their place but change their forms so that their neighbors would not fit to them anymore – but then other pieces may be able to fill the gap. Some pieces may stay at their places for decades without changing much, others will rapidly appear as dead ends and will be forgotten – until they reappear eventually in the future at another place.

The puzzle as metaphor for history is another way to imagine progress, avoiding simplicist linear conceptions of technological change. It avoids reducing technological change to new developments only. It allows to integrate the question of the old in the question of the new as David Edgerton has asked for.[23] As we have seen throughout this book, old technologies like the guidance cable idea persist and are sometimes reintroduced in the new.

---

[19] Heßler, Martina, and Heike Weber. 2019. Provokationen der Technikgeschichte, Eine Einleitung. In *Provokationen der Technikgeschichte, Zum Reflexionszwang historischer Forschung*, ed. Martina Heßler and Heike Weber, p. 8. Leiden, Boston, Singapore, Paderborn.

[20] Dickmanns, Ernst Dieter. 2002. The Development of Machine Vision for Road Vehicles in the Last Decade. *Intelligent Vehicle Symposium* 1, p. 277.

[21] Mody, Cyrus C. M. 2017. *The Long Arm of Moore's Law, Microelectronics and American Science*. Cambridge, London.

[22] Ceruzzi 2005, p. 217, Lenzen 2018, p. 39.

[23] Heßler, Weber, p. 23.

5.4 Lessons of History

Finally we should answer the question if there are any "lessons" that can be learned from history. Mahoney said that "the past can be a good place to look for the future."[24] History helps us "to know where we might be headed by establishing where we are and how we got here. The future is largely the result of our present momentum." Even surprises "wind up having a history, first because the new most often comes incrementally embedded in the old, and because in the face of the unprecedented we look for precedents."[25]

Some researchers working in the field express the hope that an understanding of "the similarities and the differences" between past and present could help to make "accurate predictions about deployment timelines in the future."[26] What did not change until today is the fact that it remains difficult to make "accurate predictions" about the deployment of autonomous vehicles as they depend on too many unknown factors. After having presented half-of-the-century research efforts we should remain cautious with promises of a fast market introduction. As we have outlined in the introduction, since over 100 years autonomous vehicles were often promised to become reality 20 years later. The entire period that was investigated in this book saw big investments, many inventions and considerable advances in full automation or autonomous driving. But despite all these efforts this type of vehicle remained an element of basic research and fiction – *fully autonomous vehicles* continue to be *permanent future technologies*.

Nevertheless, it would also be inappropriate to speak of market failure. While transnational engineering communities worked on *autonomous driving*, testing their research vehicles in laboratory yards, *automated driving functions* entered as innovations the field of real-world-applications: Starting with the invention of the Cruise Control by Ralph Teetor in 1948, more and more *partially automated* driving functions were integrated into *manual* driven automobiles. What is today on the market – and what would not exist without the enormous investments and the development efforts of the research projects described in this book – is a great number of Advanced Driver Assistance Systems (ADAS). Only in this field technological *innovation* can be observed, defined as a technology that leaves the sphere of basic research and reaches the market.

Let us come back to the request of Kurt Möser, presented in the introduction, to center historical research on the reconstruction of long historical trends. The historian Gijs Mom observed a long historical trend in the competition between the electric and the gasoline automobile: He showed that the gasoline car each year became

[24] Mahoney 2005, p. 119.

[25] Ibid., p. 120.

[26] Shladover, Steven E., Jane Lappin, Robert P. Denaro, and Bryant Walker Smith. 2018. Introduction: The Transportation Research Board's 2013 Workshop on Road Vehicle Automation. In *Road Vehicle Automation*, ed. Gereon Meyer and Sven Beiker, p. 8. Cham, Heidelberg, New York, Dordrecht, London.

more electrical, it imitated the electric car. A similar trend can be observed in the field of vehicle automation: Each year, more functions of our manually driven cars become automated. A long-term trend of an *automatization* of the driving task can be observed – without having *autonomous* vehicles on the road. And that is perhaps the only prediction that can be made: Martina Heßler emphasizes that the incremental development of several small steps of automation may lead to fully autonomous systems. A continuous delegation of tasks to technology will finally represent at a certain moment a "break", a "jump" when it culminates in an autonomous vehicle.[27] In other words, the incremental succession of assistance technologies could one day lead to the commercialization of an autonomous vehicle, and that would perhaps not even be noticed as a revolutionary event.

Bibliography

Berger, Arthur Asa. 2005. *Media Analysis Techniques*. Thousand Oaks, London, New Delhi.

Bowler, Peter. 2017. *A History of the Future, Prophets of Progress from H.G. Wells to Isaac Asimov*. Cambridge.

Ceruzzi, Paul E. 2005. Moore's Law and Technological Determinism: Reflections on the History of Technology. *Technology and Culture* 46 (3): 584–593.

Crevier, Daniel. 1993. *The Tumultuous History of the Search for Artificial Intelligence*. New York.

Dickmanns, Ernst Dieter. 2002. The Development of Machine Vision for Road Vehicles in the Last Decade. *Intelligent Vehicle Symposium* 1: 268–281.

Earnest, Lester. Stanford Cart, 2018. https://web.stanford.edu/~learnest/les/cart.html. Accessed on February 3, 2019.

Gruber, Angela, 2016. Kleiner geht's nicht, in Spiegel Online, March 26, online in: https://www.spiegel.de/netzwelt/web/moore-s-law-die-goldene-regel-der-chipherstellerbroeckelt-a-1083468.html

Gugerli, David. 2018. *Wie die Welt in den Computer kam, Zur Entstehung digitaler Wirklichkeit*. Frankfurt a.M.

Lenzen, Manuela. 2018. *Künstliche Intelligenz, Was sie kann & was uns erwartet*. München.

Lossau, Norbert. Das erste autonome Auto kostete 200.000 D-Mark. *Die Welt*, October 13, 2017.

Mahoney, Michael S. 2005. The Histories of Computing (s). *Interdisciplinary Science Reviews* 30 (2): 119–135.

Mody, Cyrus C. M. 2017. *The Long Arm of Moore's Law, Microelectronics and American Science*. Cambridge, London.

Moravec, Hans. 1988. *MIND Children, The Future of Robot and Human Intelligence*. Cambridge, London.

Mysliwetz, Birger. 1990. *Parallelrechner-basierte Bildfolgen-Interpretation zur autonomen Fahrzeugsteuerung*, Dissertation, Universität der Bundeswehr München, Fakultät für Luft- und Raumfahrttechnik. Institut für Systemdynamik und Flugmechanik.

Shladover, Steven E., Jane Lappin, Robert P. Denaro, and Bryant Walker Smith. 2018. Introduction: The Transportation Research Board's 2013 Workshop on Road Vehicle Automation. In *Road Vehicle Automation*, ed. Gereon Meyer and Sven Beiker, 1–12. Cham, Heidelberg, New York, Dordrecht, London.

Thorpe, Charles, Martial Hebert, Takeo Kanade, and Steven Shafer. 1991. Toward Autonomous Driving: The CMU Navlab, Part I, Perception. *IEEE Expert* 6 (4): 31–42.

[27] Heßler, p. 260.

Chapter 6
Timeline with Firsts, Milestones and Shifts

The 1930s and 1940s

1939

US: First public promotion of an **automated highway system** at the Futurama of General Motors, World's Fair New York.

1948

US: William Grey Walter developed the **first mobile robots** that could avoid obstacles.

The 1950s

1953

US: Vladimir Zworykin starts to work on highway automation with guidance-cables at **Radio Corporation of America (RCA).** First demonstration of the system with miniature cars.

1956

US: General Motors (GM) joins RCA in the work on highway automation. The Firebird II shown in the Motorama popularizes the guide-wire approach.

US: **AI-research** starts with the **Dartmouth Conference.**

1957

US: The Soviet Union launches **Sputnik**, the first satellite circulating in a low Earth orbit. It is followed by a second satellite that carries a dog named Laika. These events trigger the creation of ARPA.

US: First demonstration of **automated lateral control with guide-wires by RCA and GM** in Lincoln, Nebraska. First demo of an infrastructure-based driver assistance system on a real highway.

F. Kröger, *From Automated to Autonomous Driving*, Archimedes 70,
https://doi.org/10.1007/978-3-031-49881-7_6

1958

US: Foundation of the **Advanced Research Projects Agency (ARPA)** by the US-Defense Department. The agency will later fund autonomous driving research.

US: **GM tests the guidance-system** on a test track in Warren, Michigan, achieving a speed of 35 mph.

The 1960s

1960

US: **Demonstration of the guide-wire-system** at the test-track of **RCA**, Princeton, New Jersey, showing automated lane tracking, lane changing at certain locally fixed points, and car following.

England: Tests with **guidance-cables for automated longitudinal and lateral control** of automobiles are started at **Road Research Laboratory, England.**

US: Creation of the mobile robot **Stanford Cart** at **Stanford Mechanical Engineering Laboratory**, aiming to explore **remote control by vision**. Speed: 0.2 mph, later 5 mph. Unused from 1963–1966.

1961

Japan: The **Government Mechanical Laboratory** in Japan starts research on the automatization of some functions in automobile driving.

1963

Foundation of the **Stanford Artificial Intelligence Laboratory (SAIL)** with funding from ARPA.

1965

US: Acquisition of the Stanford Cart by SAIL. In 1967 research on **computer control of the cart by vision**, aiming at an autonomous vehicle. Speed: 0,8 mph.

1964

US: **Robert Fenton** at **Ohio State University** starts to work on **lateral and longitudinal control**.

US: **Johns Hopkins University** develops a mobile robot dubbed *The Beast* that is able to search for power plugs and recharge itself.

1965

Japan: **Economic Boom** Izanagi in Japan (1965–1970).

1967

Japan: **Automated steering control with a guidance cable is tested in Japan** up to a speed of 100 km/h.

1966

US: **Shakey the robot** is created at **Stanford Research Institute (SRI).** Funded by ARPA until 1972. First mobile robot that could perceive its environment, reason about its actions, plan a course, act autonomously and learn.

1967

US: Tests with two guidance-cables at Ohio State University.

1968

Japan: Tests with two guidance-cables at Government Mechanical Laboratory in Japan. Cable guidance is then abandoned in Japan.

Germany: **Continental and Siemens use guidance-cables for lateral control** with a Mercedes 250 at Contidrom, Germany. Longitudinal control is remote controlled.

France, end of the 1960s: **Peugeot uses guidance cables** to test new vehicles in Besançon and at the end of the 1970s in Sochaux.

1969

US: Bendix studies Personal Rapid Transit (PRT)-Systems.

1969

US: Intel develops the first microprocessor.

The 1970s

1970

Japan: Test of the **guideway-based Computer Controlled Vehicle System (CVS)** in Higashi-Murayama, Japan (1970–1978).

Japan: The Government Mechanical Laboratory in Japan becomes the **Mechanical Engineering Laboratory.**

1971

Germany: **MAN and Bosch** start experiments on **vehicle automation with a guidance-cable**. They also work on radar and infrared sensors.

1972

US: ARPA becomes DARPA.

1973

US: Hans Moravec uses the **Stanford Cart** at SAIL until 1980 to build an **obstacle avoider**. Velocity: 3 to 5 meters per hour. Bottom-up approach based on sensor data.

1975

Japan: First experiments on **obstacle detection with stereo vision by MEL** in Japan.

Japan: One of the first **vehicle navigation systems** is tested in Japan, the Comprehensive Automobile Traffic Control System (CACS).

England: The Road Research Laboratory in England **abandons guidance-cables** in the mid-1970s.

1977

US: The FHWA decides to **abandon the development of active-wire following systems**. OSU stops all studies on guide-wire systems. The development of active systems in the vehicle is preferred now, e. g. radar.

France: The **mobile robot HILARE** is developed by the *Laboratoire d'automatique et d'analyse de systèmes (LAAS)* in Toulouse with the aim to **sense the environ-**

ment and to avoid obstacles. Laboratory-bound with off-board computing power, it reaches a speed of 1 m/s.

Germany: The *Institut für Systemdynamik und Flugmechanik (ISF)* at **University of Bundeswehr Munich chooses machine vision as long-term research goal.**

1978

Japan: MEL develops the **first partially autonomous automobile guided laterally along a guard-rail by vision with on-board computers** at a speed of 36 km/h (**First generation Intelligent vehicle**).

Germany: In 1978 **Dickmanns** develops the **window-concept** for real-time image sequence processing.

1979

US: Foundation of the Robotics Institute of **Carnegie Mellon University** in Pittsburgh.

Germany: In 1979 **research on visual guidance of ground- and air vehicles starts at UniBwM** with a 3-axis motion simulator. In 1979/1980 an image sequence processor called BVV is developed.

Japan: Sadayuki Tsugawa publishes an IJCAI-paper about visual tracking of lateral guide rails by vertical stereo vision.

The 1980s

1980

Japan: MEL develops the second generation Intelligent Vehicle (1980–1984) with improved obstacle detection. It detects figures on the ground but has **no full lane-marking detection.** It is able to follow a preplanned path represented by a sequence of points at 10 km/h.

1981

US: **The Reagan-Administration stops the development of automatic transportation systems.**

US: CMU starts to work on the **CMU Rover** (1981–1984).

Japan: MITI starts the **Fifth Generation Computer System**-initiative with the aim to create an artificial intelligence.

1982

Germany: First dissertation on visual guidance of a ground vehicle with the simulation loop at UniBwM (Meissner).

Germany: The car-industry becomes interested in visual steering systems: Walter Zimdahl develops the **Opto-Pilot** at Volkswagen, aiming at **semi-autonomous driving by a few lines of video images:** While **steering is handled by a computer,** longitudinal control is performed by a human driver. In 1985, the system is tested at 100 km/h on a straight free road.

1983

US: As a response to Japans Fifth generation Computing-project DARPA starts the ***Strategic Computing Initiative***, aiming at full machine intelligence.

1984

US: In the context of the SCI-Initiative of DARPA the development of an **Autonomous Land Vehicle (ALV)** is started in 1984 and explored until 1988.

France: The French military starts one of the **earliest European military projects** in the field of UGVs, called *Demonstrateur autonome à rapidité de déplacement pour la surveillance (DARDS)* and aiming at a **semi-autonomous teleoperated** ground vehicle.

US: CMU develops the **Terregator**. Speed: 1 km/h.

Germany: UniBwM acquires a Mercedes Benz 508 D van dubbed ***Versuchsfahrzeug für autonome Mobilität und Rechnersehen*** **(VaMoRs)** for real-world experiments. A two-axis gaze control platform with a bifocal set of cameras is used for driving tests by vision.

Japan: MEL studies **vehicle-to-vehicle communication** with small automated vehicles in factories aiming to control a **platoon** of cars.

1985

Germany: Daimler-Benz AG starts to work on automated driving but without exploiting machine vision approaches.

1986

Europe: **Prometheus-program (1986–1994) of the European Union.** After a preparatory definition-phase, research is started in January 1988.

Germany: Daimler-Benz AG und UniBwM plan to work together in the project **Autonom Mobile Systeme**. **First test drives with VaMoRs**, using a real-time image analysis system for fully autonomous driving. In December, VaMoRs demonstrates **autonomous longitudinal and lateral guidance by machine vision** at the skidpan-test-track of Daimler in Untertürkheim. Goal is a technology-transfer to a Mercedes-Spurbus in 1988.

US: **Carnegie Mellon University** starts research with the **NavLab1, equipped with on-board computers**. In 1987 Navlab1 follows a preprogrammed path at 2.7 mph (4.3 km/h), in 1990 it reaches a speed of 20 mph.

US: The research program ***Partners for Advanced Transit and Highways (PATH)*** is started by the Californian Department of Transportation and the Institute of Transportation Studies of the University of California in Berkeley.

1987

Japan: Personal Vehicle System (PVS) with **machine vision** in Japan, successor of the guideway-based Computer Controlled Vehicle System (1970s). 1989 it runs along straight lines with 30 km/h.

US: In August, the **Autonomous Land Vehicle (ALV)** shows the first cross-country traverse with a speed of 3 km/h.

US: PATH starts experiments with **vehicle platoons** guided by magnetic markers.

Germany: VaMoRs demonstrates in November the **first fully autonomous longitudinal and lateral vehicle guidance by machine vision at high speed (96 km/h) on a free stretch of Autobahn** near Dingolfing. First publication of the method called 4-D approach for dynamic scene understanding based on image sequences.

1988

Germany: First summarizing publication on dynamic machine vision by Dickmanns of UniBwM. VaMoRs/Spurbus demonstrate stopping in front of a stationary obstacle at 40 km/h in Rastatt.

1989

Japan: The Japanese Ministry of Construction starts research on AHS.

Germany: The German Federal Ministry of Defense starts the experimental ***Program of Intelligent Mobile Unmanned Systems*** **(PRIMUS)**, aiming at autonomous navigation for military vehicles. Dornier GmbH Friedrichshafen and UniBwM cooperate during PRIMUS A (1989–1992) to equip a Mercedes GD-300 with a vision system. The vehicle drives autonomously at a speed of 5 km/h while avoiding obstacles. With a better LADAR it achieves 25 km/h in open terrain and 60 km/h on a country road.

Germany: **Volkswagen** works on a multiple-vehicle **platoon**, called **convoy driving**.

The 1990s

1990

US: Creation of the Intelligent Vehicle and Highway Society of America (**IVHS America**).

US: CMU uses a **neural network** approach for the **Navlab 2 army** vehicle.

US: **Demo I program** (1990–1992) aiming at teleoperated military vehicles.

Japan: JSK works on an information system for drivers based on **inter-vehicle communication** (*Super Smart Vehicle System).*

1991

Japan: Advanced Safety Vehicle-initiative of the Ministry of Transport in Japan, focusing on **assistance systems**. In the US, the shift to assistance systems can only be observed in 1997 with the Intelligent Vehicle Initiative (IVI).

Germany: The VITA-van of Daimler Benz, equipped with the vision system and software of UniBwM, demonstrates **vision based lane following**, convoy driving and Stop&Go (Prometheus-demo in Torino, Italy).

France: The French National Institute for Research in Computer Science and Control (INRIA) starts a research project on an **automated taxi service**, called *Développement d'un engin de déplacement automatique en libre service (DEDALS).*

US: Congress signs the Intermodal Surface Transportation Efficiency Act (ISTEA) requiring a **prototype of a fully automated roadway in operation by 1997**.

1992

US: **First platooning experiment of PATH** with automatic lateral control *or* automatic longitudinal control.

US: **Demo II program** (1992–1996) focusing on **scout missions** with teams of four vehicles.

1993

France: The **French project Praxitèle** works on a **new public transport system** with automated and shared electrical vehicles. In 1995 automatic parallel parking, lane changing and obstacle detection are demonstrated.

UniBwM and Daimler acquire and equip two Mercedes 500 SEL with the second-generation vision system based on transputers for the Prometheus-demo in 1994.

1994

Germany: **VaMoRs demonstrates transition into convoy driving, lane changing and Stop&go** with the second generation vision system of UniBwM.

US: **CMU abandons the neural network / machine learning approach** of ALVINN, substituting it with RALPH.

US: Creation of the National Automated Highway System Consortium (NAHSC), aiming to conduct an AHS-program.

US: **First public demo of an AHS-system** in history: PATH shows a platooning demo with four Lincoln Town Car in San Diego.

France: First ITS congress is held in Paris.

Europe: **Final demo of PROMETHEUS** in October in Paris. **First demonstration of autonomous driving on a public highway:** Two Mercedes 500 SEL, the *Versuchsfahrzeug für autonome Mobilität PKW* (VaMP) of UniBwM and VITA-II of Daimler demonstrate about a thousand kilometers of **autonomous free-lane driving, autonomous transition into convoy driving and autonomous lane changes at speeds up to 130 km/h in public three-lane traffic on Autoroute-1** near the Airport Charles de Gaulle.

1995

US: In July **CMU started the test-drive *No Hands Across America*** from Pittburgh to San Diego (2800 miles) with a Pontiac Trans Sport (NavLab 5). **Only lateral control is performed autonomously**, longitudinal control is handled by a human driver.

Japan: First Japanese **AHS-demonstration** in Tsukuba with 19 vehicles, initiated by the Ministry of Construction.

US/Germany: Memorandum of Understanding (MoU) between the US-American and the German Ministries of Defense; in its framework the AUTONAV project was performed from 1998 till 2003 (see below).

Germany: **Fully autonomous long-distance test drive of VaMP** in November from Munich to Odense, Denmark (95% driven autonomously ~1600 km), showing **autonomous lane keeping and autonomous performance of lane changes, autonomous longitudinal control and collision avoidance**. Max speed in fully autonomous driving mode was 175 km/h.

1996

Germany: End of the cooperation between Daimler and UniBwM. Funding of the project *Intelligente Fahrzeug-Funktionen* (IFF) with VaMoRs by the Bundesamt für Wehrtechnik und Beschaffung (BWB), bound to the participation of UniBwM in the German-US AUTONAV agreement.

Using the vision system of UniBwM, Dornier and Deutsche Aerospace AG transform a **Wiesel 2 tank** during the **PRIMUS C-program** (1996–1999) into an autonomous robot. From 1997–2001 the Wiesel is part of AUTONAV.

Germany: VaMoRs can turn autonomously into an unknown crossroad, **using active gaze control for situation assessment during the approach**.

Japan: Second Japanese AHS-demo with 11 vehicles on Joshinetsu Expressway. **Shift to a combination of infrastructure-based and vehicle-based approaches:** lateral control with magnetic markers, speed control with LCX-cables. Lane detection with computer vision, longitudinal distance controlled with LIDAR.

1997

US: **Demo'97 in San Diego: Demonstration of an Automated Highway System in North America** with eight Buick LeSabre.

US: *Intelligent Vehicle Initiative* of the US Department of Transportation, **focusing on assistance systems.** Similar to the ASV-initiative of 1991 in Japan.

US/Germany: **Autonomous Vehicle Navigation Control System (AUTONAV)-agreement** (1997–2003) aiming at autonomous vehicles for on- and off-road navigation. Technology transfer between the Demo III program in the US and Intelligente Fahrzeug-Funktionen (IFF) in Germany. Development of the **third-generation vision system of UniBwM**, called Expectation-based, Multi-focal Saccadic Vision (EMS), aiming at scout-type vision without in-advance information about the environment. This was to be transferred onto the Wiesel-2 tank by Dornier.

France: The **CyCab, a two-seater without steering wheel**, is presented to the public in France.

France: Development of an **automated highway lane and automated low-speed vehicles** (up to 60 km/h) for semi-urban areas in the **project *La route automatisée*** in France.

Japan: JSK tests infrared based **cooperative driving system** with four cars in Japan.

1998

Italy: ***Mille Miglia in Automatico*, semi-automated test run (2000 km) of University of Parma,** Italy in June with a Lancia Thema 2000 called ARGO. **Lateral control is realized by computer vision**, longitudinal control by a human driver.

US: Demo III-AUTONAV-project (1998–2003); a cooperation between the US and Germany for transfer of know-how. It aimed at the capability of autonomous navigation on networks of minor roads including cross-country sections and night mobility.

The 2000s

2000

Japan: Cooperative driving demo of MITI and JSK with five vehicles in Tsukuba. **System independent from road infrastructure,** but without machine vision, only inter-vehicle communication, GPS and LIDAR.

Japan: Smart Cruise '21-Demo of the Japanese Ministry of Transport and the Ministry of Construction, in Tsukuba, showing both, **infrastructure- and vehicle based systems.**

France: INRIA and Yamaha develop the **CyberCab, guided by a wire in the ground**. Five vehicles are shown at INRIA Roquencourt.

2001

Germany: UniBwM test-track Neubiberg: Final demo with VaMoRs of the national IFF-project as part of AUTONAV. Autonomous navigation in a network of minor roads, including off-road sections with unknown negative obstacles (ditches).

PRIMUS D (2001–2003): The capability to detect and classify ditches and trenches is transferred onto the tracked vehicle Wiesel-2.

US: Congress defines as goal that one third of military ground vehicles should be able to drive autonomously in 2015. Trigger for DARPA-Challenges.

2003

Final demo of the AUTONAV project with the real-time, full-frame stereo system of Sarnoff / PVS on a single standard board for insertion in one of the four COTS-processor systems of EMS-vision.

Bibliography

Adams, James L. 1961. *Remote Control with Long Transmission Delays,* PhD in Mechanical Engineering. Stanford University.

Adam, John A. 1986, January. Aerospace and Military. *IEEE Spectrum.*

Albert, Dan. 2018. *Are We There Yet?: The American Automobile Past, Present, and Driverless.* New York.

Albus, James S., and Alexander M. Meystel. 2001. *Engineering of Mind, An Introduction to the Science of Intelligent Systems*. New York.

Albus, James, and Anthony Barbera. 2006. 4D/RCS Reference Model Architecture for Unmanned Ground Vehicles. In *Intelligent Vehicle Systems: 4 D/RCS Approach*, ed. Raj Madhavan, Elena R. Messina, and James Albus, 1–26. New York.

Anderson, Edward. 1996. Some Lessons from the History of Personal Rapid Transit (PRT), August 4, 1996. Conference Paper, online in: http://staff.washington.edu/jbs/itrans/history.htm. Accessed July 8, 2019.

AP Archive, Driverless Car, Crowthorne / UK, May 31, 1971. http://www.aparchive.com/metadata/youtube/c9e8b5995bdf4e21bbe1daaea76a018b.

Auer, Ashley, Shelley Feese, and Stephen Lockwood. 2016. History of Intelligent Transportation Systems, Performed by U.S. Department of Transportation, Intelligent Transportation Systems Joint Program Office, Report No. FHWA-JPO-16-329, Produced by Booz Allen Hamilton, May 2016.

Author Unknown. 1958. Highway of the Future. *Electronic Age (RCA-Journal)* 17 (1): 12–14.

———. 1970. Bendix Works on Automatic Vehicle Project. *Ann Arbor News*, June 19, 1970, online in: https://aadl.org/aa_news_19700619-bendix_works_on_automatic_vehicle_project. Accessed in June, 13, 2019.

———. 1972. Firm Devises Driverless Vehicle Controls. *Ann Arbor News*, June 30, 1972, online in: https://aadl.org/aa_news_1972060-firm_devises_driverless_vehicle_controls. Accessed on June 18, 2019.

———. 1991. Take Me, I'm Yours, The Autobiography of SAIL, 1991, online in: http://infolab.stanford.edu/pub/voy/museum/pictures/AIlab/SailFarewell.html. Accessed on April 10, 2019.

———. 2007. More Attacks, Mounting Casualties. *The Washington Post*, September 30, 2007, online in: https://www.washingtonpost.com/wp-dyn/content/graphic/2007/09/28/GR2007092802161.html. Accessed July 7, 2020.

———. 2009. The Inside Story of the SWORDS Armed Robot "Pullout" in Iraq. *Popular Mechanics*, October 1, 2009. https://www.popularmechanics.com/technology/gadgets/a2804/4258963/.

F. Kröger, *From Automated to Autonomous Driving*, Archimedes 70, https://doi.org/10.1007/978-3-031-49881-7

Automobile Inspection and Registration Information Administration, 自動車保有台数の推移 (軽自動車を含, (eng.: Changes in the number of car owners [including mini vehicles]), online in: http://www.airia.or.jp/publish/statistics/ub83el00000000wo-att/hoyuudaisuusuiihyou.pdf. Accessed November 9, 2017.

Baille, Gérard, Philippe Garnier, Hervé Mathieu, and Pissard-Gibollet, Roger. 1999. *Le cycab de l'INRIA Rhône-Alpes, Rapport de recherche*. INRIA.

Barbaza, Pierre. 1984. Profession Pilote Essayeur. *Echappement*, 90.

Baten, Stefan, Michael Lützeler, Ernst Dieter Dickmanns, Robert Mandelbaum, and Peter J. Burt. 1998. Techniques for Autonomous, Off-Road Navigation. *IEEE-Intelligent Systems and Their Applications* 13 (6): 57–65.

Bel Geddes, Norman. 1940. *Magic Motorways*. New York.

Beltran, Alain, and Pascal Griset. 2007. Des véhicules intelligents pour assurer la sécurité routière? *Code Source, L'hebdomadaire des 40 ans de l'INRIA*, No 36, November 5, 2007, p. 1, online in: https://www.inria.fr/sites/default/files/2019-11/Code_Source_1998-2007.pdf. Accessed October 8, 2020.

Bender, James G., Robert E. Fenton, and Karl W. Olson. 1971. An Experimental Study of Vehicle Automatic Longitudinal Control. *IEEE Transactions on Vehicular Technology* VT-20 (4): 114–123.

Bender, James G., and Robert E. Fenton. 1969. A Study of Automatic Car Following. *IEEE Transactions on Vehicular Technology* VT-18 (3): 134–140.

Berger, Arthur Asa. 2005. *Media Analysis Techniques*. Thousand Oaks, London, New Delhi.

Bertozzi, Massimo, Alberto Broggi, Gianni Conte, and Alessandra Fascioli. 1997. Obstacle and Lane Detection on ARGO, *Proceedings of IEEE Intelligent Transportation Systems conference 1997*, online in: http://www.ce.unipr.it/people/broggi/publications/iv97.pdf.

Bertozzi, Massimo, Alberto Broggi, Massimo Cellario, Alessandra Fascioli, Paolo Lombardi, and Marco Porta. 2002. Artificial Vision in Road Vehicles. *Proceedings of the IEEE* 90 (7): 1258–1269.

Bertozzi, Massimo, Alberto Broggi, and Alessandra Fascioli. 2000. Vision-Based Intelligent Vehicles: State of the Art and Perspectives. *Robotics and Autonomous Systems* 32 (1): 1–16.

———. undated. The Evolution of Vision-based Unmanned Ground Vehicles: the VisLab Perspective (undated, without publication source), in: http://citeseerx.ist.psu.edu/viewdoc/download?doi=10.1.1.73.7759&rep=rep1&type=pdf. Accessed July 20, 2020.

Bertozzi, Masssimo, Alberto Broggi, and Alessandra Fascioli. 2006. VisLab and the Evolution of Vision-Based UGVs. *Computer* 39 (12): 31–38.

Bertozzi, M., A. Broggi, et al. 2007. Vision Technologies for Intelligent Vehicles, 2007, p. 6. online in: http://www.ce.unipr.it/people/broggi/publications/ivia2007.pdf. Accessed May, 5, 2021.

Bidwell, J.B. 1959. Vehicles and Drivers – 1980, General Motors Research Laboratories, Presented at SAE Annual Meeting, Detroit.

Bishop, Richard. 2001. Whatever Happened to Automated Highway Systems (AHS)? Originally published in the August–September 2001 issue of Traffic Technology International, online in: https://staff.washington.edu/jbs/itrans/bishopahs.htm. Accessed on July 3, 2019.

Bishop, Richard, Kevin Dopart, and Steven Shladover. 2020. Commemorating the 20th Anniversary of the National Automated Highway Systems Consortium Demo '97 in San Diego, Powerpoint-Presentation, online in: https://path.berkeley.edu/sites/default/files/demo97foravs17v6.pdf. Accessed March 30, 2020.

Bloom, Justin L., and Shinsuke Asano. 1981. Tsukuba Science City: Japan Tries Planned Innovation. *Science,* New Series 212 (4500): 1239–1247.

Blosseville, J.M., and Saïd Mammar. 2000. LaRA Update, the French Program La Route Automatisée, Powerpoint-Presentation, 2000. http://aramis.iup.univ-evry.fr:8080/~smam/publications/publis2000/lara_2000.pdf. Accessed October 12, 2020.

Bommel, Nathan. 2015. Toyota testing self-driving cars 'since the 1990s'. *USA Today*, October 6, 2015. https://eu.usatoday.com/story/money/cars/2015/10/06/toyota-driverless-cars-automated-highway-driving/73448028/. Accessed July 13, 2019.

Bowler, Peter. 2017. *A History of the Future, Prophets of Progress from H.G. Wells to Isaac Asimov*. Cambridge.

Boyer, Robert, Elsie Charron, Ulrich Jürgens, and Steven Tolliday. 1998. Introduction: Between Imitation and Innovation: The Transfer and Hybridization of Productive Models in the International Automobile Industry. In *Between Imitation and Innovation, The Transfer and Hybridization of Productive Models in the International Automobile Industry*, ed. Robert Boyer, Elsie Charron, Ulrich Jürgens, and Steven Tolliday, 1–20. New York.

Braid, Deborah, Alberto Broggi, and Gary Schmiedel. 2006. The TerraMax Autonomous Vehicle, Field Report. *Journal of Field Robotics* 23 (9): 693–708.

Braisted, Paul. 1963. *Study of a Predictor for Remote Control Systems Operating with Signal Transmission Delays, PhD in Mechanical Engineering*. Stanford University.

Brauckmann, M.E., C. Goerick, J. Groß, and T. Zielke. 1994. *Towards All Around Automatic Visual Obstacle Sensing for Cars, Proceedings of International Symposion on Intelligent Vehicles*, 79–84. Paris.

Broggi, Alberto, Masssimo Bertozzi, Alessandra Fascioli, Lo Bianco, Corrado Guarino, and Aurelio Piazzi. 1999a. The ARGO autonomous vehicle's vision and control systems. *International Journal of Intelligent Control and Systems* 3 (4): 409–441.

Broggi, Alberto, Massimo Bertozzi, and Alessandra Fascioli. 1999b. ARGO and the MilleMiglia in Automatico Tour. *IEEE Intelligent Systems and Their Applications* 14 (1): 55–64.

Broggi, Alberto, Paolo Grisleri, Chris Yakes, and Chris Hubert. 2007. The Oshkosh-VisLab Joint Efforts on UGVs: Architecture, Integration, and Results. Platform Innovations and System Integration for Unmanned Air, Land and Sea Vehicles, 146th NATO Workshop on Applied Vehicle Technology Panel, Florence 2007, pp. 16/1–14, online in: http://www.ce.unipr.it/people/broggi/publications/natoavt146.pdf.

Broggi, Alberto. 2008. History of AHS in Italy and future issues. *Proceedings of the 2008 IEEE International Conference on Vehicular Electronics and Safety*, Columbus, OH, USA, September 22–24, 2008, pp. 11–12.

Brooks, Rodney. 1986. A Robust Layered Control System for a Mobile Robot. *IEEE Journal of Robotics and Automation* 2 (1): 14–23.

Brüdigam, Claus. 1994. *Intelligente Fahrmanöver sehender autonomer Fahrzeuge in autobahnähnlicher Umgebung*, Dissertation. Universität der Bundeswehr München, Fakultät für Luft- und Raumfahrttechnik.

Bundeswehr, Die Laufbahn der Offiziere (M/W/D). 2020. https://www.bundeswehr.de/de/ueber-die-bundeswehr/dienstgrade-laufbahnen-bundeswehr/laufbahn-offiziere. Accessed Mai 8, 2020.

Burns, Lawrence D. 2018. *Autonomy: The Quest to Build the Driverless Car – And How It Will Reshape Our World* (Together with Christopher Shulgan). London.

California PATH, Platoon Demonstration, Demo '97-Flyer. http://www.path.berkeley.edu/sites/default/files/my_folder_76/NAHSC_Brochure- Demo%26Groups-971.pdf, pp. 14–17. Accessed June 17, 2018.

———, Automated Driving Mini-Demonstration. http://www.path.berkeley.edu/sites/default/files/my_folder_76/NAHSC_Brochure-Demo%26Groups-971.pdf, p. 34. Accessed June 17, 2018.

Capps, Robert. 2006. The 50 Best Robots of All Time. *WIRED*, January 1, 2006, online in: https://www.wired.com/2006/01/robots-3/. Accessed October 1, 2019.

Carafano, James Jay, and Andrew Gudgel. 2008. The Pentagon's Robots: Arming the Future. In *Competitive Technologies for National Security*, ed. James Jay Carafano, Andrew Gudgel, and Alane Kochems, 27–32. Washington, DC: The Heritage Foundation. online in: https://www.heritage.org/node/15537/print-display. Accessed on June, 24, 2020.

Cardew, K.H.F. 1970. The Automatic Steering of Vehicles: An Experimental System Fitted to a DS 19 Citroen Car, Driver Aids and Abilities Section, Road Research Laboratory, Crowthorne, Berkshire, RRL report LR 340 (1970), online in: https://trl.co.uk/sites/default/files/LR340.pdf. Accessed June 22, 2019.

Carey, James, and John Quirk. 2009. The History of the Future. In *Communication as Culture: Essays on Media and Society*. New York, London.

Cassel, David. 2017. Remembering Shakey, the First Intelligent Robot. *The New Stack*, March 5, 2017, online in: https://thenewstack.io/remembering-shakey-first-intelligent-robot/.

Ceruzzi, Paul E. 2005. Moore's Law and Technological Determinism: Reflections on the History of Technology. *Technology and Culture* 46 (3): 584–593.

Chalmers, Johnson. 1982. *MITI and the Japanese Miracle: The Growth of Industrial Policy, 1925–1975*. Stanford.

Chatila, Raja. 1996. Autonomous Mobile Robots in France: Some Projects and Scientific Approaches. In *Autonome Mobile Systeme 1996*, ed. Günther Schmidt and Franz Freyberger, 2–7. München.

Champeny-Bares, Lee, Syd Coppersmith, and Kevin Dowling. 1991. *The Terregator Mobile robot*. Pittsburgh: Field Robotics Center, The Robotics Institute, Carnegie Mellon University.

Chand, Naresh. 2016. Reconnaissance, Surveillance, Target Acquisition and Precision Technologies. SP's Land Forces, No. 5, 2016, online in: http://www.spslandforces.com/story/?id=427&h=Reconnaissance-Surveillance-Target-Acquisition-and-Precision-Technologies. Accessed June 24, 2020.

Chatani, Yoshiyuki. 2008. Antennas for the Bullet Train. In *Mobile Antenna Systems Handbook*, ed. Kyohei Fujimoto, 3rd ed., 417–425. Boston, London.

Cheon, Sanghyun. 2003. *An Overview of Automated Highway Systems (AHS) and the Social and Institutional Challenges They Face, UC*. Berkeley: University of California Transportation Center. online in: https://escholarship.org/uc/item/8j86h0cj, p. 23. Accessed on March, 28, 2020.

Chun, Wendell H., and Todd M. Jochem. 1994. Unmanned Ground Vehicle Demo II: Demonstration A. *Unmanned Systems* 12 (1): 14–20.

Chun, Wendell, J. Walt Faulconer, and Steven Munkeby. n.d. UGV Demo II Reuse Technology for AHS. *Proceedings of the Intelligent Vehicles '95 Symposium*, pp. 247–252.

CNRS, La science de la machine intelligente, Interview with Georges Giralt, interviewed by Patrick Danès, *La lettre du LAAS*, N° 37, October 2008, p. 58, online in: https://www.laas.fr/LettreDuLaasStatic/content/download/793/4835/version/3/file/La-Lettre-n37–5.pdf.

Committee on Japan (ed.). 1997. *Maximizing U.S. Interests in science and technology relations with Japan, Committee on Japan framework statement and report of the competitiveness task force*. Washington, D.C., online in: https://www.nap.edu/read/5850/chapter/1#iii. Accessed January 20, 2018.

Congressional Research Service, Defense Primer, U.S. Policy on Lethal Autonomous Weapons Systems, December 19, 2019.

Continental AG, Controlled by Electronics: Continental Launched Its First Driverless Vehicle 50 Years Ago, Press Release September 11, 2018, The "Ghost Car", online in: https://www.continental.com/en/press/press-releases/controlled-by-electronics-144932. Accessed September 25, 2020b.

———, Contidrom Electronic test, online in: https://www.continental.com/de/presse/pressemitteilungen/50-jahre-fahrerloses-fahren-144612. Accessed July 4, 2019.

———, 50 Jahre fahrerloses Fahren: Pressematerial, online in: https://publicarea.admiralcloud.com/p/a49d3d8ba92f3cdbfa864f. Accessed September 25, 2020a.

———, The "Ghost Car", online in: https://www.continental.com/en/company/history/the-ghost-car-145316. Accessed September 25, 2020b.

Crevier, Daniel. 1993. *The Tumultuous History of the Search for Artificial Intelligence*. New York.

Daimler, The PROMETHEUS project launched in 1986: Pioneering autonomous driving, September 20, 2016a, online in: https://media.daimler.com/marsMediaSite/ko/en/13744534.

———, Das Projekt PROMETHEUS ab 1986: Vorreiter des autonomen Fahrens, September 20, 2016b, online in: https://media.daimler.com/marsMediaSite/de/instance/ko.xhtml?oid=13744534.

Darrach, Brad. 1970. Meet Shaky, the First Electronic Person. *LIFE Magazine*, November 20, 1970, pp. 58A–58D, 61, 63–64, 65–66, 68.

Davies, Alex. 2021. *Driven, The Race to Create the Autonomous Car*, New York, London, Toronto, Sydney, New Delhi.

Davies, Alex, Driven, *The Race to Create the Autonomous Car*, New York, London, Toronto, Sydney, New Delhi 2021.

Defense Advanced Research Projects Agency, Next Generation Explosive Ordnance Disposal Remote Controlled Vehicle (NGEODRCV), Common Systems Architecture, Draft version 5, Arlington 2001.

Delcker, Janosch. 2018. The man who invented the self-driving car. *Politico*, July 19, 2018, online in: https://www.politico.eu/article/delf-driving-car-born-1986-ernst-dickmanns-mercedes/.

Department of Defense, Unmanned Ground Vehicle Master Plan, July 1992. https://apps.dtic.mil/dtic/tr/fulltext/u2/a277757.pdf. Accessed June 25, 2020.

Department of Defense, Office of the Secretary of Defense. 2000. *Joint robotics program master plan*. Washington, DC.

Department of Defense, Directive, N° 3000.09, Autonomy in Weapon Systems, November 21, 2012, p. 2, online in: https://www.esd.whs.mil/Portals/54/Documents/DD/issuances/dodd/300009p.pdf. Accessed March 3, 2020.

De Vries, Marten Johannes, Pushkin Kachroo, and Kaan Ozbay, eds. 1998. Intelligent transportation systems. *Proceedings of the Society of Photo-Optical Instrumentation Engineers (SPIE)*, vol. 3207., 15–17 October 1997. Pittsburgh.

Dickmanns, Ernst Dieter. 1980. Untersuchungen und mögliche Arbeitsschritte zum Thema "Künstliche Intelligenz: Rechnersehen und -steuerung dynamischer Systeme".

———. 1986. Computer Vision in Road Vehicles – Chances and Problems. *ITCS-Symposion on "Human Factors Technology for Next-Generation Transportation Vehicles"*, Amalfi, Italy.

———. 1987. PROMETHEUS, 11 170 Integrated Approaches, State of the art review, March 1987.

———. 1988a. An Integrated Approach to Feature Based Dynamic Vision. *Proceedings of the Computer Society Conference on Computer Vision and Pattern Recognition*, Ann Arbor, pp. 820–825.

———. 1988b. 4-D-Dynamic Scene Analysis with Integral Spatio-Temporal Models. *Proceedings of the Fourth International Symposium on Robotics Research, Santa Cruz 1987*, ed. Robert C. Bolles and Bernard Roth, 311–318. Cambridge: Robotics Research.

———. 1993a. Visueller Autobahn/Autopilot/Copilot auf Transputerbasis. *Hochschulkurier* 22: 4–5.

———. 1993b. Interview, Der Lotse geht an Bord. *Hochschulkurier*, pp. 6–7.

———. 1994. The 4D-Approach to Dynamic Machine Vision. *Proceedings of the 33rd Conference on Decision and Control*, pp. 3770–3775.

———. 1997a. *Dynamic Vision for Perception and Control of Motion*. Heidelberg.

———. 1997b. Vehicles Capable of Dynamic Vision. *IJCAI'97 Proceedings of the Fifteenth International Joint Conference on Artificial Intelligence*, Vol 2, pp. 1577–1592.

———. 1997c. Improvements in Visual Autonomous Road Vehicle Guidance 1987–1994. In *Visual Navigation: From Biological Systems to Unmanned Ground Vehicles*, ed. Y. Aloimonos, 375–404.

———. 2002. The Development of Machine Vision for Road Vehicles in the Last Decade. *Intelligent Vehicle Symposium* 1: 268–281.

———. Forschungsbericht 1.10.1998–30.9.2002. https://dokumente.unibw.de/pub/bscw.cgi/999528. Accessed May 21, 2020a.

———. Contributions to Visual Autonomous Driving, Part I, 2015, online in: http://dyna-vision.de/main/Text_PDF/2015%20ContribPart%20I%20to%20Visual%20Auton.%20Driving%20EDD.pdf. Accessed October 6, 2020.

———. 2016. Summary of contributions to Vision-based Autonomous Vehicles. http://www.dyna-vision.de/main/Review2016%20of%20ContribVisBasAutVeh.html.

———. H.2.3 Van VaMoRs, online in: http://www.dyna-vision.de/main/H.2.3%20Van%20 VaMoRs.html. Accessed October 18, 2018.

———. May a Pair of "Eyes" be Optimal for Vehicles Too?, Draft-Version 6.12.2019.

———. May a Pair of 'Eyes' Be Optimal for Vehicles Too ?, *Electronics*, 2020, 9, 759, published on May 5, 2020 online in: https://www.mdpi.com/2079-9292/9/5/759.

Dickmanns, Ernst Dieter, and A. Zapp. 1985. Guiding Land Vehicles Along Roadways by Computer Vision. *Proceedings Congrès Automatique 1985*, AFCET, Toulouse, pp. 233–244.

———. 1986. A Curvature-Based Scheme for Improving Road Vehicle Guidance by Computer Vision. *Mobile Robots, SPIE Proceedings*, vol. 727, 161–168. Cambridge, MA.

———. 1987. Autonomous high speed road vehicle guidance by computer vision. *Proceedings of the International Federation of Automatic Control (IFAC)* 20 (5, Part 4): 221–226.

Dickmanns, Ernst Dieter, V. Graefe, and W. Niegel. Abschlußbericht, Definitionsphase PROMETHEUS, Pro-Art, 1.10.86 bis 14.11.87, Universität der Bundeswehr München.

Dickmanns, Ernst Dieter, and Volker Graefe. 1988a. Sehendes Auto fährt Weltrekord. *Hochschulkurier* 14: 17–23.

———. 1988b. Dynamic monocular machine vision. *Machine Vision and Applications* 1: 223–240.

———. 1988c. Sehendes Auto fährt Weltrekord. *Hochschulkurier* 14: 17–23.

Dickmanns, Ernst Dieter, R. Behringer, D. Dickmanns, T. Hildebrandt, M. Maurer, F. Thomanek, and J. Schiehlen. The seeing passenger car VaMoRs-P. *Proceedings of the Intelligent Vehicles '94 Symposium*, pp. 68–73.

Dowling, Kevin, Robert Guzikowski, Jim Ladd, Henning Pangels, Sanjiv Singh, and William Whittaker. 1990. Navlab. An Autonomous Navigation Testbed. In *Vision and Navigation. The Carnegie Mellon Navlab*, ed. Charles E. Thorpe, 273–275. Boston/Dordrecht/London.

Dudek, Gregory, and Michael Jenkin. 2010. *Computational Principles of Mobile Robotics*. Cambridge, New York, Melbourne, Madrid: Cambridge University Press. (first edition: 2000).

Dunne, Jim. 1966. This 140-m.p.h. Test Car is Driven by Remote Control. *Popular Science*, pp. 82–83.

Earnest, Lester. Stanford Cart, 2018. https://web.stanford.edu/~learnest/les/cart.html. Accessed on February 3, 2019.

Edgerton, David. 2008. *The shock of the old, Technology and global history since 1900*. London.

Eliot, Lance. 2018. Introduction to Driverless Self-Driving Cars: The Best of the AI Insider, (no place of publication mentioned).

Ellwein, Thomas, Achatz von Müller, and Harro Plander. 1974. *Hochschule der Bundeswehr zwischen Ausbildungs- und Hochschulreform*, Opladen.

Engineering and Technology History Wiki, Milestones: SHAKEY: The World's First Mobile Intelligent Robot, 1972, last modified on May 24, 2017. http://ethw.org/Milestones:SHAKEY:_The_World's_First_Mobile_Intelligent_Robot,_1972. Accessed on December 5, 2019.

Fallon, Michael. 2019. *Self-Driving Cars, the New Way Forward*. Minneapolis.

Farivar, Cyrus. 2016. Dallas deployment of robot bomb to kill suspect is "without precedent". Ars technica, July 8, 2016, online in: https://arstechnica.com/tech-policy/2016/07/is-it-ok-to-send-a-police-robot-to-deliver-a-bomb-to-kill-an-active-shooter/. Accessed June 24, 2020.

Faure, Guy. 2007. Le METI et la science industrielle au Japon. In *L'Empire de l'intelligence: Politiques scientifiques et technologiques du Japon depuis 1945*, ed. Jean-François Sabouret. Paris, online in: http://books.openedition.org/editionscnrs/4945. Accessed on January 26, 2018.

Feitz, Anne. 2019. Premiers coups de frein sur la voiture autonome. *Les Echos*, March 26, 2019, online in: https://www.lesechos.fr/tech-medias/intelligence-artificielle/premiers-coups-de-frein-sur-la-voiture-autonome-1003606. Accessed on July 3, 2019.

Fenton, Robert E., Robert L. Cosgriff, Karl W. Olson, and Lyle M. Blackwell. 1968. One Approach to Highway Automation. *Proceedings of the IEEE* 56 (4): 556–566.

Fenton, Robert E., and Karl W. Olson. 1969. The Electronic Highway. *IEEE Spectrum*, pp. 60–66.

Fenton, Robert E. 1970. Automatic Vehicle Guidance and Control – A State of the Art Survey. *IEEE Transactions on Vehicular Technology* VT 19 (1): 153–161.

Fenton, Robert E., and Robert J. Mayhan. 1991. Automated Highway Studies at the Ohio State University – An Overview. *IEEE Transactions on Vehicular Technology* 40 (1): 100–113.

Fenton, Robert E. 1994. IVHS/AHS: Driving into the Future. *IEEE Control Systems Magazine* 14 (6): 13–20.

Firgau, Walter. Ohne Fahrer auf richtigem Kurs. *Süddeutsche Zeitung*, September 6, 1973.

Flory, Leslie E. 1961. Electronic Control of Highway Vehicles. *Journal of the IEEE* 7 (77): 271–273.

Flory, Leslie E., et al. 1962. Electronic techniques in a system of highway vehicle control. *RCA Review* 23: 293–310.

Foasso, Cyrille. Hilare autonomous mobile robot. https://artsandculture.google.com/asset/hilare-autonomous-mobile-robot-laboratory-of-analysis-and-architecture-of-systems/UwFV2X6hPn0NjA.

Future of Life Institute. Autonomous Weapons: An Open Letter from AI & Robotics Researchers, online in: https://futureoflife.org/open-letter-autonomous-weapons/?cn-reloaded=1. Accessed March 5, 2020.

Gage, Douglas W. 1995. UGV History 101: A Brief History of Unmanned Ground Vehicle (UGV) Development Efforts, *Unmanned Systems, Special Issue on Unmanned Ground Vehicles* 13 (3): 2–7.

Garçon, Anne-Françoise. 2012. *L'Imaginaire et la pensée technique, Une approche historique, XVIe-XXe siècle*. Paris.

Gardels, K. 1960. Automatic car controls for electronic highways, General Motors Research Laboratories, *Rep. GMR-276*. Warren, MI.

General Motors. 1940. Futurama, Exhibition brochure, USA.

———. 1953. An Automatically Guided Automobile Cruised Along a One-Mile Check Road at General Motors Technical Center today.... Press release, USA.

———. 2020. Firebird II, "Three-Zero-Four", The Gas Turbine Family Car, General Motors latest "Laboratory on Wheels", p. 13, in: The Old Car Manual Project, Brochures, GM Corporate and Concepts, http://www.oldcarbrochures.com/static/NA/GM%20Corporate%20and%20Concepts/1956_GM_Firebird_II/1956%20Firebird%20II-12-13.html. Accessed on April 23, 2020.

Gershgorn, Dave. 2016. Police Used Bomb Disposal Robot to Kill a Dallas Shooting Suspect. *Popular Science*, July 8, 2016, online in: https://www.popsci.com/police-used-bomb-disposal-robot-to-kill-dallas-shooting-suspect/. Accessed June 24, 2020.

Giles, C.G., and J.A. Martin. 1961. *Cable Installation for Vehicle Guidance Investigations in the New Research Track at Crowthorne, Report RN/40 57/CGG*. Crowthorne: Road Research Lab.

Giralt, Georges, R.P. Sobek, and Raja Chatila. 1979. A multi-level planning and navigation system for a mobile robot; a first approach to HILARE. *Proc. 6th IJCAI*, Tokyo, Japan.

Giralt, Georges. 1990. Robots et intelligence des machines. *Les Cahiers du MURS* n° 22, 4ème trimestre, pp. 27–47.

Giralt, Georges, Raja Chatila, and Marc Vaisset. 1990. An Integrated Navigation and Motion Control System for Autonomous Multisensory Mobile Robots. In *Autonomous Robot Vehicles*, ed. Ingemar J. Cox and Gordon T. Wilfong, 420–443. New York, Berlin, Heidelberg, London, Paris Tokyo, Hong Kong.

Gleitsmann, Rolf-Jürgen, Rolf-Ulrich Kunze, and Günther Oetzel. 2009. *Technikgeschichte*, Konstanz.

Graefe, Volker, and Klaus-Dieter Kuhnert. 1992. Vision-Based Autonomous Road Vehicles. In *Vision-Based Vehicle Guidance*, ed. Ichiro Masaki, 1–29. New York.

Gregor, R., M. Lützeler, M. Pellkofer, K.H. Siedersberger, and Ernst Dieter Dickmanns. 2002. EMS-Vision: A Perceptual System for Autonomous Vehicles. *IEEE Transactions on Intelligent Transportation Systems* 3 (1): 48–59.

———. EMS-Vision: A Perceptual System for Autonomous Vehicles, pp. 1–6.

Gruber, Angela, 2016. Kleiner geht's nicht, in Spiegel Online, March 26, online in: https://www.spiegel.de/netzwelt/web/moore-s-law-die-goldene-regel-der-chipherstellerbroeckelt-a-1083468.html

Gugerli, David. 2018. *Wie die Welt in den Computer kam, Zur Entstehung digitaler Wirklichkeit.* Frankfurt a.M.

Hagen, Hans. 1974, April 19. *Lastwagen ohne Fahrer*. VDI-Nachrichten.

Harayama, Yuko, and Laurent Genod. 2007. La politique scientifique et technologique au Japon: une perspective historique. In *L'Empire de l'intelligence: Politiques scientifiques et technologiques du Japon depuis 1945*, ed. Jean-François Sabouret, 23–49. Paris, online in: http://books.openedition.org/editionscnrs/4942. Accessed January 20, 2018.

Hartford, Bill. 1967, April. You 'Fly' this Car with a Stick. *Popular Mechanics*, pp. 87–89 and p. 207.

Hartwich, Erwin, and Walter Zimdahl. 2012. Regelung von Straßenfahrzeugen: Ein Zeitzeugenbericht aus dem Institut von Prof. W. Oppelt, TH Darmstadt. *at-automatisierungstechnik* 60 (6): 338–343.

Hawkins, Andrew J. 2019. Waymo's Driverless Car: Ghost-Riding in the Back Seat of a Robot Taxi. *The Verge*, December 9, 2019, online in: https://www.theverge.com/2019/12/9/21000085/waymo-fully-driverless-car-self-driving-ride-hail-service-phoenix-arizona. Accessed September 3, 2020.

Hebert, Martial, Charles E. Thorpe, and Anthony Stentz. 1997. Introduction. In *Intelligent Unmanned Ground Vehicles, Autonomous Navigation Research at Carnegie Mellon*, ed. Martial H. Hebert, Charles E. Thorpe, and Anthony Stentz, 1–15. New York.

Henderson, Harry. 2006. *Modern Robotics, Building Versatile Machines*. New York.

Herrmann, Andreas, Walter Brenner, and Rupert Stadler. 2018. *Autonomous Driving: How the Driverless Revolution will Change the World.* Bingley.

Heßler, Martina. 2019. Technik und Autonomie, Kulturhistorische Bemerkungen zu einem komplexen Verhältnis. In *Autonome Systeme und Arbeit*, ed. Hartmut Hirsch-Kreinsen and Anemari Karacic, 247–274. Bielefeld.

Hirose, Takeshi. 1989. Position Prediction of Automobiles – White Line Detection (in Japanese). *Proceedings of the 28th SICE Annual Conference* 1: 23–24.

保坂明夫, 青木啓二, 津川定之, 自動運転(第2版):システム構成と要素技術, 森北出版 (Translation of the book title in English: Hosaka, Akio, Keiji Aoki, Sadayuki Tsugawa. 2019. *Automated Driving, Systems and Technologies*, 2nd ed. Tokyo: Morikita Publishing).

Houston Metro, Demo '97-Flyer. http://www.path.berkeley.edu/sites/default/files/my_folder_76/NAHSC_Brochure- Demo%26Groups-971.pdf, p. 45. Accessed June 17, 2018.

Ijichi, Tomohiro. 2012. Innovation Policies and Reforms in Japan. In *The Dynamics of Regional Innovation, Policy Challenges in Europe and Japan, Series on Innovation and Knowledge Management*, ed. Yveline Lecler, Tetsuo Yoshimoto, and Takahiro Fujimoto, vol. 10, 41–71. Singapore.

Ingraham, Joseph C. 1960. Electronic Roads Called Practical; New System of Guiding Cars Safely on Highways Is Shown at Princeton Fruit of 7 Years' Study R.C.A. and G.M. Jointly Conducted It – Full Use Seen 15 Years Away. *The New York Times*, June 6, 1960.

INRIA, Rapport d'activité 1997, Action Praxitele. https://raweb.inria.fr/rapportsactivite/RA-Dev97/PRAXITELE/. Accessed October 11, 2020.

Ioannou, Petros A., ed. 1997. *Automated Highway Systems*. New York, London.

Ishii, Takemochi, Masakazu Iguchi, and Masaki Koshi. 1976. CVS: Computer-Controlled Vehicle System, translated by Akira Yamashita. In *Personal Rapid Transit III*, ed. D.A. Gary, M.J. Gary, A.L. Kornhauser, and W.L. Garrard, 77–83. Minneapolis: University of Minnesota. online in: http://staff.washington.edu/jbs/itrans/cvs%20prt%20paper%20with%20illustrations.pdf. Accessed July 7, 2019.

Ishiyama, Takeshi. Introduction of JARI, 1st Asia Automobile Institute Summit, 26–27 November 2012, Tokyo. http://www.jari.or.jp/Portals/0/resource/uploads/AAI_Summit_01-JARI.pdf. Accessed November 3, 2017.

ITS Japan, About ITS, 2010. http://www.its-jp.org/english/about_e/. Accessed January 30, 2018.

Jacobsen, Annie. 2015. *The Pentagon's Brain, An Uncensored History of DARPA, America's Top Secret Military Research Agency*. New York.

Jacomy, Bruno. 1990. *Une histoire des techniques*. Paris.

Jamthe, Sudha. 2017. 2030 The Driverless World: Business Transformation from Autonomous Vehicles, (independently published).

Jochem, Todd, Dean Pomerleau, Bala Kumar, and Jeremy Armstrong. PANS: A Portable Navigation Platform. *Proceedings of the Intelligent Vehicles '95. Symposium*, pp. 107–112, also in: https://www.cs.cmu.edu/~tjochem/nhaa/navlab5_details.html. Accessed on August 3, 2018.

Jochem, Todd, and Dean Pomerleau. No Hands Across America General Information (undated.), in: https://www.cs.cmu.edu/~tjochem/nhaa/general_info.html. Accessed on August 3, 2018.

———. *No Hands Across America Journal*. http://www.cs.cmu.edu/~tjochem/nhaa/Journal.html. Accessed on August 3, 2018.

Johnson, Ann. 2009. *Hitting the Brakes, Engineering Design and the Production of Knowledge*. Durham and London.

Juhlin, Oskar. 1997. *Prometheus at the wheel – Representations of Road Transport Informatics* (Doctoral thesis, Institute of Tema Research, Linköping University, Sweden), Linköping Studies in Arts and Science No. 162, Linköping, see also: https://www2.bibl.liu.se/liupubl/disp/disp97/arts162s.htm.

Julia, Luc. 2019. *L'intelligence artificielle n'existe pas*. Paris.

Kanade, Takeo. 1990. Foreword. In *Vision and Navigation: The Carnegie Mellon Navlab*, ed. Charles E. Thorpe, xiii–xiv. Boston.

Kanaris, Alexander, Petros Ioannou, and Fu-Sheng Ho. 1997. Spacing and Capacity Evaluations for Different AHS Concepts. In *Automated Highway Systems*, ed. Petros A. Ioannou, 125–167. New York, London.

Kato, S., et al. 2002. Vehicle Control Algorithms for Cooperative Driving with Automated Vehicles and Inter-vehicle Communications. *IEEE Transactions on Intelligent Transportation Systems* 3 (3): 155–161.

Kay, Jennifer S. 1997, January. *STRIPE: Remote Driving Using Limited Image Data*, PhD thesis. Carnegie Mellon University, Computer Science Department.

Kelly, Fred C., ed. 2002. *Miracle at Kitty Hawk, The Letters of Wilbur & Orville Wright*. New York.

Kilbon, Kenyon. 1960, Autumn. Tomorrow's Thruway is Here Today! *Electronic Age*, pp. 26–29.

Knop, Carsten. 2018. Ein Deutscher war Tesla 30 Jahre voraus. *Frankfurter Allgemeine Zeitung*, February 1, 2018, online in: https://www.faz.net/aktuell/wirtschaft/autonomes-fahren-ein-deutscher-war-tesla-30-jahre-voraus-15424550.html.

König, Wolfgang. 2009. *Technikgeschichte, Eine Einführung in ihre Konzepte und Forschungsergebnisse*. Stuttgart.

Krishnan, Armin. 2009. *Killer Robots. Legality and Ethicality of Autonomous Weapons*. Surrey/Burlington.

Kröger, Fabian. 2012. Fahrerlos und unfallfrei. Eine frühe automobile Technikutopie und ihre populärkulturelle Bildgeschichte. In *Technology Fiction, Technische Visionen und Utopien in der Hochmoderne*, ed. Uwe Fraunholz and Anke Woschech, 93–114. Bielefeld.

———. 2014. Die Zukunft, die nie eintrat, Über die Nützlichkeit filmischer Zukunftsbilder des autonomen Fahrens für die Prospektive. *Technology Assessment – Theory and Practice (TATuP)*, N° 1/2014, Verkehrszukünfte – Visionen jenseits aktueller öffentlicher Aufmerksamkeit, pp. 3–6.

———. 2015. Das automatisierte Fahren im gesellschaftsgeschichtlichen und kulturwissenschaftlichen Kontext. In *Autonomes Fahren, Technische, rechtliche und gesellschaftliche Aspekte*, ed. M. Maurer, J.C. Gerdes, B. Lenz, and H. Winner. 41–67. Heidelberg.

———. 2016a. Automated Driving in its Social, Historical and Cultural Contexts. In *Autonomous Driving, Technical, Legal and Social Aspects*, ed. M. Maurer, J.C. Gerdes, B. Lenz, and H. Winner, 41–68. Heidelberg.

———. 2016b. Le long cheminement de l'imaginaire cinématographique de la voiture autonome. In *Choc de Mobilités*, ed. Mathieu Flonneau and Stéphane Levesque, 55–67. Paris.

Kuan, D., and U.K. Sharma. 1987. Model Based Geometric Reasoning for Autonomous Road Following. *Proceedings of the IEEE Int. Conf on Robotics and Automation*, Raleigh, NC, USA, March 31 – April 3, 1987, pp. 416–423.

Kuan, D., and G. Phipps. 1987. A-Shuan Hsueeh, Autonomous Land Vehicle Road Following. *Proceedings of the First Int. Conf. on Computer Vision*, London England, June 1987, pp. 557–566.

Kurfess, Thomas R. 2005. *Robotics and Automation Handbook*. Boca Raton, London, New York, Washington D.C.

Kurman, Melba. 2017. Electronic Highway of the Future. *Mechanix Illustrated*. March 10, 2017, online in: https://mechanixillustrated.technicacuriosa.com/2017/03/10/electronic-highway-of-the-future/. Accessed May 25, 2021.

Laboratoire Central des Ponts et Chaussées (LCPC), LA ROUTE AUTOMATISÉE, Un scenario périurbain, Brochure, May 2001.

Lanson, Florian. 2010. Les Systèmes de Transport Intelligents au Japon, Rapport d'Étude, Ambassade de France au Japon, Service pour la Science et la Technologie.

Latour, Bruno. 1992. *Aramis ou l'amour des techniques*, Paris.

Laugier, Christian, S. Sekhavat, F. Large, J. Hermosillo, and Z. Shiller. 2001. Some steps towards autonomous cars. *IFAC Symposium*, 9–17. Japan.

Laugier, Christian. 2002. Towards "Automated Road", Some Projects and New Technologies, Keynote ITSC 2002 Singapore, online in: http://emotion.inrialpes.fr/laugier/Research/itsc02.pdf. Accessed October 12, 2020.

———. 2019. A Journey into the History of Automated Driving, Invited pioneer's talk. *IROS 2019 – IEEE/RSJ International Conference on Intelligent Robots and Systems.*, Nov 2019, Macau, China, in: https://hal.inria.fr/hal-02428196/document. Accessed November 11, 2020.

Laurgeau, Claude. 2009. *Le siècle de la voiture intelligente*. Paris.

Lecler, Yveline. 1991. La dynamique du progrès scientifique et technique au Japon. In *Gestion 2000, Management & Prospective*, No. 4, August–September 1991, pp. 149–178.

Lenzen, Manuela. 2018. *Künstliche Intelligenz, Was sie kann & was uns erwartet*. München.

Lipson, Hod, and Melba Kurman. 2017. *Driverless, Intelligent Cars and the Road Ahead*. Cambridge, London.

Lossau, Norbert. Das erste autonome Auto kostete 200.000 D-Mark. *Die Welt*, October 13, 2017.

Lutz, Randolph B. 1966, April. *The Motor Vehicle of the Future*. Michigan Department of State Highways, Technical Report Number 2.

Lützeler, M., and Ernst Dieter Dickmanns. 2000. EMS-Vision: Recognition of Intersections on Unmarked Road Networks. *Proceedings of the IEEE Intelligent Vehicles Symposium 2000*, pp. 302–307, cit. version found in: https://dyna-vision.de/wp-content/uploads/2020/10/2000-EMS-Vision-RecogIntersectionsOnUnmarkedRoadNetworks.pdf.

Lukács, Gabriella. 2010. *Scripted Affects, Branded Selves: Television, Subjectivity, and Capitalism in 1990s*. Japan, Durham and London.

Madhavan, R., T. Hong, and E. Messina. 2006. Temporal Registration of Sensed Range Images for Autonomous Navigation. In *Intelligent Vehicle Systems: 4 D/RCS Approach*, ed. Raj Madhavan, Elena R. Messina, and James Albus, 173–198. New York.

Mahoney, Michael S. 1988, April–June. The History of Computing in the History of Technology. *Annals of the History of Computing* 10 (2): 113–125.

Mahoney, Michael S. 2005. The Histories of Computing (s). *Interdisciplinary Science Reviews* 30 (2): 119–135.

Main, Douglas. 2013. Robots to the Rescue. *Popular Science*, April 8, 2013, online in: https://www.popsci.com/trp-sponsored-article-slideshow/. Accessed July 7, 2020.

MAN Bus & Trucks, Controlled by an invisible hand, January 28, 2020, online in: https://www.mantruckandbus.com/en/moments/1971.html. Accessed September 25, 2020.

Mann, Martin. 1958. The Car That Drives Itself. *Popular Science*, No. 172, May 1958, pp. 75–79 and 226–227.

Martinez, Augustin. 2008. La centrale technologique microélectronique, in: *Lettre du LAAS*, N° 37, October 2008, p. 39, online in: https://www.laas.fr/LettreDuLaasStatic/content/download/793/4835/version/3/file/La-Lettre-n37-5.pdf.

Massot, Marie-Hélène. 2000. Praxitèle: un concept, un service et une expérimentation, Bilan d'un prototype. *Transports Environment Circulation* 59: 25–32.

Matsumoto, Shuntetsu. 1971. Highway Communication System Using Guidance Cable. *IEEE Transactions on Vehicular Technology* VT-20 (1): 12–16.

Matthies, Larry H. 1998. Robotic Perception for Autonomous Navigation of Mars Rover. *1998 Symposium on Frontiers in Engineering*, September 1998, pp. 1–6.

Maurer, M., R. Behringer, S. Fürst, F. Thomanek, E.D. Dickmanns. 1996. A Compact Vision System for Road Vehicle Guidance. *Proceedings of 13th International Conference on Pattern Recognition*, pp. 313–317.

McCarthy, John, Marvin Minsky, Nathan Rochester, Claude Shannon. 1955. A Proposal for the Dartmouth Summer Research Project on Artificial Intelligence.

McCarthy, John. 2019. Computer Controlled Cars, (according to the author, the paper was written around 1968 and revised in the 1970s), online in: http://jmc.stanford.edu/commentary/progress/cars.pdf. Accessed October 1, 2019.

McGrath, Michael E. 2018. *Autonomous Vehicles: Opportunities, Strategies, and Disruptions* (independently published).

McKay, Shawn, Matthew E. Boyer, Nahom M. Beyene, Michael Lerario, Matthew W. Lewis, Karlyn D. Stanley, Randall Steeb, Bradley Wilson, and Katheryn Giglio. 2020. *Automating Army Convoys, Technical and Tactical Risks and Opportunities*. Santa Monica: RAND Corporation. in: https://www.rand.org/content/dam/rand/pubs/research_reports/RR2400/RR2406/RAND_RR2406.pdf. Accessed on June 24, 2020.

McMillan, Stuart I., and Jason G. McPhee. 2014. *An Analysis of the First Fifteen Years of the Department of Defense Framework for Unmanned Ground Systems*, Thesis, Naval Postgraduate School, Monterey, California.

Meissner, Hans-Georg. 1982. *Steuerung dynamischer Systeme aufgrund bildhafter Informationen*, Dissertation UniBwM/LRT.

Merrin, William. 2019. *Digital War, A Critical Introduction*. Oxon/New York.

Meyer, Wolfgang. 1968. Ein Auto ohne Fahrer rast mit 120 Sachen über die Beton-Piste. *Bild*, September 13, 1968.

Meystel, Alex. 1991. *Autonomous Mobile Robots, Vehicles with Cognitive Control*. Singapore.

Ministry of Land, Infrastructure, Transport and Tourism, Road Bureau, ITS Homepage, ITS Handbook Japan 2000, Section 4 ITS deployment progress in Japan, Toward practical use of the Smart Cruise System, 2000. http://www.mlit.go.jp/road/ITS/2000HBook/chapter4/4-19e.html. Accessed December 12, 2017.

Ministry of Land, Infrastructure, Transport and Tourism, Road Bureau, ITS Homepage, Comprehensive Plan for ITS in Japan, Chapter 1 Significance of Promoting ITS, 2008. http://www.mlit.go.jp/road/ITS/5Ministries/chap1.html. Accessed December 12, 2017.

Ministry of Land, Infrastructure, Transport and Tourism, Road Bureau, ITS Homepage, Archive, ITS Handbook Japan 2008, AHS (Advanced Cruise-Assist Highway System), 2008. http://www.mlit.go.jp/road/ITS/1998HBook/chapter3/3-3e.html. Accessed December 12, 2017.

Ministry of Land, Infrastructure, Transport and Tourism, Study Group for the Promotion of ASV, Seeking Even Greater Traffic Accident Reductions through Vehicle Advancements, October 2017. http://www.mlit.go.jp/jidosha/anzen/01asv/resourse/data/asv6pamphlet-e.pdf. Accessed July 13, 2017.

Minx, Eckard, and Rainer Dietrich. 2015. *Autonomes Fahren, Wo wir heute stehen und was noch zu tun ist*, 38–55. Ladenburg.

Möser, Kurt. 2012. Innovationskulturen um das Automobil?, In *Innovationskulturen um das Automobil, Von gestern bis morgen, Stuttgarter Tage zur Automobil- und Unternehmensgeschichte,* ed. Rolf-Jürgen Gleitsmann, Jürgen E. Wittmann, Wiss. Schriftenreihe der Mercedes-Benz Classic Archive, Vol. 16, 25–50. Mühlacker/Mühlhausen.

Mom, Gijs. 1997. History of Tomorrows Automobile, In Geschiedenis van de auto van morgen, Mom, Gijs, Technische Universiteit Eindhoven, p. 7–27.

———. 2004. *The Electric Vehicle, Technology and Expectations in the Automobile Age.* Baltimore, London.

Moravec, Hans. 1980, March. *Obstacle Avoidance and Navigation in the Real World by a Seeing Robot Rover*, Ph.D. thesis, Computer Science Department, Stanford University.

———. 1982. The CMU rover. *AAAI Proceedings*, pp. 377–380.

———. 1988. *MIND Children, The Future of Robot and Human Intelligence.* Cambridge, London.

———. 1990. The Stanford Cart and the CMU Rover. In *Autonomous Robot Vehicles*, ed. Ingemar J. Cox and Gordon T. Wilfong, 407–419. New York, Berlin, Heidelberg, London, Paris Tokyo, Hong Kong.

———. 1999. *Robot: Mere Machine to Transcendent Mind.* Oxford.

Morris-Suzuki, Tessa. 1994. *The Technological Transformation of Japan: From the Seventeenth to the Twentieth-First Century.* Cambridge/New York/Melbourne: Cambridge University Press.

Morris, Jefferson. 2006. iRobot's Dyer Sees Missions Multiplying for PackBot. *Aerospace Daily & Defense Report* 217 (36).

Moye, William T. 1997, May. *The Genealogy of ARL.*

Murphy, Robin R. 2000. *Introduction to AI Robotics.* Cambridge, London: The MIT Press.

Musso, Pierre, Laurent Ponthou, and Éric Seuillet, ed. 2007. Fabriquer le futur 2, L'imaginaire au service de l'innovation. Paris.

Mysliwetz, Birger. 1990. *Parallelrechner-basierte Bildfolgen-Interpretation zur autonomen Fahrzeugsteuerung,* Dissertation, Universität der Bundeswehr München, Fakultät für Luft- und Raumfahrttechnik. Institut für Systemdynamik und Flugmechanik.

Nagano, Hiroshi. 2012. Science and Technology Policy in Japan, In *The Dynamics of Regional Innovation, the Policy Challenges in Europe and Japan.* ed. Yveline Lecler, Tetsuo Yoshimoto, and Takahiro Fujimoto. Singapore.

NAHSC. 2018a. What is An Automated Highway System?, Demo '97-Flyer, in: http://www.path.berkeley.edu/sites/default/files/my_folder_76/NAHSC_Brochure-Demo%26Groups-971.pdf. Accessed June 17, 2018.

———. 2018b. Where the research meets the road, Demo '97-Flyer, in: http://www.path.berkeley.edu/sites/default/files/my_folder_76/NAHSC_Brochure-Demo%26Groups-971.pdf. Accessed June 17, 2018.

National Research Council. 2002. *Technology Development for Army Unmanned Ground Vehicles.* Washington DC.

Nebraska State Historical Society. 2019. Online Archive, online in: https://nebraskahistory.pastperfectonline.com/search?utf8=✓&search_criteria=%22automatic+driving%22&searchButton=Search. Accessed on June 18, 2019.

Nilsson, Nils J. 2010. *The Quest for Artificial Intelligence, a History of Ideas and Achievements.* New York: Cambridge University Press.

Nocks, Lisa. 2007. *The Robot: The Life Story of a Technology.* Westport.

Nora, Simon, and Alain Minc. 1978. *L'Informatisation de la Société: rapport à M. le Président de la République.* Paris: La Documentation Française.

Nowacki, G. 2012. Development and Standardization of Intelligent Transport Systems. *International Journal on Marine Navigation and Safety of Sea Transportation* 6 (3): 403–411.

Nozaki, Taketoshi. 1976. Characteristics of an Automatic Automobile Steering System Using the Information of a Course Deviation at Some Distance Ahead of the Automobile (article in Japanese). *Transactions of the Society of Instrument and Control Engineers* 12 (6): 687–693. online in: https://www.jstage.jst.go.jp/article/sicetr1965/12/6/12_6_687/_pdf/-char/ja. Accessed on June 27, 2019.

Ohio State University. Alternative technologies, Demo '97-Flyer, in: http://www.path.berkeley.edu/sites/default/files/my_folder_76/NAHSC_Brochure-Demo%26Groups-971.pdf, pp. 47–54.

Ohmine, Hiroyuki. 2008. Antennas for ITS. In *Mobile Antenna Systems Handbook*, ed. Kyohei Fujimoto, 427–457. Boston, London: Artech House.

Ohnsman, Alan. 2018. At Waymo, It's Launch Time for Google's Biggest Moonshot, *Forbes*, November 13, 2018, online in: https://www.forbes.com/sites/alanohnsman/2018/11/30/10-million-miles-later-waymo-ceo-says-its-self-driving-cars-are-ready-for-prime-time/#222ddf675e7e.

Okimoto, Daniel I., and Thomas P. Rohlen, eds. 1988. *Inside the Japanese System: Readings on Contemporary Society and Political Economy*. Stanford.

Oshima, Y., E. Kikuchi, M. Kimura, and S. Matsumoto. 1965. Control System for Automatic Automobile Driving. In *Proceedings of the International Federation of Automatic Control (IFAC), Tokyo Symposium on Systems Engineering for Control System Design*, ed. Kankuro Kaneshige and Keisuke Izawa, 347–357.

Özgüner, Umit, Keith A. Redmill, and Alberto Broggi. 2004. Team TerraMax and the DARPA Grand Challenge: A General Overview. *Intelligent Vehicles Symposium*, 232–237. University of Parma.

Özgüner, Ümit, Tankut Acarman, and Keith Alan Redmill. 2011. *Autonomous Ground Vehicles*. Boston, London.

Packer, Jeremy. 2008. Automobility and the Driving Force of Warfare: From Public Safety to National Security. In *The Ethics of Mobility, Rethinking Place, Exclusion, Freedom and Environment*, ed. Sigurd Bergmann and Tore Sager, 39–64. Hampshire, Burlington.

Parent, Michel. 1995. Le future de l'automobile, Présentation INRIA-IRISA, December 3, 1995. https://www.irisa.fr/manifestations/seminaires-1999/3dec99/parent/sld007.htm. Accessed October 12, 2020.

Parent, Michel, and Pascal Daviet. 1996. Automated Urbain Vehicles: Towards a Dual Mode PRT (Personal Rapid Transit). *Proceedings of IEEE International Conference on Robotics and Automation*, Minneapolis, Minnesota, April 22–28, 1996.

Parent, Michel, Yang, Ming. Cybercars: A Complement to Mass Transport, Powerpoint-Presentation (without date), p. 10, online in: https://slideplayer.com/slide/3354708/

Paromtchik, Igor E., and Christian Laugier. 1997. Automatic Parallel Parking and Returning to Traffic Maneuvers. *Proceedings of the 1997 IEEE/RSJ International Conference on Intelligent Robot and Systems, Innovative Robotics for Real-World Applications, IROS '97, Grenoble*, France, Vol. 3, pp. V21–V23, in: http://emotion.inrialpes.fr/~paromt/infos/papers/paromtchik:laugier:video:iros:1997.pdf.

Paromtchik, Igor, and Christian Laugier. 1996a. Motion Generation and Control for Parking an Autonomous Vehicle. *Proceedings of the IEEE International Conference on Robotics and Automation*, 3117–3122. Minneapolis.

——— 1996b. Autonomous Parallel Parking of a Nonholonomic Vehicle, 1996. *Proc. of the IEEE Intelligent Vehicles Symp.*, Tokyo, Japan, September 19–20, 1996, pp. 13–18, online in: http://emotion.inrialpes.fr/~paromt/infos/papers/paromtchik:laugier:iv:1996.pdf.

Pearson, Jordan. 2016. A 60s Japanese Cartoon Predicted Our Autonomous Car Dilemma. *Motherboard*, September 30, 2016, online in: https://motherboard.vice.com/en_us/article/wnx4ay/60s-japanese-cartoon-predicted-autonomous-car-dilemma-self-driving-crash-blame-astro-boy. Accessed September 20, 2018.

Plet, Cécile. 2001. Michel Parent, "Les voitures entièrement automatiques ? Vous en verrez dans 5 ans." *Internetactu.net,* February 22, 2001, online in: http://www.internetactu.net/2001/02/22/michel-parent-les-voitures-entirement-automatiques-vous-en-verrez-dans-5-ans/. Accessed October 8, 2020.

Podnar, Gregg W. 1985a, February. The NEPTUNE Mobile Robot. *Annual Report, Technical Report CMU-RI-TR-86-04*, Robotics Institute, Carnegie Mellon University, pp. 123–125.

———. 1985b, February. The URANUS Mobile Robot. *Annual Report, Technical Report CMU-RI-TR-86-04*. Robotics Institute, Carnegie Mellon University, pp. 127–129.

Pomerleau, Dean A. 1993. *Neural Network Perception for Mobile Robot Guidance*. New York.

Prätorius, Gerhard. 1993. *Das PROMETHEUS-Projekt, Technikentstehung als sozialer Prozeß*. Wiesbaden.

Prometheus – Proart (ed.), Rapport de fin de contrat de recherche, 1994, December.
Quigg, Doc. 1960, June 7. Reporter Rides Driverless Car. *The Press-Courier*.
Rajamani, Rajesh, Han-Shue Tan, Boon Kait Law, and Wei-Bin Zhang. 2000. Demonstration of Integrated Longitudinal and Lateral Control for the Operation of Automated Vehicles in Platoons. *IEEE Transactions on Control Systems Technology* 8 (4): 695–708.
Rescher, Nicholas. 1998. *Predicting the Future: An Introduction to the Theory of Forecasting*. New York.
Reuters, U.K. 1960. New Road Research Track Opened at Crowthorne, October 25, 1960, in: https://reuters.screenocean.com/record/547949. Accessed June, 25 2019.
Reynolds, John. 2001. Cruising into the Future. *Telegraph*, May 26, 2001, online in: https://www.telegraph.co.uk/motoring/4750544/Cruising-into-the-future.html. Accessed June 12, 2019.
Rieu, Alain-Marc. 2007. Le Japon comme "société de connaissance": quelles leçons pour la France?, Politiques de recherche, innovation et réformes institutionnelles depuis la crise de l'énergie des années 1970. In *L'Empire de l'intelligence: Politiques scientifiques et technologiques du Japon depuis 1945*, ed. Jean-François Sabouret, 185–216. Paris.
Rillings, Jim. 1997, February. *The National Automated Highway System: Cooperative Program*. National Automated Highway System Consortium.
Rives, P., and L. Marce. 1985, October. Use of Moving Vision Sensors to Control Robots in an Unknown Universe. *Proceedings of the 5th International Conference on Robot Vision and Sensory Controls*, 165–176. Amsterdam.
Roland, Alex, and Philip Shiman. 2002. *Strategic Computing, DARPA and the Quest for Machine Intelligence*. Cambridge, London.
Roth, Joshua Hotaka. 2017. The Shared Road: Cars, Pedestrians and Bicyclists in Japan. In *Cars, Automobility and Development in Asia: Wheels of Change*, ed. Arve Hansen and Kenneth Bo Nielsen, 134–151. London/New York.
Rousselot, Michel. 2001. Préface. In *La route automatisée, Un scenario périurbain*, ed. LCPC. Paris.
Sabouret, Jean-François. 2007. Introduction, Japon: Le grand secret. In *L'Empire de l'intelligence: Politiques scientifiques et technologiques du Japon depuis 1945*, ed. Jean-François Sabouret, 9–21. Paris.
SAE International. 2019. SAE J3016, Levels of Driving Automation, in: https://www.sae.org/news/press-room/2018/12/sae-international-releases-updated-visual-chart-for-its-"levels-of-driving-automation"-standard-for-self-driving-vehicles. Accessed July 2, 2019.
Saito, Mitsuo. 2000. *The Japanese Economy*. Singapore, New Jersey, London, Hong Kong.
Schaub, Günter, Alfred Pfändner, and Christoph Schäfer. 2004. PRIMUS: Autonomous Navigation in Open Terrain with a Tracked Vehicle. *Unmanned Ground Vehicle Technology VI*, ed. Grant R. Gerhart, Chuck M. Shoemaker, and Douglas W. Gage, vol. 5422, 156–165. Proceedings of SPIE.
Schefter, Jim. 1985, October. Look, Ma! No Driver. *Popular Science*, pp. 94–95, 114–116.
Schiehlen, Joachim. 1995. *Kameraplattformen für aktiv sehende Fahrzeuge, VDI-Fortschritt-Berichte Nr. 514*, Dissertation, Universität der Bundeswehr München, Fakultät für Luft- und Raumfahrttechnik, Institut für Systemdynamik und Flugmechanik.
Schmidt, Rodney. 1971. *A Study of the Real-Time Control of a Computer-Driven Vehicle*. PhD in Electrical Engineering. Stanford University.
Schmidt, Herbie. 2020. Audi steckt beim autonomen Fahren zurück. In *Neue Zürcher Zeitung*. April 28, 2020, online in : https://www.nzz.ch/mobilitaet/auto-mobil/autonomes-fahren-stufe-3-audi-verzichtet-im-a8-auf-staupilot-ld.1553933. Accessed May 24, 2020.
Schneider, Jerry. 2014. Description of the Japanese Computer-Controlled Vehicle System (CVS) (derived largely from a report entitled Automated Guideway Transit: An Assessment of PRT and Other New Systems, U.S. Congress, Office of Technology Assessment, June 1975), November 11, 2014, online in: http://staff.washington.edu/jbs/itrans/cvs1.htm. Accessed July 7, 2019.
Schodt, Frederik L. 2007. *The Astro Boy Essays: Osamu Tezuka, Mighty Atom, and the Manga/Anime Revolution*. Berkeley: Stone Bridge Press.

Schwartz, Ingo. 1999, April. *Primus: An Autonomous Driving Robot, Proceedings of the SPIE*. Vol. 3693. Orlando: AeroSense Session on Unmanned Ground Vehicle Technology.

Schwartz, Samuel I. 2018. *No One at the Wheel: Driverless Cars and the Road of the Future*. New York.

Sekine, Taro. 2015. The Image of Automobile Safety. In *International Association of Traffic and Safety Sciences, Interdisciplinary Wisdom of IATSS.*, 2015, online in: http://www.iatss.or.jp/common/pdf/en/publication/commemorative-publication/iatss40_theory_06.pdf, p. 62. Accessed November 9, 2017.

Sfez, Lucien. 2002. La technique comme fiction, *Revue européenne des sciences sociales*, XL-123, published December 3, 2009, online in: http://journals.openedition.org/ress/612. Accessed August 23, 2020.

Shladover, Steven E., Advanced Vehicle Control Systems. 1990. *SAE Technical Paper* 901129: 103–112.

Shladover, Steven E. 1992. The California PATH Program of IVHS Research and Its Approach to Vehicle-Highway Automation. *Proceedings of the Intelligent Vehicles '92 Symposium*, Detroit, pp. 347–352.

———. 1997. Reasons for Operating AHS Vehicles in Platoons. In *Automated Highway Systems*, ed. Petros A. Ioannou, 11–27. New York, London.

Shladover, Steven. 2007. PATH at 20 – History and Major Milestones. *IEEE Transactions on Intelligent Transportation Systems* 8 (4): 584–592.

Shladover, Steven E. 2008. AHS Research at the California PATH Program and Future AHS Research Needs. *Proceedings of the 2008 IEEE International Conference on Vehicular Electronics and Safety*, Columbus, September 22–24, 2008, pp. 4–5.

Shladover, Steven E., Jane Lappin, Robert P. Denaro, and Bryant Walker Smith. 2018. Introduction: The Transportation Research Board's 2013 Workshop on Road Vehicle Automation. In *Road Vehicle Automation*, ed. Gereon Meyer and Sven Beiker, 1–12. Cham, Heidelberg, New York, Dordrecht, London.

Shladover, Steven E., Jane Lappin, Robert P. Denaro, and Bryant Walker Smith. 2018. Introduction: The Transportation Research Board's 2013 Workshop on Road Vehicle Automation. In *Road Vehicle Automation*, ed. Gereon Meyer and Sven Beiker, 1–12.

Shoemaker, Charles. 2006. Development of Semi-Autonomous Robotic Ground Vehicles, DoD's Ground Robotics Research Programs: Demo I through Demo III. In *Intelligent Vehicle Systems: 4 D/RCS Approach*, ed. Raj Madhavan, Elena R. Messina, and James Albus, 283–315. New York.

Siedersberger, Karl-Heinz. 2003. *Komponenten zur automatischen Fahrzeugführung in sehenden (semi-)autonomen Fahrzeugen*, Dissertation. Universität der Bundeswehr München.

Simoudis, Evangelos. 2017. *The Big Data Opportunity in Our Driverless Future*. Menlo Park.

Sperling, Daniel. 2018. *Three Revolutions: Steering Automated, Shared, and Electric Vehicles to a Better Future*. Washington, DC.

Spofford, John R., Raymond D. Rimey, and Steve H. Munkeby. 1997. Overview of the UGV / Demo II program, online in: https://pdfs.semanticscholar.org/72fa/74930bb02e33e2c7d887f56 1dd78d4bf4271.pdf. Accessed on April 8, 2020.

Statista. 2020. By 2030, one in 10 vehicles will be self-driving globally, Press release, February 6, 2020, https://mailchi.mp/statista/autonomous_cars_20200206?e=145345a469.

Statistics Bureau, Ministry of Internal Affairs and Communications, Chapter 29, Disasters and Accidents, Road Traffic Accidents 1924–2005, online in: http://www.stat.go.jp/data/chouki/zuhyou/29-11.xls. Accessed November 9, 2017.

Stone, Wesley L. 2005. The History of Robotics. In *Robotics and Automation Handbook*, ed. Thomas R. Kurfess, 1–12. Boca Raton, London, New York, Washington, DC.

Strachan, Jamie Robert. 2009. *NOVA: Nottingham Offroad Vehicle Architecture*, PhD thesis. University of Nottingham, in: http://eprints.nottingham.ac.uk/10842/1/Microsoft_Word_-_Complete_Thesis_-_issue_M.pdf.

Sukthankar, R. RACCOON: A Real-time Autonomous Car Chaser Operating Optimally at Night. *Proceedings of the Intelligent Vehicles '93 Symposium*, pp. 37–42.

Tamaney, L.S. Real Time Intelligent Control of a Mobile Robot, NATO Advanced Research Workshop on Mobil Robot Implementation, Porto, Portugal, 28 Sept. – 2 Oct. 1987.

Taniguchi, M., et al. 1991. The Development of Autonomously Controlled Vehicle, PVS. *Proceedings of the 1991 Vehicle Navigation and Information Systems Conference*, pp. 1137–1141.

Tezuka, Osamu. Astro Boy (1), Complete Collection of Tezuka Osamu, Kodansha 2009 (手塚 治虫, 鉄腕アトム (1), 手塚治虫文庫全集, 講談社 2009).

Team Navlab. Carnegie Mellon University, Demo '97-Flyer, in: http://www.path.berkeley.edu/sites/default/files/my_folder_76/NAHSC_Brochure-Demo%26Groups-971.pdf, p. 46.

Temple, David W. 2015. *Motorama, GM's Legendary Show and Concept Cars*. North Branch.

Tessier, Catherine. 2015. Autonomie: enjeux techniques et perspectives. In *Drones et killer robots: faut-il les interdire?*, ed. Ronan Doaré, Didier Danet, Gérard de Boisboissel, Rennes.

The Congress of the United States, Congressional Budget Office, High-Tech Highways: Intelligent Transportation Systems and Policy, October 1995, online in: https://www.cbo.gov/sites/default/files/104th-congress-1995-1996/reports/High-techHighways.pdf. Accessed March, 23, 2020.

Thielman, Sam. 2016. Use of Police Robot to Kill Dallas Shooting Suspect Believed to be First in US History. The Guardian, July 8, 2016, online in: https://www.theguardian.com/technology/2016/jul/08/police-bomb-robot-explosive-killed-suspect-dallas. Accessed June 24, 2020.

Thomanek, F., Ernst Dieter Dickmanns, and Dirk Dickmanns. Multiple Object Recognition and Scene Interpretation for Autonomous Road Vehicle Guidance. *Proceedings of the Intelligent Vehicles '94 Symposium*, pp. 231–236.

Thorpe, Charles, Martial H. Hebert, Takeo Kanade, and Steven A. Shafer. 1988. Vision and Navigation for the Carnegie-Mellon Navlab. *IEEE Transactions on Pattern Analysis and Machine Intelligence* 10 (3): 362–373.

Thorpe, Charles E. 1990. Chapter 1: Introduction. In *Vision and Navigation: The Carnegie Mellon Navlab*, ed. Charles E. Thorpe, 1–8, Boston.

Thorpe, Charles, Martial Hebert, Takeo Kanade, and Steven Shafer. 1991a. Toward Autonomous Driving: The CMU Navlab, Part I, Perception. *IEEE Expert* 6 (4): 31–42.

———. 1991b. Toward Autonomous Driving: The CMU Navlab, Part II, System and Architecture. *IEEE Expert* 6 (4): 44–52.

Thorpe, Chuck. 1998. Mobile Robots and Smart Cars. In *Field and Service Robotics*, ed. Alexander Zelinsky, 1–5. Berlin, New York.

Thorpe, Chuck, Todd Jochem, and Dean Pomerleau. 2020. Automated Highways and the Free Agent Demonstration, online in: https://www.ri.cmu.edu/pub_files/pub2/thorpe_charles_1997_1/thorpe_charles_1997_1.pdf. Accessed September 11, 2020.

———. 1998. The 1997 Automated Highway Free Agent Demonstration. *Proceedings of Conference on Intelligent Transportation Systems 1997*, IEEE, S. 496–501.

Thorpe, Chuck. 2018. An oral history conducted in 2010 by Peter Asaro and Selma Šabanović, Indiana University, Bloomington Indiana, for Indiana University and the IEEE, in: http://ethw.org/Oral-History:Chuck_Thorpe#Interview. Accessed Mai 7, 2018.

Thrun, Sebastian, et al. 2006. Stanley: The Robot that Won the DARPA Grand Challenge. *Journal of Field Robotics* 23 (9): 661–692. online in: https://onlinelibrary.wiley.com/doi/epdf/10.1002/rob.20147. Accessed Mai 7, 2018), also In The 2005 DARPA Grand Challenge, the Great Robot Race ed. Buehler, Martin, Iagnemma, Karl, Singh, Sanjiv, Berlin, Heidelberg, New York 2007, pp. 1–44.

Toffler, Alvin. 1980. *The Third Wave*. New York.

Tombo, 日本で最初のテストコース 旧東村山自動車試験道路, July 2, 2009, in: http://blog.livedoor.jp/thetombo2/archives/51493535.html. Accessed December 3, 2017.

Torchinsky, Jason. 2019. *Robot, Take the Wheel: The Road to Autonomous Cars and the Lost Art of Driving*. New York.

Toyota. 2018. A 75-Year History through Text, Chapter 2: Satisfying Public Needs, Section 2: Response to Increase in Traffic Accidents, Item 5: Research and Development of a

Comprehensive Automobile Traffic Control System, in: http://www.toyota-global.com/company/history_of_toyota/75years/text/entering_the_automotive_business/chapter2/section2/item5.html. Accessed January 3, 2018.
Transport Research Board, National Research Council, National Automated Highway System Research Program, a review, Special Report 253, Washington, DC, 1998.
Tsugawa, Sadayuki, Teruo Yatabe, Takeshi Hirose, Shuntetsu Matsumoto. 1979. An Automobile with Artificial Intelligence. *Proceedings of the 6th International Joint Conference on Artificial Intelligence*, Vol. 2, San Francisco, pp. 893–895.
Tsugawa, Sadayuki, Takeshi Hirose, and Teruo Yatabe. 1984. An Intelligent Vehicle with Obstacle Detection and Navigation Functions. *Industrial Electronics Conference (IECON)* 1: 303–308.
Tsugawa, Sadayuki. 1992. Concept of Super Smart Vehicle Systems and Their Relation to Advanced Vehicle Control Systems. *Transportation Research Record* 1358: 42–49.
Tsugawa, Sadayuki, Masayoshi Aoki, Akio Hosaka, and Kaoru Seki. A Survey on Present IVHS Activities in Japan. *1996 IFAC, 13th Triennial World Congress*, pp. 7844–7849.
Tsugawa, Sadayuki. 2005. Issues and recent trends in vehicle safety communication systems. *IATSS Research* 29 (1): 7–15.
———. 2008. A History of Automated Highway Systems in Japan and Future Issues, Institute of Electrical and Electronics Engineers. *International Conference on Vehicular Electronics and Safety*, Ohio, USA, September 22, 2008, p. 2–3.
———. 2009. A Survey on Effects of ITS–related Systems and Technologies on Global Warming Prevention. In *Proceedings of the 12th IFAC Symposium on Transportation Systems*, Redondo Beach, USA, September 2–4, 2009, pp. 334–341.
———. 2011. Automated Driving Systems: Common Grounds of Automobiles and Robots. *International Journal of Humanoid Robotics* 8 (1): 1–12.
———. 2013. An Overview on an Automated Truck Platoon within the Energy ITS Project. *7th IFAC Symposium on Advances in Automotive Control, The International Federation of Automatic Control (IFAC)*, September 4–7, 2013, Tokyo, Japan, pp. 41–46.
Tsugawa, Sadayuki, Sabina Jeschke, and Steven E. Shladover. 2016. A Review of Truck Platooning Projects Foe Energy Savings. *IEEE Transactions on Intelligent Vehicles* 1 (1): 68–77.
Tuchman, Barbara. 1982. In Geschichte denken, Essays. Düsseldorf.
Turk, Mathew A., David G. Morgenthaler, Keith D. Gremban, and Martin Marra. 1988. VITS–A Vision System for Autonomous Land Vehicle Navigation. *IEEE Transactions on Pattern Analysis and Machine Intelligence* 10 (2): 342–361.
Ulmer, Berthold. 1994. VITA II – Active Collision Avoidance in Real Traffic. *Proceedings of the Intelligent Vehicles Symposium 1994*, Paris, pp. 1–6.
Ullrich, Otto. 1980. Der Charakter des Fortschritts moderner Technologien, Zur Geschichtsphilosophie der Technikgeschichtsschreibung. *Technologie und Politik*, Issue 16, Hamburg, pp. 23–51.
Universität der Bundeswehr München, Präsidialabteilung, Struktur- und Entwicklungsplan der Universität der Bundeswehr München 2015–2020, München 2015. online in: https://www.unibw.de/universitaet/universitaet/leitung/struktur-und-entwicklungsplan-unibw-m-2015-2020.pdf. Accessed May 7, 2020.
Urry, John. 2004. The 'System' of Automobility. *Theory, Culture and Society* 21 (4/5): 25–39.
US Army Material Command, The U.S. Army's Report on International Cooperative Programs, Urbana-Champaign 1998.
US Department of Transportation, Request for Applications to Establish a National Automated Highway System Consortium, Contract number DTFH61-94-X-0000, December 15, 1993.
———, The National Intelligent Transportation Systems Program: Where We've Been & Where We're Going, Report No. FHWA-JPO-97-0027, March 1997.
Uzulis, André. 2005. *Die Bundeswehr: Eine politische Geschichte von 1955 bis heute*. Hamburg, Berlin, Bonn.

Vaid, Kamal. 2018. *Selbst ist das Auto – automatisiertes und autonomes Fahren*, Die Zukunft der Mobilität, Schriftenreihe der Northern Business School zur angewandten Wissenschaft, Band 10, Hamburg.

VanderWerf, Joel, Steven Shladover, and Mark A. Miller 2004. Conceptual Development and Performance Assessment for the Deployment Staging of Advanced Vehicle Control and Safety Systems, California PATH Research Report UCB-ITS-PRR-2004-22, August 2004, online in: https://pdfs.semanticscholar.org/3476/e3f1195402060c9398e95674f7c2c6d0fa49.pdf. Accessed on July 8, 2019.

Velez-Green, Alexander. 2015. The Foreign Policy Essay: The South Korean Sentry—A "Killer Robot" to Prevent War, March 1, 2015, in: https://www.lawfareblog.com/foreign-policy-essay-south-korean-sentry—-killer-robot-prevent-war. Accessed March 3, 2020.

Vieweg, Christof. 2015, July 31. Wer hat das Roboterauto erfunden ? Die Bundeswehr! *DIE ZEIT*.

Volkswagen Newsroom, Automatically bound for the future, The history of autonomous driving at Volkswagen, August 24, 2018. online in: https://www.volkswagen-newsroom.com/en/stories/automatically-bound-for-the-future-4118. Accessed March 17, 2020.

Volvo, Volvo Cars CEO urges governments and car industry to share safety-related traffic data, Press release, April 3, 2017. online in: https://www.media.volvocars.com/global/en-gb/media/pressreleases/207164/volvo-cars-ceo-urges-governments-and-car-industry-to-share-safety-related-traffic-data. Accessed July 3, 2019.

Wadhwa, Vivek, and Alex Salkever. 2019. *The Driver in the Driverless Car: How Our Technology Choices Will Create the Future,* Oakland 2017. 2nd ed.

Waugh, Rob. 2013. How the first "driverless car" was invented in Britain, *Yahoo News UK*, July 17, 2013, online in: https://uk.news.yahoo.com/how-the-first%2D%2Ddriverless-car%2D%2Dwas-invented-in-britain-in-1960-093127757.html. Accessed September 29, 2020.

Weber, Jutta. 2013. Vorratsbomben im Himmel, Über digitalen Terror, unsichtbare Opfer und die Rhetorik der Präzision. In *High-Tech Kriege, Frieden und Sicherheit in den Zeiten von Drohnen, Kampfrobotern und digitaler Kriegsführung*, ed. Heinrich-Böll Stiftung, 31–43. Berlin.

Weber, Marc. 2014. Where To? A History of Autonomous Vehicles, May 8, 2014, in: http://www.computerhistory.org/atchm/where-to-a-history-of-autonomous-vehicles/.

Weinberger, Sharon. 2018. *The Imagineers of War: The Untold History of DARPA, the Pentagon Agency That Changed the World*. New York.

Weingroff, Richard F. 2001. Creating a Landmark: The Intermodal Surface Transportation Act of 1991, *Public Roads*, 65(3), November/December 2001, online in: https://www.fhwa.dot.gov/publications/publicroads/01novdec/istea.cfm. Accessed March 5, 2020.

Wengenroth, Ulrich. 1997. Technischer Fortschritt, Deindustrialisierung und Konsum, Eine Herausforderung für die Technikgeschichte. *Technikgeschichte* 64: 1–18.

Wetmore, Jameson M. 2003. Driving the Dream, The History and Motivations Behind 60 Years of Automated Highway Systems in America. *Automotive History Review*, Summer 2003, pp. 4–19.

———. 2020. Reflecting on the Dream of Automated Vehicles, Visions of Hands Free Driving Over the Past 80 Years. *Technikgeschichte* 87 (1): 69–94.

Wettergreen, David, Chuck Thorpe, and William Whittaker. 1993. Exploring Mount Erebus by Walking Robot. *International Conference of Intelligent Autonomous Systems*, pp. 72–81.

Winner, Hermann. 2015. PROMETHEUS – Welche Visionen wurden Realität? *VDA, Technischer Kongress*, pp. 93–124.

Wootton, J.R., and A. García-Ortiz. 1995. Intelligent Transportation Systems. In *Mathematical and Computer Modelling* 22 (4–7): 259–268.

Wünsche, Hans-Joachim. 1986. Detection and Control of Mobile Robot Motion by Real-Time Computer Vision. In *Advances in Intelligent Robotics Systems, Proceedings of the Society of Photo-Optical Instrumentation Engineers (SPIE)*, ed. Nelson Marquina, vol. 727, 100–109.

———. 1987. *Bewegungssteuerung durch Rechnersehen*, Dissertation. Universität der Bundeswehr München, Fakultät für Luft- und Raumfahrttechnik, Institut für Systemdynamik und Flugmechanik.

———. 2013. Geschichte des Automatischen Fahrens – eine kurze Zusammenfassung, Eingeladener Vortrag auf der 6. Tagung Fahrerassistenz, TÜV-SÜD, München, November 28, 2013, online in: https://www.unibw.de/lrt8/forschung-en/geschichte/automated_driving_history. Accessed October 6, 2017.

Yazigi, Monique P. 1992. Promethean Plans for the Next Generation of Cars. *New York Times*, September 13, 1992.

Zapp Alfred. 1988. *Automatische Straßenfahrzeugführung durch Rechnersehen*, Dissertation. Universität der Bundeswehr München, Fakultät für Luft- und Raumfahrttechnik, Institut für Systemdynamik und Flugmechanik.

Zimdahl, Walter. 1963. *Die Bewegung des Straßenfahrzeugs auf vorgeschriebener Bahn*, Diploma Thesis, 1963, Institut für Regelungstechnik, Technische Hochschule Darmstadt (supervised by Fritz Frank).

———. 1965. Führungsverhalten des vierrädrigen Straßenfahrzeugs bei Regelung des Kurses auf festgelegter Bahn. *Regelungstechnik, Zeitschrift für Steuern, Regeln und Automatisieren* 13 (5): 221–268. online in: https://doi.org/10.1524/auto.1965.13.112.221. Accessed October 1, 2020.

———. 1969. *Regelungstechnische Probleme bei dem Leitkabelgeführten Kraftfahrzeug*, Dissertation. Fakultät für Elektrotechnik, Technische Hochschule Darmstadt.

Zimdahl, W., I. Rackow, and T. Wilm. 1986. OPTOPILOT – ein Forschungsansatz zur Spurerkennung und Spurführung bei Straßenfahrzeugen. *VDI Berichte Nr.* 162: 49–60.

Zworykin, V.K., and L.E. Flory. 1958. Electronic Control of Motor Vehicles on the Highway. *Highway Research Board Proceedings* 37: 436–451.

Zworykin, Vladimir K. 1959. The Human Aspect of Engineering Progress. The Fourth Clerk Maxwell Memorial Lecture, July 1, 1959, Cavendish Laboratory, Cambridge. *Journal Brit. I. R. E.*, September 1959, pp. 529–544.

Video Sources

US-Chapter

Automatedhighway (YouTube-account), NAHSC Demo 97 Platoon Scenario (video title), uploaded on April 23, 2011, online in: https://www.youtube.com/watch?v=aaLPBtS8b9E. Accessed on January 25, 2020.

——— (YouTube-account), Automated Highway History (video title), uploaded on April 24, 2011, online in: https://www.youtube.com/watch?v=dyqU85D0lkA, 4:11–5:10. Accessed on April 21, 2020.

Cmurobotics (YouTube-account), History Channel 1998: Driverless Car Technology Overview at Carnegie Mellon University (video title), in: YouTube, uploaded on September 29, 2017, online in: https://www.youtube.com/watch?v=2KMAAmkz9go. Accessed on December 3, 2019.

——— (YouTube-account), Terregator – CMU – Field Robotics Circa 1983 (video title), in: YouTube, uploaded on October 13, 2016, online in: https://www.youtube.com/watch?v=beGRM9ZmWlI. Accessed on December 13, 2019.

JournalStarNews (YouTube-account), Driverless Car (video title), in: YouTube, uploaded on September 13, 2017, online in: https://www.youtube.com/watch?v=Eufcsa46gGk. Accessed on June 13, 2019.

Stanford University Libraries (YouTube-account), Shakey: Experiments in Robot Planning and Learning (1972) (video title), Shakey, An experiment in robot planning and learning, (produced by Peter Hart and Nils Nilsson, assisted by Michael Wilber, sponsored by the Advanced Research Projects Agency and the National Aeronautics and Space Administration, Research performed by the Artificial Intelligence Center at Stanford Research Institute, Menlo Park,

California), in: YouTube, uploaded on September 23, 2016, online in: https://www.youtube.com/watch?v=GmU7SimFkpU. Accessed on December 5, 2019.

Todd Jochem (YouTube-account), Navlab 84–94 (video title), Carnegie Mellon University, Navlab: Vision & Navigation 1984-194, Promotional Video, 1994, in: YouTube, uploaded on January 19, 2018, online in: https://www.youtube.com/watch?v=yfVxt-eBVLo. Accessed on December 3, 2019.

——— (YouTube-account), Navlab 2 Cross Country Navigation 1992 (video title), Official Video from CMU, 1992, in: YouTube, uploaded on January 8, 2018, in: https://www.youtube.com/watch?v=HmLafSKdCRs. Accessed on December 3, 2019.

——— (YouTube-account), Navlab 2 on Flagstaff Hill Pittsburgh People, Places, and Things (video title), TV-emission from 1994, Reporter: Lisa Beatty, published on January 7, 2018, in: https://www.youtube.com/watch?v=btJDzWt82wU. Accessed on August 3, 2018.

Europe-Chapter

British Movietone, Early Driverless Car in 1971, July 2, 2015, online in: https://www.youtube.com/watch?v=5ocvNxjN3dc. Accessed on June 16, 2019.

Dyna-vision, VaMoRs Autobahn Dingolfing, 1987, online in: http://www.dyna-vision.de/main/WebQualVideo/04%20VaMoRsAutobahn%20Dingolfing%201987.wmv. Accessed September 4, 2020.

———, VaMoRs-test run in Rastatt 1988, online in: http://dyna-vision.de/main/WebQualVideo/07%20VaMoRs%20Stat.Obstacle%20Rastatt%201988.mpeg. Accessed September 4, 2020.

———, Prometheus, Turin 1991, DB-VITA, online in: http://www.dyna-vision.de/main/WebQualVideo/09%20VisionBasedStopBehindCar%20Torino%201991.wmv. Accessed September 4, 2020.

———, Turnoff Neubiberg Taxiway. http://www.dyna-vision.de/main/WebQualVideo/23%20TurnoffNbbTaxiway%201996.wmv. Accessed September 4, 2020.

———, Turning off onto an unknown crossroad, in: http://www.dyna-vision.de/main/A.3.7%20Turning%20off%20onto%20an%20unknown%20crossroad.html.

Laraimara, Parking automatique à l'aide d'ultrasons – automated parking using ultrasounds, Uploaded on February 2, 2010, in: https://www.youtube.com/watch?v=BkHHS4uAuOk. Accessed October 12, 2020.

LAAS, Video: Robotique et Intelligence Artificielle, in: https://www.laas.fr/vod/public/films_Georges/5-robotique_IA_720p.html. Accessed November 19, 2020.

Raglan tribe (YouTube-Account), Robot Car | Raglan Tribe (Title), Uploaded 23.2.2013, https://www.youtube.com/watch?v=AlLZhcnpXYI.

Index

F. Kröger, *From Automated to Autonomous Driving*, Archimedes 70,
https://doi.org/10.1007/978-3-031-49881-7

MIX
Papier aus verantwortungsvollen Quellen
Paper from responsible sources
FSC® C105338

If you have any concerns about our products,
you can contact us on
ProductSafety@springernature.com

In case Publisher is established outside the EU,
the EU authorized representative is:
Springer Nature Customer Service Center GmbH
Europaplatz 3, 69115 Heidelberg, Germany

Printed by Libri Plureos GmbH
in Hamburg, Germany